Accreting White Dwarfs

From exoplanetary probes to classical novae and Type 1a supernovae

Online at: https://doi.org/10.1088/2514-3433/ac930c

AAS Editor in Chief

Ethan Vishniac, Johns Hopkins University, Maryland, USA

About the program:

AAS-IOP Astronomy ebooks is the official book program of the American Astronomical Society (AAS) and aims to share in depth the most fascinating areas of astronomy, astrophysics, solar physics and planetary science. The program includes publications in the following topics:

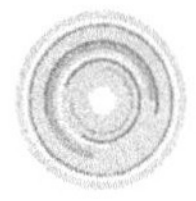
GALAXIES AND COSMOLOGY

INTERSTELLAR MATTER AND THE LOCAL UNIVERSE

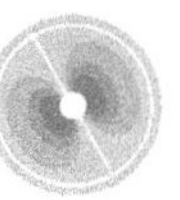
STARS AND STELLAR PHYSICS

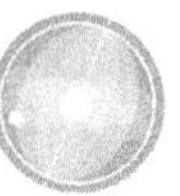
EDUCATION, OUTREACH, AND HERITAGE

HIGH-ENERGY PHENOMENA AND FUNDAMENTAL PHYSICS

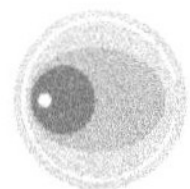
THE SUN AND THE HELIOSPHERE

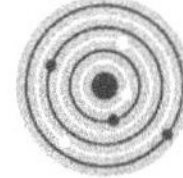
THE SOLAR SYSTEM, EXOPLANETS, AND ASTROBIOLOGY

LABORATORY ASTROPHYSICS, INSTRUMENTATION, SOFTWARE, AND DATA

Books in the program range in level from short introductory texts on fast-moving areas, graduate and upper-level undergraduate textbooks, research monographs, and practical handbooks.

For a complete list of published and forthcoming titles, please visit iopscience.org/books/aas.

About the American Astronomical Society

The American Astronomical Society (aas.org), established 1899, is the major organization of professional astronomers in North America. The membership (~7,000) also includes physicists, mathematicians, geologists, engineers, and others whose research interests lie within the broad spectrum of subjects now comprising the contemporary astronomical sciences. The mission of the Society is to enhance and share humanity's scientific understanding of the universe.

Accreting White Dwarfs

From exoplanetary probes to classical novae and Type 1a supernovae

Edward M Sion

*Department of Astrophysics and Planetary Science, Villanova University,
Villanova, PA 19085, USA*

IOP Publishing, Bristol, UK

ISBN   978-0-7503-2042-9 (ebook)
ISBN   978-0-7503-2040-5 (print)
ISBN   978-0-7503-2043-6 (myPrint)
ISBN   978-0-7503-2041-2 (mobi)

DOI   10.1088/2514-3433/ac930c

Version: 20231001

AAS–IOP Astronomy
ISSN 2514-3433 (online)
ISSN 2515-141X (print)

British Library Cataloguing-in-Publication Data: A catalogue record for this book is available from the British Library.

Published by IOP Publishing, wholly owned by The Institute of Physics, London

IOP Publishing, No.2 The Distillery, Glassfields, Avon Street, Bristol, BS2 0GR, UK

US Office: IOP Publishing, Inc., 190 North Independence Mall West, Suite 601, Philadelphia, PA 19106, USA

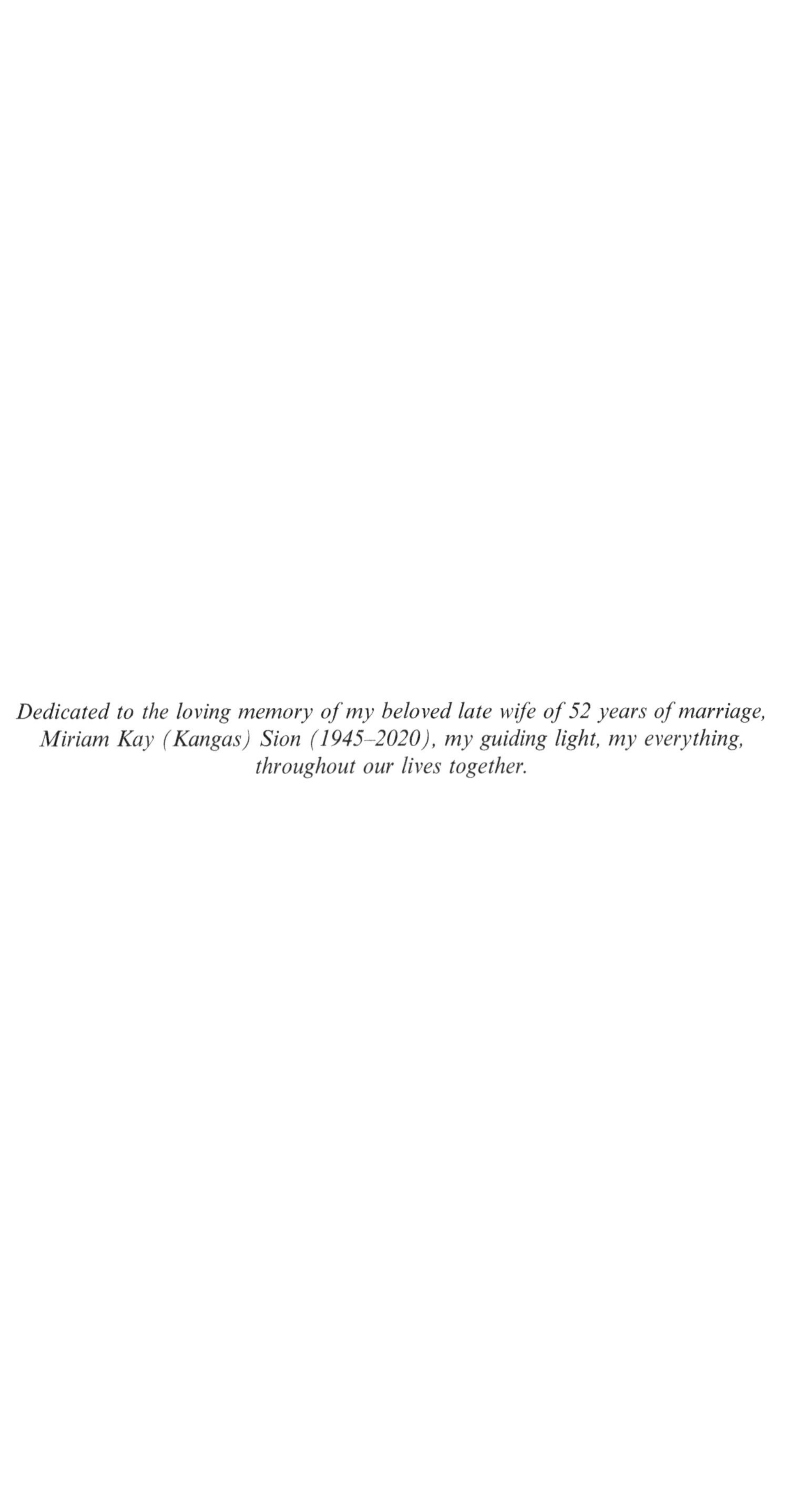

Dedicated to the loving memory of my beloved late wife of 52 years of marriage, Miriam Kay (Kangas) Sion (1945–2020), my guiding light, my everything, throughout our lives together.

Contents

Preface   xiii

Acknowledgements   xv

Author biography   xvii

1   **A Brief Introduction to the Physics of Non-accreting White Dwarfs**   1-1

1.1   A Historical Background of the Study of White Dwarfs   1-1

1.2   Electron Degeneracy   1-2

1.3   The Inverse Relation between WD Mass and Radius   1-5

1.4   The Chandrasekhar Limit   1-6

1.5   Core Compositions Versus WD Mass   1-6

1.6   Cooling Evolution   1-7

1.7   WD Atmospheres and Envelope Compositions   1-8

1.8   Gravitational and Thermal Diffusion   1-10

1.9   Radiative Acceleration   1-13

1.10   Convective Mixing and Dredge-up   1-14

1.11   Rotational Velocities of Non-accreting Isolated WDs   1-15

     References   1-17

2   **Accretion of Interstellar Gas and Dust by White Dwarfs**   2-1

2.1   Early Spectroscopic Observations of Photospheric Metals in Cool White Dwarfs   2-2

2.2   Interstellar Gas and Dust Accretion as the External Source of Metals   2-5

2.3   The Two-Phase ISM Accretion Model of Dupuis, Fontaine and Wesemael   2-7

2.4   Comparison of White Dwarf Kinematics with Maps of the Local ISM   2-10

2.5   The Demise of ISM Accretion as the Source of Metals in Cool White Dwarfs   2-14

     References   2-16

3   **Accreting White Dwarfs as Probes of the Formation, Composition and Evolution of Exoplanetary Systems**   3-1

3.1   Introduction to Exoplanetary Debris Accretion onto White Dwarfs   3-1

3.2   WD Accretion of Tidally Disrupted Asteroids   3-3

     3.2.1   Poynting–Robertson Effect   3-6

3.2.2 Yarkovsky Effect 3-8

3.2.3 YORP Effect 3-8

3.3 Comet Accretion onto a WD 3-10

3.4 Does Accretion by ISM or Detached DA + dM Binaries Still Have a Role? 3-13

3.5 Disk Structure and Accretion Rates onto WDs 3-14

3.6 Gaseous Disks 3-17

3.7 Accretion Rates 3-17

3.8 Diffusion in WDs Accreting Exoplanetary Matter 3-20

3.9 Spectroscopic Abundances and the Chemical Composition of Exoplanetary Matter Accreted by WDs 3-21

3.10 Summary 3-29

References 3-30

4 Accreting WDs in Roche Lobe-detached Post-common Envelope Main Sequence–White Dwarf Binaries **4-1**

4.1 Mass Loss Rates from Late-type Dwarfs in PCEBs 4-3

4.2 Characterizing the Population of Main Sequence–WD PCEBs 4-4

4.3 Accretion of Outflows from Detached Main Sequence Dwarfs onto WDs in PCEBs 4-5

4.4 The Ever-changing Saga of the Hot DAZ.9 WD in the PCEB Feige 24 4-5

4.5 Beyond Feige 24: The Case for Accretion of M Dwarf Winds onto Non-magnetic WDs in PCEBs 4-13

4.6 Accretion onto Magnetic WDs in PCEBs 4-20

4.7 V471 Tauri, The King of Accreting Magnetic PCEBs 4-20

4.8 Low-accretion Rate Polars (LARPS) and Pre-polars (PREPS) 4-27

4.9 Webbink and Wickramasinghe (2005) Magnetic Siphon 4-31

4.10 What Is The Origin of the High Field Magnetic Degenerates? 4-32

References 4-33

5 Accreting White Dwarfs in Symbiotic Variable Binaries **5-1**

5.1 Introduction 5-1

5.1.1 Observational Evidence of Accretion Disks around the Accreting WD? 5-1

5.1.2 The Hot Components during the Quiescent Phase 5-2

5.1.3 Observed Characteristics of the Active Phase 5-3

5.2   Theoretical Studies of the Wind Accretion in Symbiotics   5-4

    5.2.1 Far-Ultraviolet Spectroscopy of Accreting WDs in Symbiotic Variables   5-13

    5.2.2 FUSE Spectroscopy of the Hot Component in Symbiotic Variables   5-14

    5.2.3 RW Hydrae   5-17

    5.2.4 EG And   5-20

    5.2.5 CQ Dra   5-21

    5.2.6 The Weak Symbiotic System Mira AB   5-23

5.3   The Outbursts of Symbiotic Variables   5-26

    References   5-27

6   Accreting White Dwarfs in the Helium-rich AM Canum Venaticorum Systems   6-1

6.1   Introduction   6-1

6.2   Characterizing the Observed Behavior of AM CV Systems   6-3

6.3   Theoretical Interpretation of the Observed Behavior of AM CVn Outbursts   6-6

6.4   The Formation of AM CVn Binaries   6-7

6.5   The Masses, Effective Temperatures, Chemical Abundances and Rotational Velocities of Accreting WDs in AM CVn Binaries   6-8

    6.5.1 International Ultraviolet Explorer (IUE)   6-10

    6.5.2 Hubble Space Telescope   6-11

6.6   Masses of AM CVn Accreting WDs   6-18

6.7   The Heating and Cooling of Accreting WDs in AM CVn Systems   6-19

6.8   Conclusion   6-25

    References   6-26

7   The Accreting White Dwarfs in Cataclysmic Variables   7-1

7.1   Introduction to Accreting White Dwarfs in Cataclysmic Variables   7-1

    7.1.1 Historical Context   7-2

    7.1.2 Overview of Cataclysmic Variable Outbursts, Evolution and CV Subtypes   7-3

    7.1.3 Spectroscopic Detection of the Underlying CV White Dwarf   7-10

7.2   CV White Dwarf Masses   7-12

    7.2.1 Radial Velocity Curves from Emission Line Velocities for White Dwarf Masses   7-12

    7.2.2 The Photometric Eclipsing Binary CVs   7-13

7.2.3 Gravitational Redshift Masses   7-13

7.2.4 Spectroscopic Modeling Methodology for CV White Dwarfs   7-15

7.3 Masses of the Accreting Magnetic White Dwarfs in Polars and Intermediate Polars   7-16

7.4 CV White Dwarf Temperatures   7-21

7.5 Rotational Velocities of CV White Dwarfs   7-24

7.6 Chemical Abundances in CV WD Photospheres   7-29

7.7 CV WD Second Flux Components and Boundary Layer Structure: A Mystery   7-34

7.8 Accreting Pulsating CV WDs   7-38

7.9 A New Era in Research on Cataclysmic Variable Accreting WDs   7-43

7.10 Structure, Evolution, Heating and Cooling of CV WDs   7-46

7.10.1 The Physics of Compressional Heating   7-46

7.10.2 Accreting WDs: Heating and Cooling in Response to Long-Term Accretion   7-50

7.11 Classical Novae, Recurrent Novae, Symbiotic Variables: The Single Degenerate Pathway to Supernovae Type Ia?   7-58

7.11.1 The Classical Nova Explosion   7-59

7.11.2 History of Quasi-Static and Hydrodynamic Studies of Classical Novae Explosions   7-61

7.11.3 Single Degenerate Pathway to Type Ia Supernovae   7-62

References   7-65

Appendix A   **A-1**

Appendix B   **B-1**

Preface

The astrophysical process of accretion is ubiquitous in our universe. It manifests itself in many observable ways in astrophysical systems with compact ojects. The types of objects undergoing accretion range from ultra-compact binaries, protostars, massive binaries, cataclysmic variables, symbiotic variables, X-ray binaries, to the supermassive black holes at the centers of galaxies and quasars. The timescales, rates of accretion, and released energies occupy a vast range in parameter space. Whether the central accreting object in an astrophysical system is a white dwarf star, a neutron star or a black hole, many aspects and consequences of accretion physics shed light on the accretion of mass and angular momentum in all types of binaries including accreting neutron stars and black holes

The direct unambiguous detection of an underlying accreting white dwarf photosphere posed a serious challenge. This is because accretion light generated by accretion disks, boundary layers and, in magnetic systems, the accretion of magnetically-confined, luminous accretion columns of superhot plasma, overwhelms the photospheric radiation. Hence, the cataclysmic variable field was dominated by studies of the properties and light variations of accretion disks, accretion columns, X-ray emitting boundary layers, the launching of astrophysical jets and driving of accretion disk winds.

In this book, development of the accreting white dwarf field is covered, starting with the earliest detections of the underlying white dwarf accretor in the far ultraviolet and in a few cases, the optical, both of which had a few skeptics at the outset. Since the accreting white dwarfs typically radiate their peak (Planckian) fluxes in the far ultraviolet and extreme ultraviolet, the stage was set to achieve major insights into the physics (and microphysics) of how the accretion process affects the mass accreting white dwarf components themselves, using space spectroscopy.

It is ironic that if you wish to observe an accreting white dwarf, the accretion disk, as fundamentally important as it is, must be absent or be faint due to very low accretion rates that one observes only during dwarf nova quiescence or during the deep low states of nova-like variables.

In this book, I focus almost exclusively on the accreting white dwarfs. They are the central engines of their observed outbursts, either as (1) potential wells for the release of gravitational energy during accretion (dwarf nova - DNe), (2) as the sites of explosive thermonuclear runaway (TNR) shell burning in classical novae, recurrent novae, some symbiotic variables, and symbiotic novae; (3) steady shell burning (supersoft X-ray binaries), helium accreting white dwarfs (AM CVn cataclysmic binaries, Type 1a supernovae) or; (4) collapse and thermonuclear detonation if an accreting WD with a C-O core reaches the Chandrasekhar limit (Type Ia supernovae). The accreting white dwarfs serve as probes of explosive evolution, accretion physics and diffusion, spinup due to the transfer of gas with angular momentum, since they contain the thermal, chemical and rotational imprint of their long-term accretion and thermonuclear history. All of these topics are covered including single accreting white dwarfs. Few scholars ever imagined that isolated white dwarfs are actually accreting the remnant debris of tidally or collisionally disrupted asteroids, comets and even planets from the white dwarfs' extant exoplanetary systems and therefore probing the exo-geochemistry of alien planetary systems.

While I've carried out some research in theoretical astrophysics, starting with my doctoral dissertation, I am primarily a space-based observational astrophysicist, studying the beating heart (the accreting white dwarf) in a wide variety of binary system types. I have emphasized research breakthroughs that utilized specific orbiting observatories: the International Ultraviolet Explorer (IUE), the Hopkins Ultraviolet Telescope (HUT), the Orbiting and Retrievable Far and Extreme Ultraviolet Spectrometers (ORFEUS), Hubble Space Telescope (HST), and the Far Ultraviolet Spectroscopic Explorer (FUSE). I have focused on the accreting white dwarf itself and how it responds to accretion over a wide range of accretion rates, accretion histories, accretion geometries and white dwarf masses. I have attempted to cover the most recent advances in the field of accreting white dwarfs. In doing so, I have tried to write this book from first principles in such a way that advanced undergraduates, graduate students, non-specialist astrophysicists and astrophysicists who specialize in white dwarf studies, will find topics of interest both in observational and theoretical astrophysics. While this research monograph is not intended as a textbook, it could be adapted, or used as a supplementary source, in a graduate level course.

In chapters of the book where I have contributed my own research on accreting white dwarfs in the refereed literature, I have tried to minimize self-plagiarizm. However, in some instances where I spent a lot of time finding the clearest way to convey the meaning of certain concepts, especially theoretical ideas, it seemed counterproductive. Many of the citations in my book were to refereed papers that I was writing within the same time interval that I was writing this book. In those cases, it was entirely reasonable to include my write-up in both places. I also made every effort to not plagarize the work of other authors and any similarity with other works is unintentional.

Throughout the book, I use cgs units, and terminology commonly used in the astrophysical literature. For novae and supernovae, I employ the term explosions. For dwarf novae, I employ the terms outbursts and superoutbursts. I entirely avoid the term eruptions. White dwarfs are sometimes alternatively referred to as degenerate dwarfs or simply degenerates. I have minimized the use of abbreviations or acronyms.

Acknowledgements

I am deeply grateful to the founder of the Department of Astronomy and Astrophysics at Villanova University, the late Rev. Edward Felix Jenkins, O.S.A., Ph.D, to my first Chairperson, Professor George P. McCook, to the late former president of Villanova, Rev. Edmund J. Dobbin, O.S.A, Ph.D, to the Associate Dean for Sciences, the late Dr. James Markham, and to Professor Edward L. Fitzpatrick, my second Chairperson, all of whom fostered a highly supportive atmosphere for research at Villanova as the University was transitioning to an international research university.

I am deeply grateful to my undergraduate mentor at the University of Kansas Department of Physics and Astronomy, Dr. Henry G. Horak (a Ph.D recipient under Nobel Laureate S. Chandrasekhar, Univ.of Chicago), to my Ph.D mentor at the University of Pennsylvania, Dr. Samuel C. Vila, for stimulating my interest in white dwarf theory and to Dr. James Liebert (University of Arizona), above all for our lifelong friendship since our Undergraduate years at KU, and for our early collaborations in optical observational astronomy of isolated white dwarf stars.

I am indebted to Professors Sumner Starrfield and Anne Cowley for making it possible for me to spend two scientifically very stimulating years with the Department of Physics at Arizona State University, the first year on Sabbatical leave from Villanova and the second year on a leave of absence from Villanova.

It is a pleasure to thank my former Postdoctoral Fellows at Villanova, Dr. Min Huang and Dr. Fuhua Cheng and a very special thanks to my longtime Research Associate, Dr. Patrick Godon, who joined me at Villanova in 2001 and has been fully supported continuously by NASA and NSF funding to Villanova University through the present time.

My deepest appreciation is extended to Dr. Gerard Vauclair (Observatoire Midi-Pyrenees, CNRS) for sabbatical support through the Centre National de La Recherche Scientifique during my 1990-91 sabbatical leave from Villanova at the Universite de Toulouse.

I am indebted to Dr. Howard Bond for Visiting Sabbatical Support from the Space Telescope Science Institute (Hubble) during my 1990-91 sabbatical leave from Villanova.

It is a genuine pleasure to thank my very closest scientific collaborators starting with James Liebert, Sumner Starrfield, Paula Szkody, Warren Sparks, Howard Bond, Jay Holberg, George McCook, Terry Oswalt, Joanna Mikolajewska, Boris Gaensicke, Anna Pala, Gordon Hammond and Ed Guinan and to multitudes of Villanova students who have collaborated with me in research over the years including former Goldwater Scholars, Drs. Lisa Winter and Joleen Miller, Joel Urban, Dr. Ryan Hamilton, Dr. David Turnshek, Dr. John Bochanski, Dr. Kelly Kolb, Conor Larsen, Matt Fritz, Joseph McMullen, Dr. Kunegunda Belle, Dr. Adric Riedel, Dr. Ron Ballouz, Dr. Quyen Nguyen, Giannina Guzman, Craig Kolobow, Emma Carlson, Hayley Nofi and Anthony McCarthy.

I am deeply grateful to Drs. Ben Zuckerman, Jay Farihi, Paula Szkody, John Debes, Patrick Godon and Sumner Starrfield for critical readings of draft chapters and helpful constructive comments.

I am deeply grateful to the following individuals for their invaluable proof reading, editing, advice and encouragement: Marion McLauchlan, Rachel Price, Prapti Thapaliya, Emma Carlson, Harriet Shapiro, Michael E. Sion and Melanie K. Sion MD.

I am extremely grateful to the IOP Commissioning Editor Leigh Jenkins, to IOP Books Production Team member Chris Benson and Editorial Assistants, Calum Sims, Rory Weaver, and Isabelle Defillion for their kind assistance and patience during the writing of this much delayed book.

Author biography

Edward M Sion

Edward M. Sion is an American Astrophysicist who specializes on the structure and evolution of white dwarf stars, including isolated white dwarfs with no companion and white dwarfs in interacting binary systems where mass is transferred from a companion star onto the white dwarf, either through Roche lobe overflow, or through capture of stellar wind emitted by the companion star. Sion was born and raised in Wichita Kansas on January 18th, 1946.

His undergraduate education was at the University of Kansas under the tutelage of Professor Henry G. Horak (a Ph.D recipient under Nobel Laureate Subramanyan Chandrasekhar at the Univ. of Chicago). Sion did his graduate work leading to the Ph.D at the University of Pennsylvania under Professor Samuel C. Vila. His doctoral dissertation in theoretical astrophysics was titled "Mult-Modal Non-adiabatic Radial Oscillations and Pulsational stability of Hot Degenerate Dwarfs" He was hired at Villanova University in 1975 and, in 1977, Sion co-authored the Catalogue of Spectroscopically Identified White Dwarf Stars with George P. McCook. His research has been externally funded without interruption by NSF and/or NASA every year from 1978 up to the present. In 1983, Sion led a team of collaborators who developed the fundamental classification system of white dwarf stars. This system is used worldwide, and characterizes both the chemical composition class and surface temperature of every spectroscopically identified white dwarf star. In 1984, Sion uncovered empirical evidence that the hydrogen-rich DA white dwarfs transform into helium-rich DB white dwarfs when deepening helium convection, as a white dwarf cools, mixes hydrogen downward and dilutes it. In 1989, Sion was honored with the Outstanding Faculty Research Award at Villanova University. In the mid-1990s, Sion led a team of collaborators, using the Hubble Space Telescope, to unveil the physical properties of white dwarfs in explosive cataclysmic variables and how they cool and heat in response to the accretion of mass from a sun-like companion star, a process which leads to thermonuclear explosions as classical novae and, for the most massive accreting white dwarfs, Type Ia supernova explosions (the so-called "cosmological" supernovae). In 1995, Sion showed that compressional heating due to the accretion of matter by a white dwarf in a cataclysmic variable explains their post-outburst cooling response to enhanced accretion during a dwarf nova outburst. From 1996, Sion served as an Associate Editor of The Astrophysical Journal for six years. In 2007, the Government of Lebanon founded a non-profit in close association with the French Academy of Sciences, which would become their National Academy of Sciences. Sion was invited to be founding member of the Lebanese Academy of Sciences. His parents were both born in the USA but his paternal and maternal grandparents immigrated to the USA from Lebanon in 1902. Of his 642 scientific publications (270 peer-reviewed in the top five international journals), Villanova students have merited lead authorship/ co-authorship on 64 peer-reviewed papers as well as 67 poster presentations at Meetings of the American Astronomical Society. In a study conducted by Stanford University in 2019 on the scientific impact of all research scholars, he is ranked in the top 2% of all researchers in all fields of scholarly research world-wide, measured by the impact a research scholar has had on their field of research. In 2020, Sion was bestowed the Outstanding Faculty Research Mentor Award at Villanova University.

Aside from his devotion to his family and his research, Sion is a passionate student of American and European history as well as the history and archaeology of the Mediterranean region. He is also an avid sports fan of collegiate and professional basketball, American football and major league baseball.

Accreting White Dwarfs
From exoplanetary probes to classical novae and Type 1a supernovae
Edward M Sion

Chapter 1

A Brief Introduction to the Physics of Non-accreting White Dwarfs

The white dwarfs (WDs) are the terminal evolutionary state of 98% of the roughly 200 billion stars in our Milky Way. They show us what the Sun will become in the future, the history of star formation in our Galaxy, a lower limit to the age of our Galaxy as well as to the age of the universe. They also provide the white dwarf supernovae (Type Ia) that present the evidence that the expansion rate of the universe is accelerating and the existence of dark energy. They are extremely dense because they contain the mass of an average main-sequence dwarf (like the Sun) compressed into a volume the size of a rocky, metallic planet like Earth. Therefore, they offer the opportunity to study the physical behavior of matter at extreme densities not achievable in terrestrial laboratories. To better understand and appreciate the physical phenomena and processes, both non-explosive and explosive, that are the consequences of accretion onto white dwarfs, it will first prove useful to acquaint the reader with the physical processes that operate in isolated white dwarfs without accretion. An excellent starting point for this examination of the underlying physics of non-accreting white dwarfs is to first gain a historical perspective of their study.

1.1 A Historical Background of the Study of White Dwarfs

From the time that astrometric observations by Friedrich Bessel in 1844 showed that both of the bright main sequence stars Sirius and Procyon must each have "invisible" companion stars of roughly their own mass, the first actual spectrum of a WD, 40 EriB, was obtained by Williamina Paton Fleming and Edward C. Pickering at the Harvard College Observatory in 1910. It revealed spectral features of hydrogen associated with a very hot star. But for the "invisible" companions to be so hot and yet so faint, they had to be extremely small and hence unimaginably dense. At these densities, the matter is so compressed that the electrons form a separate gas from the bare nuclei and are forced to move very fast and cannot slow down (e.g., by emitting a photon. Otherwise,

doi:10.1088/2514-3433/ac930cch1   1-1

they would violate the Pauli Exclusion Principle formulated in 1925 by Wolfgang Pauli that no two electrons sharing the same four quantum numbers can occupy the same quantum state. The Pauli Principle plus the quantum-statistical theory of the degenerate electron gas published by Enrico Fermi and P. A. M. Dirac in the mid-1920s set the stage in 1930, for the work of Sir Ralph Fowler, to show that a stellar mass of electron-degenerate matter could be supported against gravitational collapse by this "quantum mechanical" pressure. His student Subhramanyan Chandrasekhar, in 1930 worked out the equations that describe the structure of a WD, combining quantum mechanics and relativity. For this work, he co-shared the Nobel Prize in Physics in 1983 with William Fowler of Caltech.

As demonstrated below (see Section 1.3), the more massive these bizarre objects are, the smaller their radii. The Pauli Principle requires this because the higher the mass of a degenerate star, the more crowded together and compact the electron gas must be in order to provide sufficient electron degeneracy pressure to support increasingly massive degenerate stars against gravitational collapse. The electron degeneracy pressure holding up the WD against gravitational collapse exceeds, by many orders of magnitude, the non-degenerate pressure of the ideal gas of ions in the core.

There are two fundamental properties of WDs which make them highly explosive. First, as seen below, the pressure of electron-degenerate matter depends only upon the density. Within the context of an accreting WD discussed in later chapters, this insensitivity to temperature of the equation of state for complete electron degeneracy underlies much of the explosive phenomena observed in accreting WDs (see Section 1.2); the classical nova explosion (Mestel 1952; Starrfield 1971), thermonuclear shell flashes in symbiotic stars and recurrent novae, the helium flash, the carbon flash, and thermal pulses on the asymptotic giant branch (AGB). (Note however that thermal pulses can occur on the AGB without electron degeneracy playing a role. Thermal pulses are driven by the Schwarzschild & Harm 1965 thin shell thermal instability.) Under degenerate conditions, when thermonuclear energy is generated in degenerate layers, there is no compensating expansion to cool and therefore quench the instability. Thus, the temperature rises rapidly, making the nuclear reactions go faster and faster until so much thermonuclear energy is quickly built up that the layer explodes. This is the underlying mechanism of classical nova explosions (CNe) discussed in Chapter 8.

Second, there is a limit to how high the mass of a WD can be before hydrostatic equilibrium can no longer be maintained; i.e., before electron degeneracy pressure can no longer support the WD against instantaneous gravitational collapse. This limiting mass of a WD is known as the Chandrasekhar limit (see Section 1.5). The instantaneous dynamical collapse of WDs that reach this limit is the underlying mechanism that triggers the explosions of Type Ia supernovae (SNe Ia), the so-called cosmological beacons (see Chapter 7).

1.2 Electron Degeneracy

Any discussion of WD structure naturally begins with a summary of electron degeneracy. First, electrons are spin-1/2 particles (fermions) whose motions are dominated by the Pauli Exclusion Principle. At high temperatures, the particle momenta and energies are greater and many quantum states are available for occupation but at lower temperatures, the fermions have less energy, and hence fewer quantum states are available. Thus, the average occupation number of each state increases. Due to the Pauli Exclusion Principle, the available levels up to some maximum energy, which is dependent on the density, are, on average, nearly filled; higher levels are, on average, nearly empty. This is the condition known as electron degeneracy.

Consider the number of particles per unit volume v with momenta in the range of p and $p + d(p)$

$$n_e(p)dp = \frac{8\pi}{h^3}p^2\,dp. \tag{1.1}$$

For a particle number density of electrons $n_e = N_e/V$, and allowing all values of particle momentum from 0 to infinity, then n_e would no longer be finite unless there is a ceiling (or threshold) momentum. For n_e to be finite, there must be a maximum value of momentum p_0 above which very few electrons can have higher values than p_0. At zero temperature, the electron gas is completely degenerate and this threshold momentum is known as the Fermi momentum. In other words, all electrons occupy energy states at or below the threshold Fermi energy, namely

$$E_f = \frac{\hbar^2}{2m_e}\left(3\pi^2\frac{N_e}{V}\right)^{2/3} \tag{1.2}$$

where again N_e is the number of electrons and V is the volume. The Fermi momentum expressed in terms of the Fermi energy is

$$p_f = (2m_eE_f)^{1/2} = \frac{h}{2}\left[\frac{3}{\pi}n_e\right]^{1/3}. \tag{1.3}$$

This assemblage of degenerate fermions is known as the Fermi sea of degenerate electrons. Ideally, the surface of the Fermi sea of degenerate electrons separates the unfilled states above the surface of the Fermi sea from the filled states in the Fermi sea. The degree of departure of the degenerate electron gas from the classical ideal equation of state is measured by a degeneracy parameter η, which is the ratio of the chemical potential μ to the Fermi energy (cf. Menzel et al. 1963)

$$\eta = \frac{\mu}{E_f}. \tag{1.4}$$

The chemical potential, μ, is the energy of the most energetic particle in a degenerate system of Fermions at zero temperature. All states with $E < \mu$ are occupied; all states with $E > \mu$ are unoccupied. When $\eta \gg 1$, indicates a highly degenerate electron gas. For $\eta \ll 1$, then the electrons behave like an ideal gas. At $T = 0$, there is complete electron degeneracy as stated earlier. The reader is cautioned that using μ as the chemical potential should not be confused with μ_e, the electron mean molecular weight or μ_t, the total mean molecular weight.

If we assume an isotropic distribution of electrons and utilize a spherical momentum distribution function in three dimensions, x, y, z, as a sphere having radius p_0 with no electrons outside the sphere with momenta $p > p_0$, then integrating over the spherical momentum distribution from $p = 0$ to $p = p_0$, we have the number of electrons per cubic centimeter

$$N_e = \frac{8\pi}{3h^3}p_0^3 \qquad (1.5)$$

or in terms of the threshold momentum,

$$p_0 = [(3h^3/8\pi)N_e]^{1/3}. \qquad (1.6)$$

The electron number density N_e relates to the density ρ as

$$N_e = \frac{1}{\mu_e}\frac{\rho}{H} \qquad (1.7)$$

where μ_e is the electron mean molecular weight $\frac{1}{\mu_e}\rho = X + \frac{1}{2}Y + \frac{1}{2}Z = \frac{1}{2}(1 + X)$ with X, Y, and Z being the usual mass fractions by weight of hydrogen, helium, and metals respectively.

The electron pressure P_e likewise follows from integrating over the spherical momentum distribution function as

$$P_e = \int\int\int_{p=0}^{p=p_0} p_x v_x \frac{z}{h^3} dp_x dp_y dp_z \qquad (1.8)$$

where v_x is the electron velocity in this spherical momentum distribution and gives the rate of momentum transport of a cubic centimeter of momentum through a square centimeter. Substituting $v_x = p_x/m$ and integrating

$$P_e = \frac{8\pi}{15mh^3}p_0^5. \qquad (1.9)$$

Using (1.5) which expresses N_e to the Fermi Momentum p_0 and (1.7) to eliminate N_e, we have the equation of state for a non-relativistic completely degenerate electron gas

$$P_e = K_1 \frac{1}{\mu_o} \rho^{5/3} \qquad (1.10)$$

with $K_1 = \frac{h^2}{20mH}\left(\frac{3}{\pi H}\right)^{2/3} = 9.91 \times 10^{12}$.

If the electrons have velocities that are predominantly relativistic, then the electron velocity component in the pressure equation is replaced with the velocity of light, namely $v_x = c(p_x/|p|)$. Carrying out the integration yields the electron pressure for the relativistic case in terms of the Fermi momentum

$$P_e = \frac{2\pi c}{3h^3} p_0^4. \qquad (1.11)$$

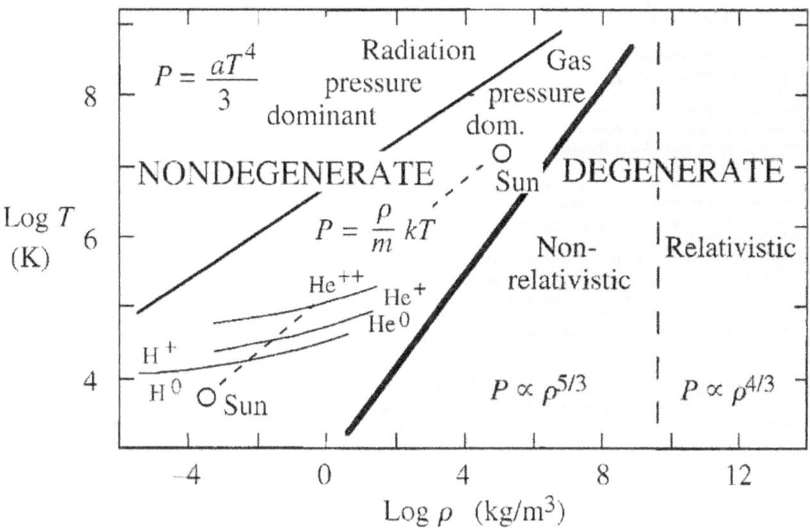

Figure 1.1. The four domains characterizing the state in a phase diagram, Temperature versus Density. (Adapted from Schwarzschild (1958). © Princeton University Press 1958.)

Hence, the relativistic electron degeneracy equation of state is

$$P = K_2 \frac{1}{\mu_e} \rho^{4/3} \tag{1.12}$$

$$K_2 = \frac{hc}{8H}\left(\frac{3}{\pi H}\right)^{1/3} = 1.23 \times 10^{12}. \tag{1.13}$$

The four domains (radiation pressure, ideal gas pressure, non-relativistic degeneracy pressure, relativistic degeneracy pressure) characterizing the state of the electron gas in the phase diagram ($\log T$ versus $\log \rho$) are shown in Figure 1.1 (from Schwarzschild 1958). The boundaries separating each domain mark where the pressure in each domain is just equal to the pressure in the adjacent domain.

1.3 The Inverse Relation between WD Mass and Radius

Starting with the hydrostatic equilibrium condition $\frac{dP}{dr} = -\frac{\rho G M_r}{r^2}$, we replace $M_r = \frac{4}{3}\pi r^3 \rho$ and assume that $P = 0$ at $r = R$. Integrating, we have

$$P(r) = \frac{2}{3}\pi G\rho^2(R^2 - r^2). \tag{1.14}$$

Taking Equation (1.10) for the non-relativistic electron degeneracy equation of state and equating it with the above expression for $P(r)$ with $r = 0$,

$$\frac{2}{3}\pi G\rho^2 R_{wd}^2 = K_1(\rho^{5/3}) \tag{1.15}$$

and replacing ρ with $\frac{M_{wd}}{\frac{4}{3}\pi R_{wd}^3}$, we see that $R_{wd} \approx M_{wd}^{1/3}$ and hence $R_{wd}^3 M_{wd} = $ constant or $M_{wd}V_{wd} = $ constant; **i.e., the larger the mass of the WD, the smaller its radius.**

1.4 The Chandrasekhar Limit

To obtain an estimate of the critical mass of a WD at the Chandrasekhar limit, the electrons will be moving close to the speed of light c, the electron degeneracy will be relativistic and therefore the appropriate equation of state is the relativistic one. Following Ostlie & Carroll (2007), the relativistic electron degeneracy pressure is given by

$$P = \left(\frac{(3\pi^2)^{1/3}}{4}\right)hc\left[\frac{Z}{A}\frac{\rho}{H}\right]^{4/3}.$$

If we equate the approximate expression (1.14) shown earlier for the central pressure with $r = 0$ at the center, to the relativistic degeneracy pressure (1.16) above, and replace the density ρ with $M_r = \frac{4}{3}\pi r^3 \rho$, then the radius of the WD cancels. Solving for the maximum WD mass (the Chandrasekhar Limit), we obtain an approximate expression for the Chandrasekhar mass

$$M_{ch} \approx \frac{3\sqrt{2\pi}}{8}\left(\frac{\hbar c}{G}\right)^{3/2}\left[\frac{Z}{A}\frac{1}{H}\right]^2. \tag{1.16}$$

For $Z/A = 0.5$, a carbon–oxygen core, then the Chandrasekhar Limit is $1.44 M_\odot$.

1.5 Core Compositions Versus WD Mass

The mass of a non-accreting isolated WD is quantitatively correlated with the initial mass of its progenitor main sequence star via the initial mass–final mass relation (IMFR). As evolution advances to the double shell burning, AGB phase, the AGB stars undergo thermal pulses, alternately in their thin He-burning and H-burning shells. Heavy mass loss occurs in the form of thermal pulse superwinds, planetary nebulae ejections that reduce their total envelope masses by a huge amount leaving the hot, exposed central core (CSPN), which is already, or soon to become, a WD. The core mass on the AGB is given approximately by the Paczynski–Uus core mass–luminosity relation (Paczynski 1970), viz.,

$$L/L_\odot \approx 6 \times 10^4 \ (M_{core}/M_\odot - 0.5).$$

Starting with the pioneering work by Volker Weidemann (1977, 2000), up to 2018, the efforts have been fraught with significant observational scatter. Most recently, Cummings et al. (2018) have used 79 WDs from 13 star clusters to derive an IMFR with greatly reduced scatter. Their IMFR covers the progenitor mass range from $0.85 M_\odot$ to $7.5 M_\odot$. Their IMFR plots are non-linear show different slopes over different progenitor mass ranges. They show that the total stellar mass loss that occurs ranges from 33% of a progenitor star's initial mass at $0.8 M_\odot$ to 83% of the initial progenitor mass at $7.5 M_\odot$.

It has generally been known from stellar evolutionary models followed to the WD stage that three different core compositions must exist, helium (He) cores, carbon–oxygen (C–O) cores and, in the most massive WDs, oxygen–neon–magnesium (ONeMg) cores. While progress in this area is model dependent, the WD mass ranges corresponding to these core compositions have been explored more recently by G. Lauffer et al. (2018). The lowest mass WDs $M_{\mathrm{wd}} < 0.4 M_\odot$ have He cores, while WDs in the mass range $0.4 < M/M_\odot < 1.05\ M_\odot$ have C–O cores with varying amounts of oxygen (via alpha particle captures). For $M_{\mathrm{wd}} > 1.05 M_\odot$, cores of ONeMg are formed. The lowest mass helium WDs had to be formed through binary evolution since the age of the galactic disk is not old enough for any low-mass single main sequence star to have had time to evolve to a helium WD. At the other mass extreme with progenitor masses $>1.05\ M_\odot$, WDs with masses larger than $\sim 1.05\ M_\odot$ may reach core temperatures high enough for stable carbon burning (Garcia-Berro et al. 1997; Siess 2007), forming either an ONe core or ONeMg core, the former if carbon burning ignites off-center and the latter core if carbon ignites at the center of the star. It is also possible that the carbon burning front may not reach the center, in which case a WD with a hybrid carbon–oxygen plus oxygen–neon core will form (Denissenkov et al. 2013).

1.6 Cooling Evolution

Isolated WD stars, without any interactions with nearby stellar neighbors or close companions, undergo a relatively simple cooling evolution, gradually leaking the thermal energy of the ions in their electron-degenerate cores into space while the atomic (and eventually molecular) opacity of their outer, non-degenerate, surface layers, blanketing the core and only 60 to 100 km thick, regulates their rate of heat loss. This cooling rate, which also depends on WD mass, determines the timescale of cooling evolution. Because of the highly efficient electron conduction, the core of a WD is essentially isothermal. In a non-accreting, isolated WD absent of any nuclear burning of residual hydrogen in its envelope, the only source of luminosity is the thermal energy of the non-degenerate nuclei in the core. Accordingly, we have

$$L = -\frac{d}{dt}\left(\frac{\frac{3}{2}kTM_{\mathrm{wd}}}{\mu_{\mathrm{t}}H}\right) \tag{1.17}$$

where M is the WD mass and T the temperature of the isothermal core. If we use a simple relation between the luminosity and the temperature in the transition region, T_{tr}, (from a "radiative zero" envelope solution) at the envelope-core boundary to replace L in (1.16), namely $L = 2 \times 10^6(T_{\mathrm{tr}}^{3.5})$, and integrate (1.16), we obtain the WD cooling time

$$\tau_{\mathrm{cool}} = \frac{\left(\frac{3}{2}kTM_{\mathrm{wd}}\right)/(\mu_{\mathrm{t}}H)}{2.5 L_{\mathrm{wd}}}. \tag{1.18}$$

Of course, when we introduce accretion onto WDs in subsequent chapters, we will see that in compact systems like cataclysmic variables, a WD has a continual heat

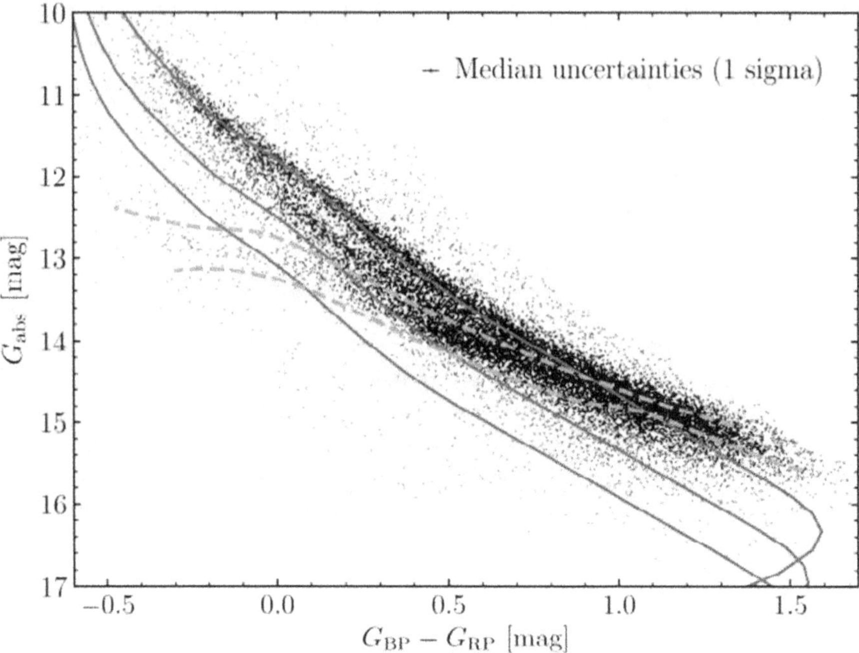

Figure 1.2. 15,000 WDs observed with Gaia within 100 pc. This data, shown as black dots, plotted on the HRD, is from the Gaia 2nd Data Release. The cooling evolutionary tracks shown in blue correspond to WD masses of 0.6, 0.9, and 1.1 Msun, respectively from the theoretical model sequences of Tremblay et al. (2019). Note the aggregation of WDs shown between orange lines. This is the first empirical evidence of the onset of core crystallization as the WDs cool below the Debye temperature. This effect has been predicted theoretically since 1968 but this represents the first observational confirmation of core crystallization in a cooling WD. (Reproduced with permission of Springer from Tremblay et al. (2019). Copyright © 2019, Springer Nature Limited.)

source shining inward from the boundary layer between the inner accretion disk and the WD photosphere as well as compressional heating (and possibly advective heating) by the accreted matter. This process resets the WD's cooling clock. If one takes advantage of the precise Gaia parallax data for thousands of WDs with Gaia-derived distances, then the cooling evolution tracks of the non-accreting WDs can be displayed on a Hertzsprung–Russell diagram (HRD) as seen in Figure 1.2.

1.7 WD Atmospheres and Envelope Compositions

Evolutionary models reveal that following the thermally pulsing AGB stage and its associated heavy mass loss, the emerging WD begins its cooling evolution with a hydrogen envelope mass of 10^{-4} $M_\odot$ below which the helium layer mass is in the range of 10^{-2} $M_\odot$ to 10^{-3} $M_\odot$. The photosphere of a WD where the continuum radiation arises and where the spectral lines form is only a few kilometers thick, comprising a mass of 10^{-16} $M_\odot$. Yet it is optical observations of this shallow layer that provide the physical properties of the atmosphere, the T_{eff}, surface gravity, log g, the projected rotational velocity, Vsini, and the chemical abundances of the atomic species.

Due to their high surface gravity, the spectra of WDs are typically mono-elemental with the lightest abundant element, either H or He, floating upward to the top. The spectra are among the simplest known, almost boring. However, these spectra yield information on several physical processes that control the flow of elements in high gravity atmospheres. Among these processes are gravitational and thermal diffusion, radiative forces levitation, possible weak wind outflows, convective mixing and dredge-up, and possible diffusion-induced nuclear burning. Each of these physical processes is briefly discussed in Sections 1.6 and 1.8–1.10. The important role of some of these processes depend on the changing temperature regimes through which the WD cools. These same processes help give rise to a puzzling zoo of spectroscopic sub classes among the H-dominated stars and the helium-dominated stars.

Following the discovery of the first spectroscopically identified white dwarf, 40Eri B (see Section 1.1), the number of such spectroscopically identified objects grew to 38 known degenerate stars by 1941. These 38 objects were tabulated by G.P. Kuiper (1941) in the Publications of the Astronomical Society of the Pacific. Mindful that white dwarf spectra bore only a vague resemblance to the spectra of normal main sequence stars, Kuiper classified each of these objects with a letter w denoting white (for white dwarf). He assigned the letters A, B, F, G as well as the spectral designation 'con' for white dwarfs revealing no line features in low resolution spectra. This provisional classification by Kuiper (1941) was further developed by Luyten (1952) who introduced upper case D for degenerate and assigned DB for the spectral type of the extremely helium-rich white dwarfs. The presently used system (Sion et al. 1983) was developed as a substantial refinement to the classifications of Greenstein (1960). The first published discussion of the evolutionary implications posed by the surprising variety of white dwarf spectra was due to Greenstein (1956, 1960). One of the many puzzling problems was the origin of the nearly pure helium DB stars which led to early published investigations of this problem by Strittmatter & Wickramasinghe (1971), Sion (1971, 1973) and Shipman (1971).

The optical spectra of WDs reveal two distinct composition classes, the DA stars (hydrogen dominated) comprising 80% of the isolated WDs and the non-DA stars (helium dominated) comprising the remaining 20%. The reason for this divide in composition was one of the major unsolved problems in stellar evolution (Greenstein 1965) and while much progress has been made, some aspects of it are still not definitively understood. Examples of the optical spectra of DA stars covering a broad range of temperature are displayed in Figure 1.3. The optical spectra of the most populous non-DA stars, the nearly pure helium DB WDs, are displayed in Figure 1.4.

The spectroscopic classification system of WDs encompassing all known spectroscopic types (Sion et al. 1983) is given in Table 1.2 along with their respective temperature ranges (Table 1.1).

PG1159 stars have high excitation spectral transitions and C and O but are He-dominated, and in some cases, e.g. H1504, C-dominated and, in at least one case, O-dominated. They are the hottest WD with T_{eff} up to 200,000 K. PG 1159 stars (McGraw et al. 1979) are formed by a born-again episode (Schoenberner 1979; Iben et al. 1983) in which a helium flash occurs during the cooling sequence (i.e., very late thermal pulse) forcing the star to evolve back to the AGB. As a result, this very late thermal pulse burns out all hydrogen left in the atmosphere of the star.

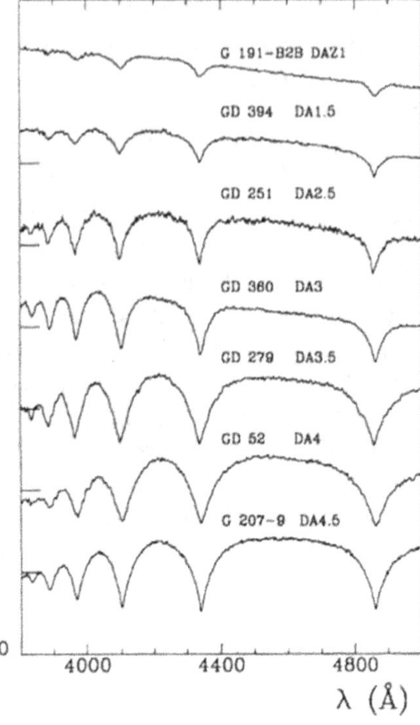
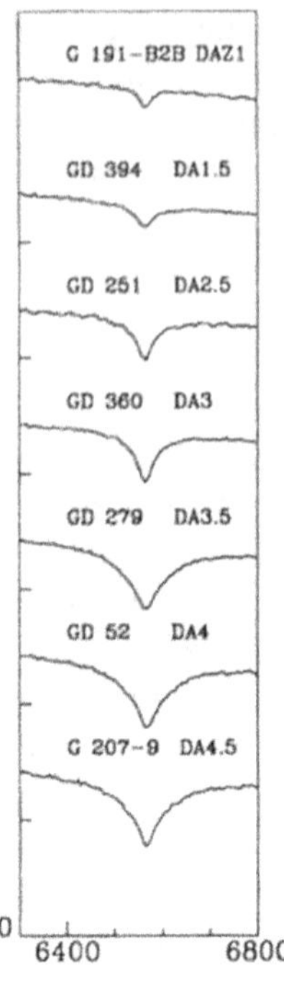

Figure 1.3. Optical spectra of stars at the hot end of the DA sequence (Table 1.1). The blue region is in the left panel, Hα in the right panel. The stars are ordered with decreasing effective temperature from top to bottom. The spectra are characterized by strong Balmer lines, generally visible up to H8 or H9. In this figure, and those that follow, the spectra are shifted vertically for clarity; the zero level of the lowermost plot is always the bottom of the figure. Plots above the lowermost one have zero points indicated by the various long tick marks. (Reproduced with permission from Wesemael et al. (1993). © 1993. The Astronomical Society of the Pacific. All rights reserved.)

The surface temperature of WDs is derived from color or synthetic spectral fits. Following the secondary letter and any additional symbols, a quantitative temperature index from 0 to 9 is used, defined by 10 times θ $\left(\theta = \frac{5040}{T}\right)$. This quantity $\frac{50,400}{T}$ given as an integer, is a temperature and luminosity indicator.

Hybrid composition WDs (e.g., DBA, DAZ, DAO, etc.) are classified by using a primary symbol for the strongest spectral features followed by secondary symbol for spectral features of other elements present. An atlas displaying examples of the optical spectra of virtually all of the spectral types on this system, including the hybrid spectra types, is presented by Wesemael et al. (1993).

1.8 Gravitational and Thermal Diffusion

Each type of ion in a WD atmosphere responds to the gravitational, electric, and radiation fields to which it is subjected by reaching a diffusive equilibrium locally. The flow of elements stops when this equilibrium is achieved and thus gives rise to a concentration

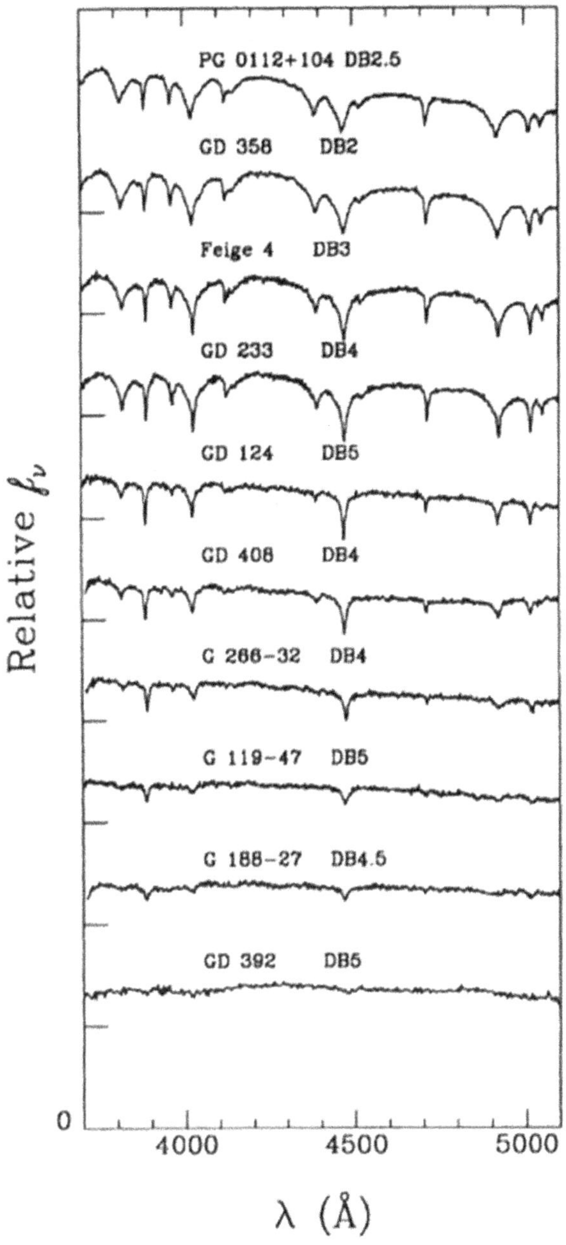

Figure 1.4. Optical spectra of classical DB stars (Table 1.4). The stars shown form an approximate temperature sequence, with the hottest one PG 0112+104, the hottest known DB star at the top and the coolest one, GD 392 a DB5 star with barely visible He I 4471 at the bottom. (Reproduced with permission from Wesemael et al. (1993). © 1993. The Astronomical Society of the Pacific. All rights reserved.)

Table 1.1. Definition of Primary Symbols: Primary Spectral Type Characteristics

DA	Only Balmer lines of H; no He I or metals present	(130,000 K–3000 K)
DB	He I lines; no H or metals present	(45,000 K–12,000 K)
DC	Continuous spectrum no lines deeper than 5, in any part of the electromagnetic spectrum	(15,000 K–3000 K)
DO	He II strong; He I or H present	(120,000 K–45,000 K)
DZ	Metal lines only; no H or He lines	(10,000 K–3000 K)
DQ	Carbon features, either atomic or molecular in any part of the electromagnetic spectrum	(25,00 0K–3000 K)

gradient term in the basic diffusion equation (see Michaud 1977). The mean diffusion velocity of an element is the average of the diffusion velocities of individual ions of that element (Montmerle & Michaud 1976; Vauclair et al. 1979). It is given by

$$\bar{V} = \bar{D}\left[\frac{-1}{c}\frac{dc}{dr} + \overline{k_p}\frac{1}{P}\frac{dP}{dr} + \overline{k_T}\frac{1}{T}\frac{dT}{dr} + mg_{rad}\right] \tag{1.19}$$

where the first term on the right-hand side is the concentration gradient, the second term is the gravitational diffusion, the third term is the thermal diffusion, and the last term takes into account the selective radiative acceleration of ions important in hot WDs. g_{rad} is the radiative acceleration discussed in Section 1.9, and m is the mass of the ion. The mean diffusion coefficient $\bar{D}$ is the summation of the diffusion coefficients of each individual ion diffusing through the background medium, namely

$$\bar{D} = \sum_i \frac{N_i D_i}{N}$$

where the D_i are the individual diffusion coefficients of each ion, N_i is the number density of each ion, and N is the total number density. The coefficients k_p and $\overline{k_T}$ bar depend on the ratio of the mass of the diffusing ion to the mean mass of the dominant background gas through which the ion is diffusing, and the mean electric charge of the medium and the electric charge of the diffusing ion.

For the diffusion of an ion in a fully ionized background gas, it is often convenient to use the Chapman & Cowling (1970) diffusion coefficient (Fontaine & Michaud 1979). In that case, the relative diffusion velocity v_g of a trace element 2 in a main constituent 1 due to gravitational diffusion alone (the pressure gradient term) is given by

$$v_g = D_{12}\left[\frac{m_2}{m_1}(1 + Z_1) - Z_2 - 1\right]\frac{\rho g}{P} \tag{1.20}$$

where D_{12} the Chapman and Cowling diffusion coefficient is given as

$$D_{12} = \frac{3(2kT)^{5/2}}{[16[(\pi m_1 m_2)/(m_1 + m_2)]^{1/2} n_1 Z_1^2 Z_2^2 e^4 A_i(2)]}$$

$$A_1(2) = \ln(1 + X_D^2)$$

$$X_D^2 = \frac{16k^2T^2\lambda_D}{(Z_1^2Z_2^2e^4)}$$

with the Debye length

$$\lambda_D = \frac{kT}{4}\pi e^2 \sum_i^{n_i}(z_i^2)^{1/2}.$$

The total relative diffusion velocity can be written as:

$$v_t = v_g(1 + f) \tag{1.21}$$

where f gives the relative importance of thermal diffusion to gravitational diffusion. Then the e-folding time of the abundance of element θ is the e-folding time:

$$\Theta = \frac{g}{4\pi G}\left(\frac{\Delta M/M}{\rho v}\right), \tag{1.22}$$

where $\Delta M/M$ is the fractional mass of the convection zone. The diffusion timescale is given by $\theta_g = \theta(1 + f)$ (Fontaine & Michaud 1979).

In Section 1.9, selective radiative forces are discussed which become important in hot WDs ($T_{\text{eff}} > 20, 000$ K), and are exerted on the atoms of a given species by bound-bound and bound-free transitions (Michaud 1970).

1.9 Radiative Acceleration

Selective radiative forces can be exerted on the atoms of a given species by bound-bound and bound-free transitions (Michaud 1970). The largest possible radiative acceleration can be transmitted to unabundant elements under the following favorable circumstances: the lines are unsaturated, and the transitions must have large f-values (transition probabilities) at or near the Planckian "peak" of the WD's flux distribution (Fontaine & Michaud 1979; Vauclair et al. 1979; Chayer et al. 1995). The pioneering calculations of radiative acceleration were carried out by Vauclair et al. (1979) and Fontaine & Michaud (1979) who demonstrated that traces of C, N, and O can be radiatively levitated through bound-bound absorption at the surface of hot WDs. They found that the line spectra of these elements can generally absorb enough momentum from the radiation field of hot WDs to oppose gravitational diffusion. Extensive calculations of radiative acceleration were presented by Chayer et al. (1995). Their computations were based upon the following expression:

$$g_i^{\text{rad}} = \frac{4\pi}{c}\frac{1}{X(A_i)}\sum_j\sum_{k>j}\int_0^\infty \kappa_i^{jk}(\nu)H_\nu d\nu \tag{1.23}$$

where c is the light speed, $X(A_i)$ is the mass fraction of element A in the ith ionization state, H_ν is the monochromatic Eddington flux ($H_\nu = F_\nu/4\pi$), and κ_i^{jk} is given by

$$\kappa_i^{jk} = \frac{N_i^j}{\rho} \frac{\pi e^2}{mc} f_i^{jk} \phi_i^{jk}(\nu)(1 - e^{-h\nu/kT}),$$

where m and e are the electron mass and charge respectively, ρ is the density, f_i^{jk} is the transition probability (oscillator strength of the bound-bound transition $j \to k$ for ions of type i, $\phi_i^{jk}(\nu)$ is the line broadening function and the factor $(1 - e^{-h\nu/kT})$ is the usual correction for stimulated emission (negative absorption) of the ion transition. For the line broadening, collisional broadening dominates in a WD over natural broadening.

On the other hand, selective radiative forces can be exerted on the atoms of a given species by bound-bound and bound-free transitions (Michaud 1970). The theory of radiative levitation has to be tested through direct numerical comparisons of observed photospheric chemical abundances of WDs and the theoretical abundances. It cannot be ruled out that weak selective ion winds are also driven which could complicate the sought after for agreement between observed and predicted abundances. For WDs with $T_{\mathrm{eff}} < 20,000$ K, radiative acceleration is unimportant.

1.10 Convective Mixing and Dredge-up

As a WD cools, convection zones develop when temperature gradients in the envelope, often due to partial ionization followed by recombination, exceed the adiabatic temperature gradient. The convection zones can deepen all the way down to the transition region between the non-degenerate envelope and the degenerate core where convective dredge-up brings gas to the surface.

In its simplest form, the criterion for stability against convection compares the actual temperature gradient, $\frac{dT}{dr}$ in a layer to the adiabatic temperature gradient $\left(\frac{dT}{dr}\right)_{\mathrm{ad}} = -\frac{g}{c_{\mathrm{p}}}$, where c_{p} is the specific heat at constant pressure, g is the acceleration of gravity and the convective blobs are assumed to rise and sink adiabatically with no heat exchange between the bubble and the surrounding gas. If $-\frac{dT}{dr} > -\left(\frac{dT}{dr}\right)_{\mathrm{ad}}$, then convective transport of energy is driven.

In the deep interior of stellar models of normal stars, it is not unreasonable to assume adiabatic convection by equating $\frac{dT}{dr}$ to $-\left(\frac{dT}{dr}\right)_{\mathrm{ad}}$. In the geometrically thin outer non-degenerate layers of a WD, the mixing length theory of convection is commonly used as a first approximation to treat convection.

The temperature excess and hence the buoyancy a convective bubble relative to its surroundings is given by the superadiabatic gradient, β, which is the difference between the actual temperature gradient and the adiabatic temperatures gradient. If beta is multiplied by $c_{\mathrm{p}}\rho$, then this gives the excess heat carried per cm^3 and multiplying by the velocity of the convective bubble yields the convective energy flux transported per cm^2 s^{-1}. Thus, this convective flux is expressed as

$$H = \beta dr c_{\mathrm{p}} \rho v \tag{1.24}$$

where dr is the initial displacement over which the bubble was warmer than its surroundings. For downward (sinking) motions of a bubble, dr changes sign which compensates for the change in sign of the velocity. Thus the kinetic energy of a bubble is given by

$$\frac{1}{2}\rho v^2 = \frac{\rho}{T}\beta\, dr\, GM_{\mathrm{r}}\left(\frac{1}{2}dr\right). \tag{1.25}$$

Now a mixing length l is introduced which is the distance over which a bubble retains its integrity before bursting and mixes with the surrounding gas. This average vertical distance can be taken to be $\bar{dr} = l/2$. Thus the expression for the convective flux becomes

$$H = c_{\mathrm{p}}\rho\left(\frac{GM_{\mathrm{r}}^{\,2}}{Tr}\right)^{1/2}\beta^{3/2}\,\frac{l^2}{4}. \tag{1.26}$$

In the geometrically thin outer non-degenerate layers of a WD, the mixing length theory of convection is commonly used as a first approximation to treat convection. Direct application of mixing length theory to interpretation of the spectra of WD is done by comparing model atmosphere line profile fits to the observed Balmer line profiles of DA stars and hybrid non-DA stars by varying the mixing length to pressure scale height ratio $\frac{l}{H_{\mathrm{p}}}$ to optimize the profile fits. This has worked remarkably for the convective layers in cool DA stars (Bergeron et al. 1995). Overall however, mixing length theory has failed to achieve similar results in other applications, especially to hotter WDs.

Beginning in the mid-1990s, a new more powerful, more physically realistic treatment of convection has emerged with multi-dimensional hydrodynamic simulations of convection in stars beginning with A type stars, the Sun, and DA WDs (Freytag et al. 1996). This new frontier in the treatment of convection is made possible with the availability of powerful, massively parallel supercomputer clusters, delivering increasingly more sophisticated treatments of convection including wider, deeper computational domains, turbulent convection and overshooting. Two different groups with different 3D hydrodynamic codes, the CO5BOLD code (e.g., Tremblay et al. 2015) and the ANTARES code (e.g., Kupka et al. 2018; Zaussinger et al. 2018) are at the forefront of these efforts.

1.11 Rotational Velocities of Non-accreting Isolated WDs

Rotational velocities of isolated, non-magnetic WDs have largely been determined from line profile fits to the sharp non-LTE cores of the Balmer lines in DA WDs (Pilachowski & Milkey 1987; Koester & Herrero 1988; Heber et al. 1997).

In Figure 1.5, the observed optical NLTE line cores of two DA WDs are shown fitted with theoretical line profiles to yield the best fitting Vsini.

The latter authors place a stringent upper limit on the rotational velocities of Vsini < 30 kms^{-1} in their sample of 13 stars. However, a number of rotational

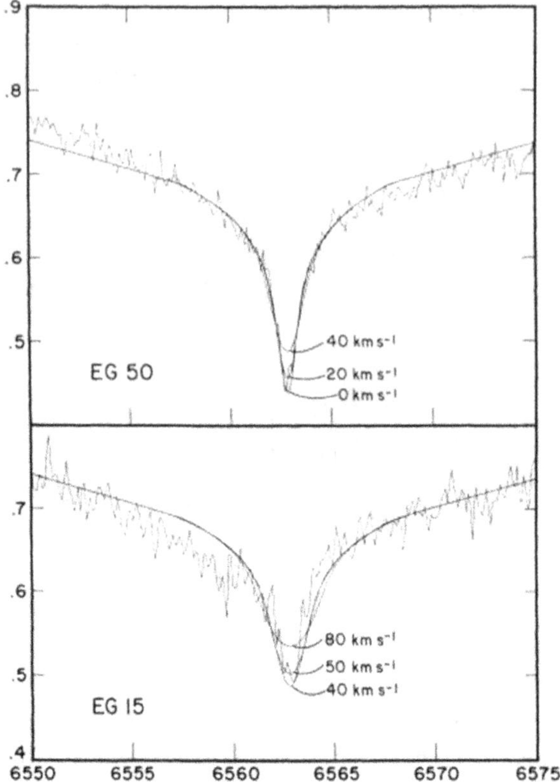

Figure 1.5. Two examples of fitting the observed profiles with theoretical profiles. The upper plot is EG 50, a star with a sharp line core indicative of a low Vsini, while the lower is EG 15, a star with a broader line core indicative of a larger Vsini. (Reproduced with permission from Pilachowski & Milkey (1987); their Figure 1. © 1987. The Astronomical Society of the Pacific. All rights reserved.)

velocities of variable DA and variable non-DA WDs have been measured from asteroseismology (Kawaler 2003). Overall, the rotational velocities of isolated WDs are remarkably small with the projected rotational velocities, Vsini < 60 kms^{-1}. This conclusion holds across virtually the entire mass distribution of DA stars with same the conclusion for isolated non-DA stars as well. These slow rotation rates are much lower than expected on the basis of the expected angular momentum conservation of their AGB progenitors. This implies large angular momentum loss during the AGB phase, spindown of the core, and efficient core–envelope coupling but the details of this process remain obscure.

For the rotational velocities of the magnetic WDs, Wickramasinghe & Ferrario (2000) find a bimodal distribution with rapid rotators having rotation periods ranging from ~700 s to several hours at one extreme of the distribution and very long rotation periods of >100 yrs and even zero rotation at the other extreme. The fast rotators may be the result of stellar mergers. The shortest rotation period they report (700 s) corresponds to Vsini $\sim$ 60 km s^{-1} which is similar to the WD rotational velocity Vsini = 65 km s^{-1} in the pre-CV V471 Tauri, which has a rotation period of 555 s. V471 Tauri, a Hyades cluster member is very likely the product of a merger in a formerly triple system (O'Brien et al. 2002).

References

Bergeron, P., Saumon, D., & Wesemael, F. 1995, ApJ, 443, 764

Carroll, B. W., & Ostlie, D. A. 2007, An Introduction to Modern Astrophysics (2nd ed.; San Francisco, CA: Pearson)

Chapman, S., & Cowling, T. G. 1970, The Mathematical Theory of Non-Uniform Gases (3rd ed.; Cambridge: Cambridge Univ. Press)

Chayer, P., Fontaine, G., & Wesemael, F. 1995, ApJS, 99, 189

Cumming, J. D., Kalirai, J. S., Tremblay, P.-E., Ramirez-Ruiz, E., & Choi, J. 2018, ApJ, 866, 21

Denissenkov, P. A., Herwig, F., Truran, J., & Paxton, B. 2013, ApJ, 772, 37

Fontaine, G., & Michaud, G. 1979, ApJ, 231, 826

Freytag, B., Ludwig, H.-G., & Steffen, M. 1996, A&A, 313, 497

Garcia-Berro, E., Isern, J., & Hernanz, M. 1997, MNRAS, 289, 973

Greenstein, J L 1956, PASP, 68, 501

Greenstein, J. L. 1960, in Stellar Atmospheres, Vol VI, Stars and Stellar Systems (Chicago, IL: Univ. Chicago Press), 703

Heber, U., Napiwotzski, R., & Reid, I. N. 1997, A&A, 323, 819

Iben, I., Kaler, J. B., Truran, J. W., & Renzini, A. 1983, ApJ, 264, 605

Kawaler, S. 2003, arXiv:astro-ph/0301539

Koester, D., & Herrero, A. 1988, ApJ, 332, 910

Kuiper, G P 1941, PASP, 53, 248

Kupka, F., Zaussinger, F., & Montgomery, M. H. 2018, MNRAS, 474, 4660

Lauffer, G., Romero, A. D., & Kepler, S. O. 2018, MNRAS, 480, 1547

Luyten, W 1952, ApJ, 116, 152

McGraw, J. T., Starrfield, S. G., Liebert, J., & Green, R. 1979, in Proc. IAU Coll. 53, White Dwarfs and Variable Degenerate Stars, ed. H. M. van Horn, & V. Weidemann (Rochester, NY: Univ. Rochester), 377

Menzel, D. H., Bhatnagar, P. L., & Sen, H. K. 1963, Stellar Interiors (New York: Wiley)

Mestel, L. 1952, MNRAS, 112, 583

Michaud, G. 1970, ApJ, 160, 641

Michaud, G. 1977, Natur, 266, 433

Montmerle, T., & Michaud, G. 1976, ApJS, 31, 489

O'Brien, S., Bond, H. E., & Sion, E. M. 2002, ApJ, 563, 971

Paczynski, B. 1970, AcA, 20, 47

Pilachowski, C. A., & Milkey, R. W. 1987, PASP, 99, 836

Schoenberner, D. 1979, A&A, 79, 108

Schwarzschild, M., & Harm, R. 1965, ApJ, 142, 855

Schwarzschild, M. 1958, Structure and Evolution of the Stars (Dover: New York)

Shipman, H 1971, BAAS, 3, 395 (Abstract)

Sion, E M 1971, BAAS, 3, 395 (Abstract)

Sion, E M 1973, Astrophys. Letters, 14, 219

Sion, E. M., Greenstein, J. L., Landstreet, J. D., et al. 1983, ApJ, 269, 253

Starrfield, S. 1971, MNRAS, 152, 307
Strittmatter, P, & Wickramasinghe, D 1971, MNRAS, 152, 47
Tremblay, P.-E., Fontaine, G., Freytag, B., et al. 2015, ApJ, 812, 19
Tremblay, P.-E., Fontaine, G., Fusillo, N., et al. 2019, Natur, 65, 202
Vauclair, G., Vauclair, S., & Greeenstein, J. 1979, A&A, 80, 79
Weidemann, V. 1977, A&A, 59, 411
Weidemann, V. 2000, A&A, 353, 647
Wickramasinghe, D., & Ferrario, L. 2000, PASP, 112, 873
Zaussinger, F., Kupka, F., Montgomery, M., & Egbers, Ch. 2018, JPhCS, 1031, 012013

Accreting White Dwarfs
From exoplanetary probes to classical novae and Type 1a supernovae
Edward M Sion

Chapter 2

Accretion of Interstellar Gas and Dust by White Dwarfs

Accretion onto a white dwarf (WD) can occur through a number of different modes; (A) accretion can occur as a WD moves through interstellar space in its galactic orbit and encounters interstellar clouds, patches of enhanced density, and the warm tenuous intercloud medium; (B) accretion can occur if the white dwarf is in close orbit with a stellar companion but the stellar companion is detached from its Roche lobe so that the only way that gas can be accreted is if the companion in the compact binary is ejecting mass via a stellar wind, stellar flares or coronal mass ejections; (C) accretion can occur in wider binaries if the companion star is a red supergiant, a red giant or subgiant, when heavy mass loss associated with cold, slow stellar winds, or thermal pulse superwinds sweep past the WD much like the hot, fast solar wind sweeps past the planets; (D) accretion can occur if a WD is surrounded by the debris of a former exoplanetary system (asteroidal debris, comets, interplanetary dust grains, exoplanets) which survived the heavy mass loss associated with the asymptotic giant branch (AGB) phase of evolution of the original main sequence star. When the WD is very hot, it may photo-evaporate and accrete an orbiting gaseous exoplanet or catastrophic accretion might occur if an exoplanet collides with the WD due to gravitational perturbations; (E) accretion can occur if a non-magnetic WD in a compact binary with a Roche lobe-filling companion accretes matter from an accretion disk or, if the WD is highly magnetic, it will accrete gas from a radially infalling stream (an accretion column) threaded along the magnetic field lines of the WD (a polar). Magnetic WDs with field strengths below those typical of polars have a truncated disk from which gas at the innermost part of the disk at the truncation radius is entrained by the magnetic field to accrete along field lines onto the WD (an intermediate polar).

In this chapter, we consider accretion mode A, the accretion of interstellar matter by WDs. Specifically, we explore the hypothesis that the source of the photospheric metals detected spectroscopically in cool WDs is due to accretion of interstellar matter. The underlying physics of interstellar accretion is discussed as well as the tests and

doi:10.1088/2514-3433/ac930cch2   2-1

challenges to which the interstellar accretion hypothesis is subjected. It is helpful before discussing interstellar accretion by white dwarfs to define the three known components of the interstellar medium. They are: a cold dense component (T < 300 K), made up of clouds of neutral and molecular hydrogen, a warm intercloud component (T $\geqslant 10^4$ K), consisting of rarefied neutral and ionized gas (WIM), and a third component made up of very hot (T $\geqslant 10^6$ K) gas that had been shock heated by supernovae occupying most of the volume of the ISM (McKee & Ostriker 1977).

2.1 Early Spectroscopic Observations of Photospheric Metals in Cool White Dwarfs

The appearance of spectral features due to absorption by any element other than the lightest ones, hydrogen or helium in the WD surface layers is unexpected because of the inexorable and rapid gravitational and thermal diffusion in high gravity degenerate stars. This causes heavy elements to sink downward out of observational detectability. Thus the presence of metals in the atmosphere of a WD signals that some process is opposing the downward diffusion and raises the tantalizing question: what is the source of the metals? Spectroscopic observations in the mid- to late-20th century primarily by Jesse L. Greenstein of Caltech at the 5.1 m Hale Reflector, James Liebert and Paul Hintzen of the Steward Observatory, Gary Wegner at Dartmouth College and their students and postdocs revealed a population of cool WDs with strong Ca II H & K, and Mg I absorption in otherwise pristine He atmospheres. These were originally classified DG, DF, DK; now known as the DZ Stars with Z denoting any metals other than carbon. The Ca II line and Mg I are the most prominent features present in their helium-dominated atmospheres, which are too cool to show neutral helium absorption. This was followed by the detection of the Ca II H & K lines in pure H WD atmospheres (classified as DAZ degenerates). The spectrum of the first DA WD with photospheric metal absorption, the DAZ star, G74-7 (Lacombe et al. 1983) is displayed in Figure 2.1. Note the Ca II H & K lines in an otherwise pure H atmosphere. The first DB WD known to show absorption features of metals was the DBZ star, GD40 (Wickramasinghe et al. 1975; Shipman et al. 1977; Klein et al. 2010). The optical spectrum of GD40 is shown in Figure 2.2. In order to identify the source of these metals, high gravity model atmospheres, and detailed models of WD envelope structure are required.

Realistic models of the structure of the envelopes of WDs, using the most sophisticated input micro-physics, first emerged by Karl-Heinz Boehm (1970) at the University of Washington, Hugh Van Horn (1970) at the University of Rochester, Detlev Koester (1972) at the University of Kiel, Annie Baglin & Gerard Vauclair (1973) at Inst. d'Astrophysique, Paris; D'Antona & Mazzitelli (1975) in Frascati; and Gilles Fontaine & Hugh Van Horn (1976) at the University of Rochester. At roughly the same time, a number of authors first addressed the process of spectral evolution of WDs, with the focus on the possible origin(s) of the nearly pure helium DB WDs (Strittmatter & Wickramasinghe 1971; Shipman 1972; Sion 1971, 1973). These investigations addressed whether cooling WDs could change their spectral types through the action of different physical processes like convective mixing and dilution, accretion and or even destruction of accreted hydrogen via slow nuclear burning.

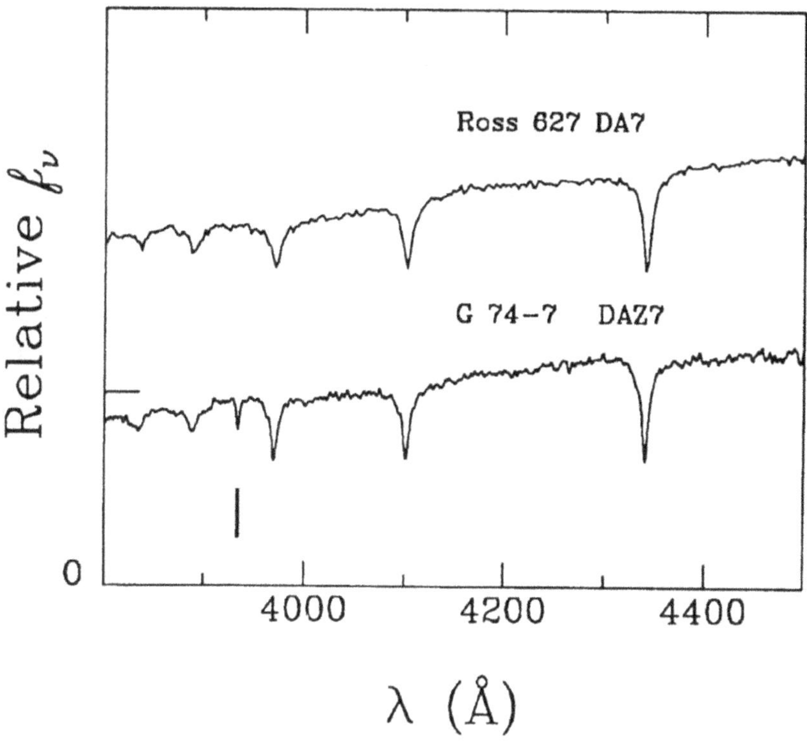

Figure 2.1. The DAZ WD G74-7, the first WD showing metals in a nearly pure hydrogen atmosphere with Ca II K shown at 3933.663Å, displayed along side the DA star, Ross 627. The Ca II H component is hidden by the Balmer line H Epsilon. (Reproduced with permission from Wesemael et al. (1993), their Figure 5. © 1993. The Astronomical Society of the Pacific. All rights reserved.)

By the end of the 1970s, the first numerical determinations of element diffusion in the envelopes of WDs were carried out independently by Vauclairet et al. (1979) and Fontaine & Michaud (1979). Both helium-rich and hydrogen-rich WDs develop convection zones, primarily due to partial ionization and recombination as they undergo their cooling evolution between ~60,000 K and the surface temperature of the coolest known degenerates ~4000 K. As convection zones deepen, the question naturally arose whether the photospheric metals could be dredged up by convection.

It was soon realized that the bottom of WD convection zones at maximum depth never extend deep enough to dredge-up primordial metals from the deep envelope and core–envelope interface (Fontaine & van Horn 1976). Indeed, the diffusion timescales are always shorter than the evolutionary cooling timescales. Even if metals were dredged up by convection, then without a replenishing source, the metals would diffuse deep into the interior leaving a pristine atmosphere composed of the lightest element, either H or He. In other words, if diffusion were acting alone, all of the metals would disappear from the surface layers. It became clear that the source of the metals (other than carbon) seen in the atmospheres of cool WDs had to be *external* (Vauclair et al. 1979; Fontaine & Michaud 1979). The one known

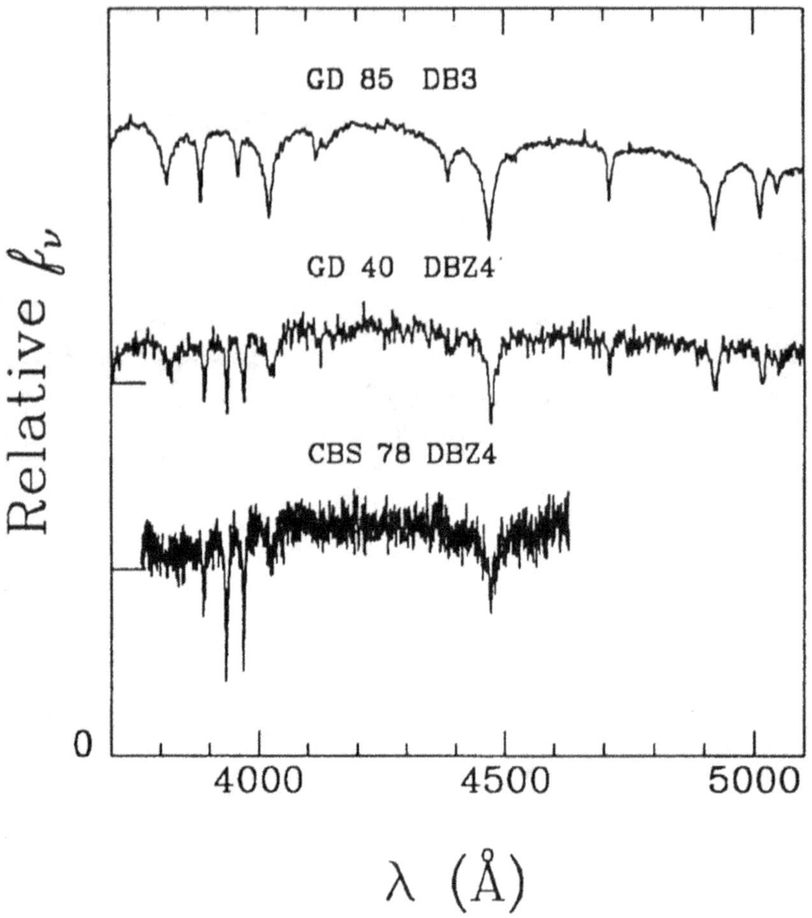

Figure 2.2. Optical spectrum of the first known DBZ star, GD 40, and the DBZ star CBS 78, with Ca II but no trace of hydrogen in their spectra shown for comparison with a pure DB spectrum, the DB3 star, GD85. The shape of the continuous energy distribution of CBS 78 has been established by assuming that it is similar to that of GD 40. (Reproduced with permission from Wesemael et al. (1993), their Figure 9. © 1993. The Astronomical Society of the Pacific. All rights reserved.)

exception to the source of surface metals being external are the helium-dominated DQ WDs (with the Swan bands of molecular carbon) where the source of the surface carbon is due to the upward convective mixing of carbon when its diffusion tail interacts with the diffusion tail of helium extending downward at the bottom of the helium convection zone (Pelletier et al. 1986 and references therein).

Further progress with WD envelope physics required ever more robust calculations of diffusion coefficients, diffusion velocities and diffusion timescales of different elements in WDs. Paquette et al. (1986) advanced our understanding of diffusion in WDs by computing the collision integrals for static screened Coulomb potentials which they approximated with analytic fits. This enabled the calculation of more accurate diffusion coefficients, diffusion velocities and diffusion timescales.

Their results were independently verified using detailed models of molecular dynamics and Monte Carlo computations that were conducted by Fontaine (1987). Thus, the work by Paquette et al. (1986) explicitly took into account screening effects in the evaluation of the transport properties of WD plasmas. They found that the diffusion timescales of different elements differ little from each other, and do not depend in any simple way on the atomic masses of the diffusing ions. Paquette et al. (1986) solidly confirmed earlier results that the diffusion timescales obtained are much smaller than evolutionary timescales. Hence, the need for physical mechanisms was indicated that would compete against diffusion in order to account for the metal abundances in the spectra of cool WDs.

The fundamental importance of the complex interplay between diffusion and accretion in WDs, particularly the peculiar chemical abundance patterns, led to further refinements and detailed comparisons with the results of Paquette et al. (1986). Recently, Heinonen et al. (2020) carried out new computations of diffusion coefficients and thermal diffusion to compare their approach with the work of Paquette et al. (1986). Basically, they used a modified method to calculate the collision integrals that is more appropriate for a partially ionized, partially degenerate plasma for the H-rich and He-rich envelopes of cool WDs. Heinonen et al. (2020) found that absolute diffusion timescales for silicon and calcium differ by more than a factor of two from those calculated by Paquette et al. (1986) in cool, He-rich DZ WD envelopes. When they compared their numerical results for the relative diffusion timescales of two of the most commonly accreted elements, silicon and calcium, with the work of Paquette et al. (1986), they found that their relative diffusion timescales of Si and Ca differed by a factor of three or more, compared with the calculations of Paquette et al. (1986).

2.2 Interstellar Gas and Dust Accretion as the External Source of Metals

With all of the above advances in the physics and structure of WD envelopes, a realistic theoretical framework for accretion and its complex interplay with diffusion and the reservoir of outer convection zones was established. The stage was now set for detailed investigations of the interplay between diffusion, convection, accretion and radiative forces levitation and their role in the spectral evolution of WDs. The most obvious external process for the presence of metals in the otherwise pristine helium atmospheres of a cool WDs was accretion of interstellar grains and gas as the WD, in its galactic orbit, encounters interstellar matter in the form of cold dense clouds, patches of enhanced density and the hot, tenuous intercloud medium. In this case, the WD convection zone serves as a reservoir for the accreted material so that the time that it takes for accreted metals to diffuse out the bottom of the convection zone, defines the diffusion timescale. Since the helium-rich WDs have deeper, more dense convection zones than the hydrogen-rich WDs, their diffusion timescales are much longer (i.e., the diffusion times are much shorter in DA stars than in DB or DZ stars at the same temperature). In general, a very rough estimate of the diffusion timescale, τ_D, is the time it takes for one metal atom to move along one local pressure scale height, H_p.

To determine how accretion of interstellar matter interacts with diffusion and other envelope physical processes, it was essential to compute the expected accretion rates as accurately as possible. According to this accretion scenario for the origin of metals in cool WDs, accretion of interstellar matter in a sufficiently dense cloud should occur hydrodynamically via the Bondi, Hoyle, Lyttleton mechanism (Bondi & Hoyle 1944), namely

$$\dot{M}_{bh} = \frac{4\pi G^2 M_{wd}^2 \rho_\infty}{(V_{rel}^2 + c_s^2)^{1.5}} \tag{2.1}$$

where M_{wd} is the WD mass, ρ_∞ is the density of the undisturbed interstellar medium (ISM), V_{rel} is the relative velocity of star and gas, and c_s is the sound speed. This accretion formula is applicable to a homogeneous ISM of cold gas, but it ignores many effects on the rate of accretion: the mix of dust grains with gas, inhomogeneities in the ISM, the radiation field of the WD, or a stellar magnetic field. However, Hunt (1971, 1979) and Shima et al. (1985) carried out hydrodynamic simulations which roughly confirm the results obtained with the Bondi–Hoyle formula. That is, Equation (2.11) gives an order of magnitude estimate of the accretion rate as long as the flow is hydrodynamical. The Bondi–Hoyle accretion rate is commonly referred to as the "fluid" rate. A major objection to the Bondi–Hoyle picture raised by Greenstein (1951) is that a tail shock must form that destroys particle momenta perpendicular to the accretion axis (wake accretion) while the hot intercloud ISM may not have sufficient density for such a tail shock to form.

If we take an average mass WD ($0.6\dot{M}$) moving through space at 60 km s^{-1}, a typical thin disk total space velocity (Sion et al. 1988) and a density of one atom/cm^3, the resulting accretion rate from the cold ISM is $\sim 10^{-15} M_\odot$ yr^{-1}. If on the other hand, a tail shock cannot form and the Bondi–Hoyle "fluid" rate is inapplicable, then accretion by direct collision of infalling particles with the WD, sometimes referred to as the "Eddington" rate occurs at a much lower value given by

$$\dot{M}_{ed} = \frac{2\pi G M_{wd} R_{wd} \rho_\infty}{V} \tag{2.2}$$

where again V is the relative velocity of star and gas. For the same values of M_{wd}, V and a particle number density of 1 H atom/cm^3, the resulting accretion rate is roughly $\dot{M}_{ed} \sim 10^{-19} M_\odot$ yr^{-1}.

The so-called accretion–diffusion picture became standard, whereby detectable amounts of metals could be maintained in the atmospheres of cool WDs by accretion–diffusion equilibrium. This required that the rate of accretion of grains and gas had to maintain a balance between sinking and replenishment by accretion. However, the fly in the ointment was this: if accretion of interstellar matter was the accepted explanation for the presence of metals in cool WDs (DZ, DZA, DAZ), then how could the essentially pure helium atmospheres of the DB WDs ($\sim 20\%$ of all WDs) avoid accretion of interstellar hydrogen? Indeed, the very existence of the DB helium atmospheres with He/H number density ratios of He/H $\geqslant 10^4$ (i.e., H mass fractions $\leqslant 10^{-5}$) with such a high degree of helium purity defied explanation. Attempts were made to find physical mechanisms that could explain the absence of accreted H while allowing the accretion of heavy metals in helium-dominated WD atmospheres. Michaud & Fontaine (1979) proposed that a hot corona and the strength of the electric field of the helium atmosphere of a DZ or DB star could repel ionized H preventing it from accreting while lacking sufficient strength to prevent metals like Ca, Mg, and Si, that were locked in evaporated grains, from accreting. Still another mechanism put forward initially by Illarionov & Sunyaev (1975) in the context of a rapidly rotating neutron star magnetosphere batting away infalling matter, was applied to WDs by Wesemael & Truran (1982). In their model, metal-laden grain accretion occurs onto a slowly rotating, weakly magnetic WD while H is batted by a propeller that arises due to the vastly different Larmor radii and charge to mass ratios of the accreting H atoms compared to the metals. The efficiency of this process increases the larger the charge to mass ratio of the accreting particles. Furthermore, the operation of this screening mechanism depends on the stellar luminosity being high enough to ionize the infalling hydrogen at the magnetospheric boundary, selectively screening out the H, but with a low enough luminosity not to evaporate the infalling grains at that surface (Wesemael & Truran 1982). Their propeller model did yield some approximate agreement with the observed abundance patterns of heavy element abundances of a number of cool, helium-rich WDs, but nevertheless there were numerous observed cases where their model-predicted abundances and the observed abundances were not in agreement. Moreover the efficiency of this screening mechanism would decline at lower surface temperatures where cool He-rich degenerates are found still with no detectable hydrogen. Despite the problem of how interstellar hydrogen can be prevented from accreting onto helium-rich cool WDs (i.e., $T_{\mathrm{eff}} < 20,000$ K), including the DB stars, the process of accretion from the interstellar medium persisted as the most widely held explanation for the presence of metals in their atmospheres.

2.3 The Two-Phase ISM Accretion Model of Dupuis, Fontaine and Wesemael

With the theoretical framework for accretion and its complex interplay with diffusion into the reservoir of outer convection zones having been established, the next major advance was spawned: the construction of the first rigorous time-dependent evolutionary models of WDs undergoing accretion and diffusion. Two landmark papers were published by Dupuis et al. (1992, 1993) of the Montreal Group who proposed a Two-Phase Accretion Process onto WDs. According to this scenario, the WD undergoes an episode of high accretion as it passes through a dense cold cloud a few parsecs in size for approximately 10^6 years during which the rate of accretion reaches an equilibrium with diffusion but for the vast majority of the time, of order 5×10^7 years, the star is moving through the hot, tenuous, intercloud medium where the particle density and the accretion rate are much lower, after which it enters a dense cloud, where the accretion rate is high again and the metal abundances in the outer layers approach an equilibrium value between accretion and diffusion.

For accretion of metals from an interstellar cloud onto a helium-rich WD, the metal abundances, including their relative abundance ratios would be essentially solar upon arrival at the top of the WD atmosphere. However, when the WD just emerges from the interstellar cloud, the action of diffusion would cause downward diffusion of the mix of metals at different diffusion velocities for each ion while any trace of hydrogen would diffuse upward to the top of the helium-rich atmosphere. With the passage of time, one expects that the total abundance of metals will decrease as the WD moves through the much lower density, hot tenuous intercloud medium while the relative abundances of the accreted elements would manifest the relative diffusion velocities of each metal. For a cool helium-rich WD of mass $M_{\mathrm{wd}} < 0.6 M_\odot$ and hence a deeper convection zone than a high mass WD, the diffusion time for a metal may be comparable to the time between encounters with dense clouds while its atmosphere still remembers the previous cloud encounter (Dupuis et al. 1992, 1993; Koester & Wilken 2006).

Given the patchy structure of the local interstellar medium, the two-phase accretion model correctly envisioned that a WD would accrete at a very low rate in the intercloud medium and at a much higher accretion rate when it encountered an interstellar cloud or patch of enhanced density. Dupuis et al. (1992) focused on cool WDs with $T_{\mathrm{eff}} < 20,000$ K, for which radiative forces and thermal diffusion are negligible. Therefore, they considered that the diffusion was governed only by the terms due to gravitational diffusion, electric fields, and the concentration gradient. Under these assumptions, a general expression for the diffusion equation was written for a plasma consisting of two ionic species and valid for any degree of degeneracy. The diffusion equation is given by,

$$V_{12} = D_{12}(1 + \gamma)\left[-\frac{\partial \ln c_2}{\partial r} + \left(\frac{A_1 Z_2 - A_2 Z_1}{Z_1 + \gamma Z_2} \right)\frac{m_{\mathrm{p}} g}{kT} + \left(\frac{Z_2 + Z_1}{Z_1 + \gamma Z_2} \right)\frac{\partial \ln p_i}{\partial r} \right]. \quad (2.3)$$

V_{12} is the relative diffusion velocity of species 2 with respect to species 1, D_{12} is the diffusion coefficient, $c_2 = n_2/(n_1 + n_2)$ where n_1 and n_2 are the number densities of species 1 and species 2, $\gamma = n_2/n_1$, A_i is the atomic weight of species 1 and 2, m_{p} is the proton mass, k is Boltzmann's constant, T is the temperature, $p_i = p_1 + p_2$ is the ionic pressure, Z_1 is the average charge of the species 1, Z_2 is the average charge of the species 2, r is the radial coordinate and g is the local gravitational acceleration.

The charges of diffusing ions and thus their ionization states must be taken into account. The ionization state of a specific metal must employ a modified version of the Saha ionization equation that is valid for partial degeneracy and includes the average charge of the diffusing ions and of the background gas through which it is diffusing (see Dupuis et al. 1993; Muchmore 1984 for details).

To follow the time-dependent evolutionary WD envelope diffusion models, the equation of continuity must be solved given by

$$\frac{\partial}{\partial t}(\rho X_2) = -\frac{1}{r^2}\frac{\partial}{\partial r}(r^2 \rho X_2 V_2) \quad (2.4)$$

where ρ is the density, X_2 the mass fraction of element 2, and V_2 is the diffusion velocity of element 2 relative to the center of mass. The diffusion velocity V_2 relative to the center of mass is given by

$$V_2 = \frac{V_{12}}{1 + A_2\gamma/A_1} \tag{2.5}$$

where the symbols in this equation are defined in Equation (2.3).

By integrating Equation (2.4) over the volume V whose upper boundary is the WD surface and lower boundary by an arbitrary radius r, and using Gauss's theorem, Equation (2.4) becomes

$$\frac{\partial}{\partial t}\int_V \rho X_2 d^3r = 4\pi r^2 \rho X_2 V_2. \tag{2.6}$$

Equation (2.6) is the boundary condition imposed at the base of the convection zone that requires that the conservation of mass be satisfied. Equation (2.4), together with Equations (2.5) and (2.6) and the boundary conditions, completely defines the time-dependent diffusion problem (Dupuis et al. 1992). When applied at the base of a superficial convection zone, Equation (2.6) becomes

$$\frac{\partial}{\partial t}X_2 = \frac{4\pi r^2 \rho V_2}{\Delta M_c}X_2 \tag{2.7}$$

where ΔM_c is the mass of the convection zone. Equation (2.7) follows from the fact that X_2 can be taken out of the integral on the left-hand side of Equation (2.7) because convection rapidly mixes and thus homogenizes the abundances.

As long as the concentration gradient is negligible at the base of the convection zone and the abundance of the diffusing element is trace, V_2 depends only on local conditions and the exact solution to Equation (2.7) is

$$X_2 = X_2(0)\exp(-t/\tau_D). \tag{2.8}$$

where $X_2(0)$ is the initial abundance in the convection zone and τ_D is the diffusion time scale given by

$$\tau_D = \frac{\Delta M_c}{4\pi r^2 \rho |V_2|} \tag{2.9}$$

where V_2 is negative when element 2 sinks toward the center of the star. Equation (2.8) is accurate when the metal distribution is far from diffusive equilibrium (diffusive equilibrium occurs when the concentration gradient exactly cancels the gravitational settling terms: the last two terms in Equation (2.3)). Even if Equation (2.3) is valid only at the base of the convection zone, it has been frequently applied by others in the radiative layers below the surface convection zone, thus yielding approximate diffusion rates for different metals.

While the two-phase accretion model seemed like a plausible scenario for the source of metals from interstellar accretion onto WDs, serious problems with this model were identified, the most important of which was the lack of cold dense clouds within the so-called Local Bubble where the DZ WDs are found. The local bubble is a large cavity that presumably formed due to the sweep of an ancient supernova blast wave. That there is no cold ISM material in the local bubble, is proven by the lack of detection of any significant Na I absorption lines. While the two-phase accretion model-predicted metal abundances with respect to helium that were found to be in quantitative agreement with the observed abundances in a number of cool WDs, there are no cold dense clouds within the local bubble. In addition, there was also the problem of understanding within the context of interstellar accretion, the wide range of derived abundances and the mix and diversity of the metal species along with the still vexing problem of how twenty percent or more of all of the WDs, i.e., the helium-rich degenerates, avoid interstellar hydrogen contamination

2.4 Comparison of White Dwarf Kinematics with Maps of the Local ISM

Aannestad & Sion (1985) tested for the first time with the then available maps of the local ISM, the interstellar accretion hypothesis that the source of the metals in DZ degenerates was accretion of interstellar gas and grains. They looked at the spatial distribution and kinematics of the DZ stars relative to known local regions of interstellar clouds. The spatial distribution of all but four of 17 DZ, DAZ, and DZA WDs suggested a possible association with Tinbergen's Patch and related hydrogen contours. They found that over half of the metal-line cool degenerates appear to have passed through regions of enhanced density less than 10^6 years ago. However, their analysis lacked full space motions due to the lack of radial velocities for the DZ stars, thus their calculations included only two-thirds of the motion, and hence no obvious correlation was found between deviations from cosmic relative metal-abundance ratios and the time elapsed since the presumed local cloud accretion episode. Their study also lacked a better knowledge of the three-dimensional structure of the local ISM. Therefore, their results must be regarded as inconclusive.

A more definitive study was carried out by Aannestad et al. (1993). They obtained radial velocities for 14 DZ WDs and a DAZ star by securing high resolution spectra with the Multiple Mirror Telescope (MMT) (Sion et al. 1990). The quality of the spectra for the 14 DZ stars is exemplified by the four DZ degenerates displayed in Figure 2.3.

Radial velocities were derived from the positions of the H and K lines of Ca II. A pressure shift correction was included, and the resulting mean gravitational redshift was 35 km s^{-1}. Typical line profiles of Ca II H and K fitted with single and double Lorentzians are shown in Figure 2.4.

Using three vector components of the space motions U, V, and W, they projected their spatial distribution and vector components onto, and perpendicular to, the galactic plane shown in Figures 2.5(a) and (b). The 14 DZ degenerates and one DAZ degenerate, G74-7, are shown in Figures 2.5(a) and (b) are identified numerically as follows: 1. vMa 2, 2. G74-7, 3. G245-58 4. GD40 5. G77-50 6. G99-44 7. G105-B2B 8. L745-46A 9. Glll-054 10. G163-028, 11. G200-039 12. G180-05 13.G139-013 14. GD 40 15. G157-35.

They compared the present and past spatial distribution of the WDs with the distribution of local interstellar matter. If the metallic-line WDs are accreting their

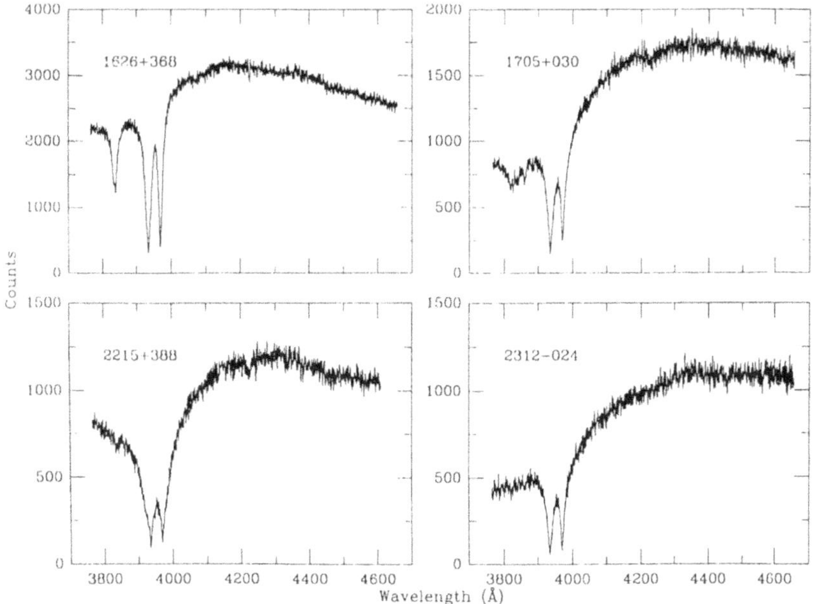

Figure 2.3. Selected DZ stars that display strong Ca II H and K lines. (Reproduced with permission from Sion et al. (1990). © 1990 The Astrophysical Journal. All rights reserved.)

metals from the interstellar medium, one would expect a spatial correlation between the stars and the local gas. However, they concluded that only very few of the DZ WDs could possibly have just accreted or presently be accreting metals from local clouds. They also considered several scenarios for interstellar accretion that may have taken place within the last million years, assuming that the local medium has not substantially changed. Their results show that Ca accretion from local interstellar matter in the past cannot have proceeded at a rate high enough to explain the observed photospheric abundances of Ca, at least in the cool metallic-line degenerates.

The major result of Aannestad et al. (1993) was *not* the lack of an obvious correlation between the spatial distribution and kinematics of the DZ stars relative to the most detailed maps of clouds and patches in the local interstellar medium, when full space motions of the DZ sample were included. Rather, Aannestad et al. (1993) demonstrated conclusively and numerically that the DZ stars situated in the local bubble cannot have acquired their accreted calcium from interstellar accretion. They considered the accretion history of the DZ stars within the bubble and followed the approach of Alcock & Illiaronov (1980) to treat grain evaporation and accretion. In their work, they utilized a silicate–graphite model to represent the dust grains and adopted grain sizes of 0.010 56 to 0.25 microns plus an independent addition of 0.005 micron grains. With the assumption of a solar abundance of Ca relative to silicon with grains having an entirely reasonable olivine composition, and noting that Ca is greatly depleted in the gas phase of the ISM (Savage & Mathis 1979) by as much as a factor of $\sim 10^4$, this ensured that

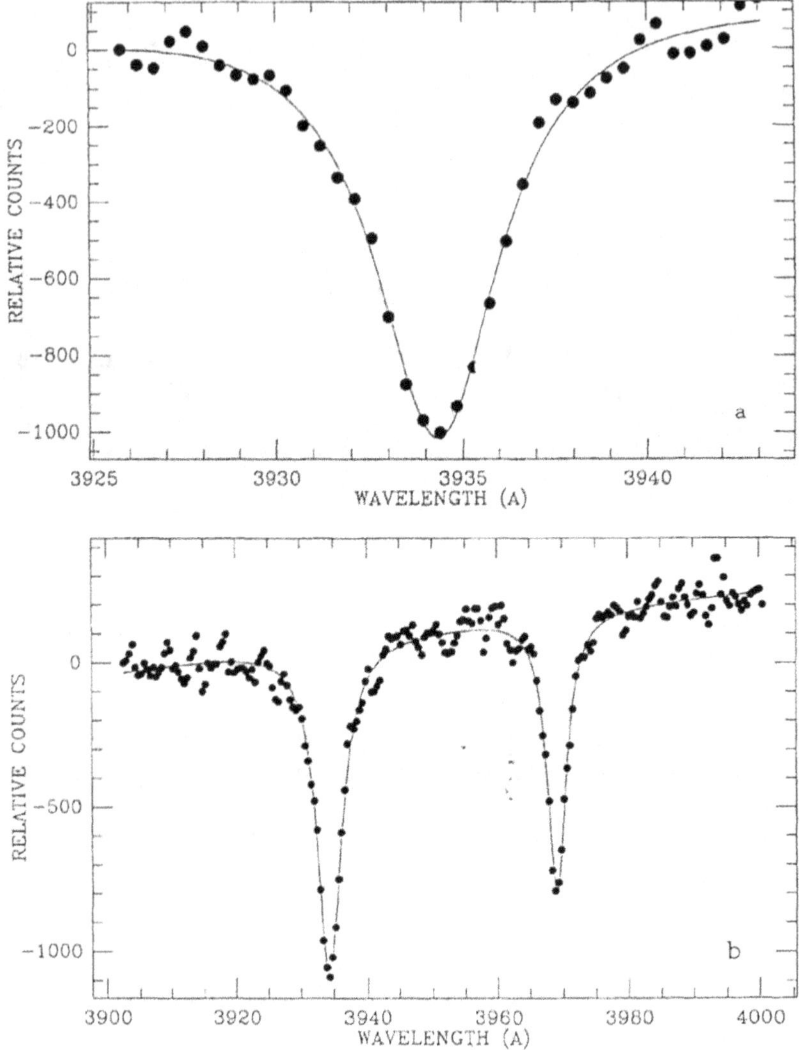

Figure 2.4. (a) A Lorentzian line plus a quadratic continuum fitted to the observations of the K line in the spectrum of the DZ WD, G111-54. (b) Two Lorentzian lines and a quadratic continuum fitted to the observations of the K and H lines in the spectrum of G111-54. (Reproduced with permission from Aannestad et al. (1993). © 1993 The Astrophysical Journal. All rights reserved.)

essentially all of the calcium is tied up in the grains. The grain density was chosen to be 3.3 g cm^{-3}, the density of silicate rocks and the condensation temperature of the silicate grains 1200 K.

Following Alcock & Illiaronov (1980), if a grain evaporates while within ionized gas and the evaporation distance $R_{evap} < R_A$ (the accretion radius), an estimate for the supersonic grain accretion rate is

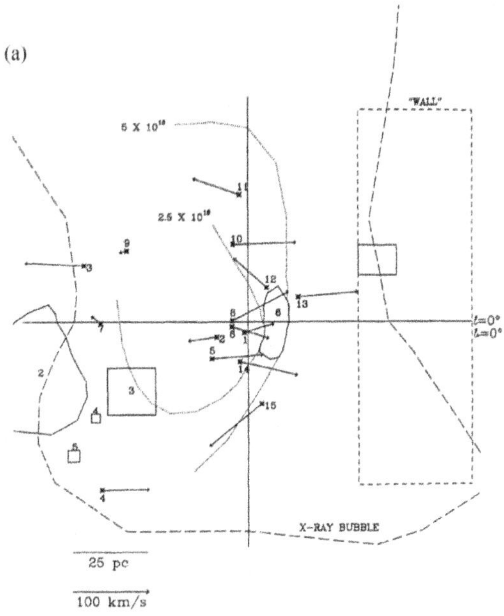

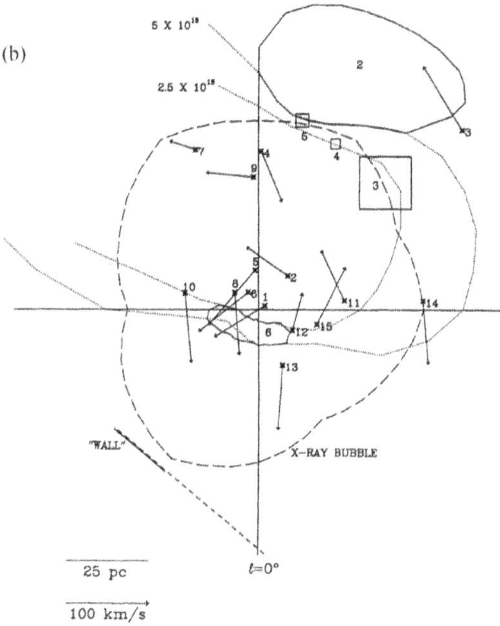

Figure 2.5. (a) A projection on to the galactic plane of the positions of the 14 DZ WDs and one DAZ, and neutral interstellar material at a time 10^6 yr ago. Also shown are the projected velocities of the WDs relative to the local ISM and their mean relative motion over 10^6 yr. (b) Same as in (a), except a projection perpendicular to the galactic plane. (Reproduced with permission from Aannestad et al. (1993). © 1993 The Astrophysical Journal. All rights reserved.)

$$\frac{dN(a_i)}{dt} = \pi R_{\mathrm{evap}}(a_i)\left[R_{\mathrm{evap}}(a_i) + R_{\mathrm{A}} \right] V_{\mathrm{T}} n(a_i) \tag{2.10}$$

where

$$R_{\mathrm{A}} = \frac{2GM_{\mathrm{wd}}}{(V_{\mathrm{T}})^2}$$

is the accretion radius, V_{T} is the relative velocity of star and gas, $n(a_i)$, is the number density of Ca-carrying grains of radius a_i. If $R_{\mathrm{evap}} \geqslant R_{\mathrm{A}}$ and the gas is still ionized at the accretion radius. Therefore, the fluid accretion rate applies given by

$$\frac{dN(a_i)}{dt} = \pi R_{\mathrm{A}}^2 V_{\mathrm{T}} n(a_i). \tag{2.11}$$

This fluid rate of grain accretion establishes an upper limit constraint to the accretion rate for a non-magnetic DZ degenerate. However, the surface temperatures of all of the DZ stars in the Aannestad et al. (1993) sample are sufficiently low that $R_{\mathrm{I}} < R_{\mathrm{evap}}$ and in some instances $R_{\mathrm{I}} \sim 0$ (i.e., $R_{\mathrm{I}} < R_{\mathrm{star}}$). Hence, the grains are released in neutral gas that accretes at the non-interacting particle rate, essentially the Eddington rate. It is important that an accretion rate as low as ($\sim$10 g/s/H atom), accreted calcium is unable to account for even the lowest observed Ca abundances in the DZ sample, which is Ca/He $\sim 10^{-11}$. Thus, Aannestad et al. (1993) showed quantitatively that the interstellar accretion hypothesis completely fails for the DZ stars in the local bubble out to a distance of 75 pc. This conclusion was strongly reinforced by the work of Farihi et al. (2010) using the large sample of DZ and DAZ stars from the Sloan Digital Sky Survey (SDSS), discussed in the next chapter (Section 3.7 in Chapter 3).

Since interstellar accretion could not explain the calcium abundances in DZ degenerates (Aannestad et al. 1993) within the local bubble that they inhabit, the only reasonable alternative for an external source of metals is circumstellar material. Eighteen years before the work of Aannestad et al. (1993), in an overlooked comment of singular prescience, Wickramasinghe et al. (1975) stated in regard to the origin of accreted calcium in the DBZ star, GD40, that "the strength and breadth of [CaII] H & K [in GD40] rule out an interstellar origin of these lines."

2.5 The Demise of ISM Accretion as the Source of Metals in Cool White Dwarfs

Nevertheless, the case for interstellar accretion onto WDs was given new life with the discovery of a new component of the interstellar medium made possible with the detection of soft X-ray emission by Snowden and collaborators (Snowden et al. 1998) dubbed the hot component of the interstellar medium (hereafter HIM). New details on the structure of this hot component were revealed by Redfield & Linsky (2002, 2004). They reported that the sight lines to many nearby stars have distinct patches with

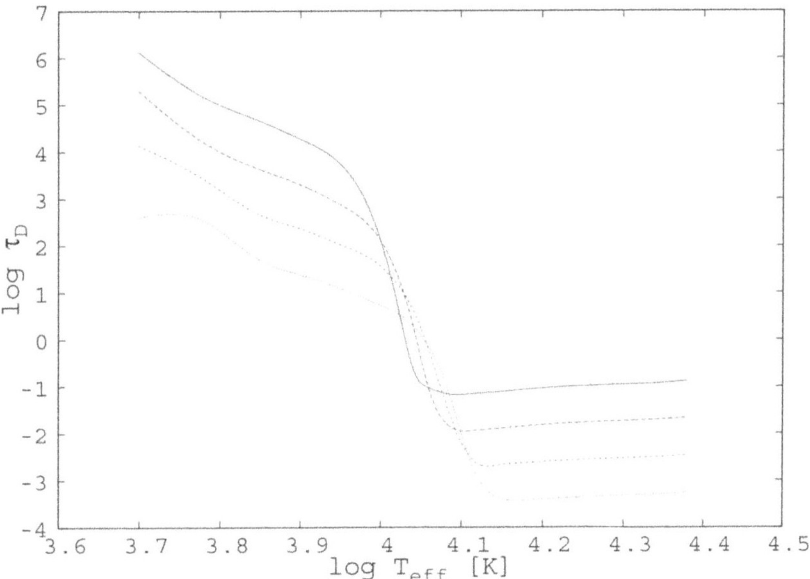

Figure 2.6. Diffusion timescales (in years) of Ca versus T_{eff}. The solid line indicates $\log g = 7.5$, then following $\log g = 8.0, 8.5, 9.0$. (Reproduced with permission from Figure 2.2 in Koester & Wilken (2006). © ESO.)

length scales between 0.1 pc to 11 pc and densities of 0.11 cm^{-3}. They found that the ISM density varies by two orders of magnitude on the scale of a few pc.

Despite the failure of interstellar accretion and accretion–diffusion equilibrium to explain the source of metals in the DZ WDs due to the absence of cold dense clouds in the Local Bubble, there is a spectroscopic subgroup of WDs, the DAZ stars, where interstellar accretion could possibly explain the calcium lines and other metals. A case was made by Koester & Wilken (2006) that the source of metals in the DAZ degenerates could be accretion from the HIM component of the interstellar medium. Koester & Wilken (2006) compiled a sample of 38 DAZ stars ranging in temperature from 25,000 K to 5190 K, a temperature range where radiative levitation is virtually inoperable. They computed new diffusion timescales for Ca, Mg, Fe, and Si in DA WDs using ML2 and a mixing length of 0.6 pressure scale heights. The diffusion timescale for Ca as a function of temperature is displayed in Figure 2.6 for DAZ stars having $\log g = 8.0, 8.5$, and 9.0.

From the absorption lines of accreted metals, they computed the corresponding accretion rates. They found that even though the density of the partially ionized plus neutral, hot, tenuous ISM is $N_h \sim 0.1 - 0.2$ cm^{-3}, the ISM accretion hypothesis could be viable if the accretion occurs at the hydrodynamic rate given by Bondi & Hoyle (1944) formula. In the DAZ stars, the diffusion timescales are so short that accretion had to have occurred in the recent past or is presently occurring.

Koester & Wilken (2006) found consistency with the findings of Redfield and Linsky that on average there are one to three such patches along each line of sight.

This means that a WD with a thin disk space velocity of 30 kms^{-1} could travel 1 pc in about 33,000 yr, which would enable accretion–diffusion equilibrium to be established and maintain the metal abundances in continuous accretion over thousands of years. They proposed this scenario as a viable explanation for the observed range of accretion rates and the derived metal abundances seen in the DAZ stars. Unfortunately, there exists strong, observational evidence that the Koester & Wilkens (2006) explanation of the metals in DAZ stars is wrong.

Subsequent to the work on Koester and Wilken (2006), various observations of DAZs (e.g. Zuckerman et al. 2007) revealed that the ISM model could not be correct. As I will discuss in Chapter 3 the WD photospheric abundances are not solar as would be expected for ISM accretion. Instead the abundances typically looked chondritic or terrestrial-like (e.g. Zuckerman et al. GD362). Moreover, if ISM material is being accreted, then why are all the ISM volatiles (e.g. carbon and oxygen) far below cosmic abundances? Since Koester and Wilken (2006) focused only upon the four refractory elements Ca, Mg, Fe, and Si, without consideration of volatiles, then it might be possible to draw the conclusion that accretion of ISM gas and dust was occurring. In Chapter 3, we will explore in considerable detail, the multiple arguments showing that accretion of tidally disrupted asteroids is indeed occurring.

References

Aannestad, P. A., Kenyon, S., Hammond, G., & Sion, E. M. 1993, ApJ, 105, 1033

Aannestad, P. A., & Sion, E. M. 1985, AJ, 90, 1832

Alcock, C., & Illarionov, A. 1980, ApJ, 235, 541

Baglin, A., & Vauclair, G. 1973, A&A, 27, 307

Boehm, K.-H. 1970, ApJ, 162, 919

Bondi, H., & Hoyle, F. 1944, MNRAS, 104, 273

D'Antona, F., & Mazzitelli, I. 1975, A&A, 44, 253

Dupuis, J., Fontaine, G., Pelletier, C., & Wesemael, F. 1992, ApJS, 82, 505

Dupuis, J., Fontaine, G., & Wesemael, F. 1993, ApJS, 87, 345

Farihi, J., Barstow, M. A., Redfield, S., Dufour, P., & Hambly, N. C. 2010, MNRAS, 404, 2123

Fontaine, G., & Van Horn, H. 1976, ApJS, 31, 467

Fontaine, G., & Michaud, G. 1979, ApJ, 231, 826

Fontaine, G. 1987, in Strongly Coupled Plasma Physics (Berlin: Springer), 161

Greenstein, J. L. 1951, in Proc. Topical Symp., Commemorating the 50th Anniversary of the Yerkes Observatory and Half a Century of Progress in Astrophysics, ed. J. A. Hynek (New York: McGraw-Hill), 526

Heinonen, R. A., Saumon, D., Dailgault, J., et al. 2020, ApJ, 896, 2

Hunt, R. 1971, PhD thesis, Univ. Cambridge

Hunt, R. 1979, MNRAS, 188, 83

Illarionov, A. F., & Sunyaev, R. A. 1975, A&A, 39, 185

Klein, B., Jura, M., Koester, D., Zuckerman, B., & Melis, C. 2010, ApJ, 709, 950

Koester, D. 1972, A&A, 16, 459

Koester, D., & Wilken, D. 2006, A&A, 453, 1051

Lacombe, P., Liebert, J., Wesemael, F., & Fontaine, G. 1983, ApJ, 272, 660

McKee, C. F., & Ostriker, J. P. 1977, ApJ, 218, 148

Muchmore, D. 1984, ApJ, 278, 769

Paquette, C., Pelletier, C., Fontaine, G., & Michaud, G. 1986, ApJS, 61, 197

Redfield, S., & Linsky, J. 2002, ApJS, 139, 439

Redfield, S., & Linsky, J. 2004, ApJ, 613, 1004

Savage, B. D., & Mathis, J. S. 1979, ARA&A, 17, 73

Shima, E., Matsuda, T., Takeda, H., & Sawada, K. 1985, MNRAS, 217, 367

Shipman, H. L. 1972, ApJ, 177, 723

Sion, E. 1971, BAAS, 3, 395

Sion, E.M., Kenyon, S.J., & Aannestad, P.A. 1990, ApJS, 72, 707

Sion, E. M. 1973, ApJL, 14, 219

Sion, E. M., Fritz, M. L., McMullin, J. P., & Lallo, M. D. 1988, AJ, 96, 251

Sion, E. M., Hammond, G. L., Wagner, R. M., Starrfield, S. G., & Liebert, J. 1990, ApJ, 362, 691

Snowden, S. L., Egger, R., Finkbeiner, D. P., Freyberg, M. J., & Plucinsky, P. P. 1998, ApJ, 493, 715

Strittmatter, P., & Wickramasinghe, D. T. 1971, MNRAS, 152, 47

Van Horn, H. 1970, ApJ, 160, 53

Vauclair, G., Vauclair, S., & Greenstein, J. L. 1979, A&A, 80, 79

Wesemael, F., & Truran, J. W. 1982, ApJ, 260, 807

Wesemael, F., Greenstein, J. L., Liebert, J., et al. 1993, PASP, 105, 761

Wickramasinghe, D. T., Hintzen, P., Strittmatter, P. A., & Burbridge, E. M. 1975, ApJ, 202, 191

Zuckerman, B., Koester, D., Melis, C., Hansen, B., & Jura, M. 2007, ApJ, 671, 872

Accreting White Dwarfs
From exoplanetary probes to classical novae and Type 1a supernovae
Edward M Sion

Chapter 3

Accreting White Dwarfs as Probes of the Formation, Composition and Evolution of Exoplanetary Systems

3.1 Introduction to Exoplanetary Debris Accretion onto White Dwarfs

As recently as the 1920s, some college level textbooks in astronomy included the possibility that the source of energy making the sunshine is the accretion of meteoroids or comets. As a large influx of meteoroids or comets plummet into the gravitational potential well of the Sun in this scenario, gravitational energy is released as sunlight. This scenario was first proposed by William Thompson of Scotland, later to be known as Lord Kelvin. While this idea preceded the theory of mass–energy equivalence and the theory of thermonuclear fusion reactions, comets and meteoroids do fall into the Sun. Until thermonuclear fusion was recognized as the only known source of energy over geological timescales that could sustain the Sun's luminosity virtually unchanged for the 4.6 billion years of the solar system's existence, meteoroid and comet accretion was among the leading ideas about what powers the Sun.

Indeed, the very formation of the Sun and its system of planets involved mass growth and heating by accretion of planetesimals versus destruction of planetesimals by collisions in a survival of the fittest competition. However, the kind of isolated star that would seem the very least likely to undergo accretion of exoplanetary matter would be a white dwarf (WD) star for the obvious reason that even if its main-sequence progenitor had a system of orbiting bodies, any such exoplanetary system would be vaporized or swept away when the parent star expanded to become a red giant, thus engulfing those objects in close orbit. It was expected that any exoplanetary matter in wider orbits during the asymptotic giant branch (AGB) phase of stellar evolution would be swept away by the momentum flux of the pulsed heavy mass outflows due to ordinary winds and thermal pulse superwinds. Or so we thought!

It came as a great surprise that accretion of extrasolar comets and extrasolar tidally disrupted asteroids is certainly occurring onto WDs and that this material is consistent with being the remains of an extant planetary system formed at the same time that the main-sequence progenitor of the WD was born. It is useful as a guide for further reading of exoplanetary accretion onto WDs to review the various chemical compositions of the asteroids and comets in our solar system.

If the parent planetesimal melted due to the intense heating by short-lived radionuclides, then the process of differentiation occurred. Its primordial (primitive) composition would undergo geochemical modifications and gravitational diffusive separation by molecular weight, with the heavy elements such as Fe and Ni diffusing downward, the lower weight elements and minerals like silicates floating toward the top of the parent body. In this geochemical melt, elements and minerals would group into families with chemical affinities for different element groups. The lowest density minerals have a chemical affinity for oxygen-bearing minerals and are known as lithophiles (e.g., Al, Ca, Sr, Rb, U, Th, Ba), the intermediate density minerals, commonly found in ore deposits, have a chemical affinity for sulfur-bearing minerals (e.g., Cu, Zn, Pb, Ag, Hg) while the siderophiles have a chemical affinity for iron and nickel (e.g., Ru, Ge, Ga, Ir). When this differentiated parent body suffers a collision, or is tidally disrupted, the fragments will have the composition of the layer of the disintegrated parent body from which they were ejected. When this fragment (meteoroid) accretes onto a WD, the vaporization and lateral spreading leaves a spectroscopic signature. From the analysis of the observed spectroscopic absorption line profiles with model line profiles, chemical abundances from the composition of the accreting bodies will be revealed. In this way, accreting WDs have provided the first known chemical compositions of exoplanetary bodies, whether fragmented extrasolar asteroids, extrasolar comets, or even fragmented or photo-evaporated exoplanets.

Since the terrestrial meteorites are the result of asteroids that have been shattered by collision, the types of meteorites resulting from these collisions shed light on similarities and differences to be found for the compositions of tidally disrupted asteroids and comets in an exoplanetary system. Therefore, in Figure 3.1, the comprehensive classification by J. Wittke and T. Bunch, of the different types of meteorites by composition are presented. At the top of Figure 3.1 are the most primitive (primeval) meteorites, the carbonaceous chondrites subtyped by their water content, still pristine in composition and never having been processed by differentiation. Glassy inclusions called chondrules (hence chondritic) are embedded in a typically silicate-rich matrix. All other stony meteorites having no chondrules are called achondrites and their compositions are less primitive. Their compositions have been altered by differentiation and they have originated from the collisional shattering of a differentiated parent body. Meteorites which are stony-iron typically contain inclusions of olivine embedded in an iron-nickel metallic matrix. These objects originated from the transition region between the rocky mantle and iron-nickel core of a differentiated parent body. Finally, the fragments of the iron-nickel core of a shattered parent meteorite planet give rise to the iron meteorites.

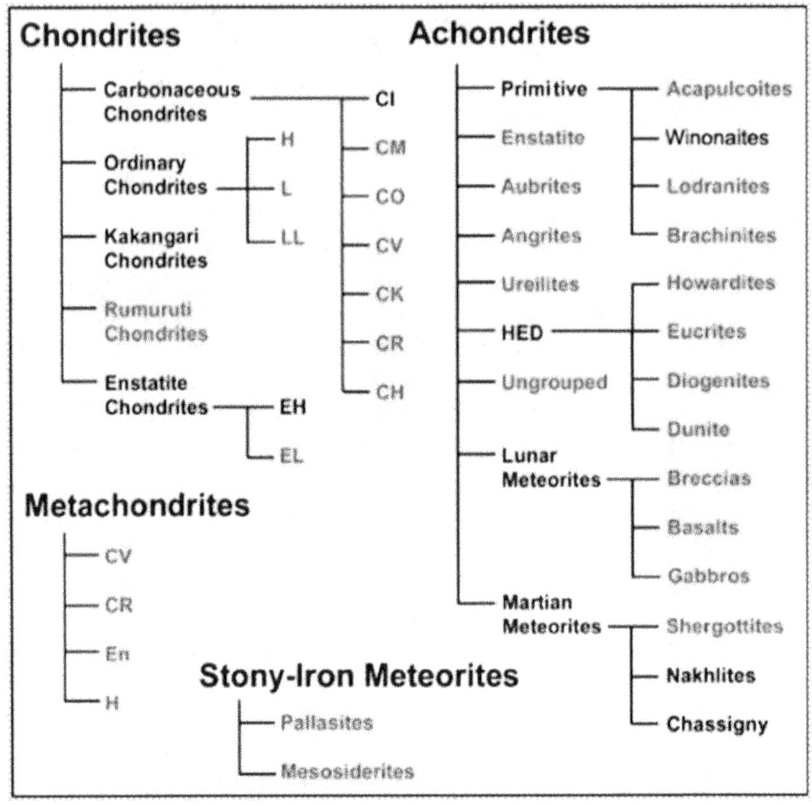

Figure 3.1. Classification of meteorites by degree of differentiation from the primitive meteorites (carbonaceous chondrites at the top with increasing degree of differentiation downward). (Reproduced with permission from Meteorite Classifications © J. H. Wittke & T. E. Bunch.)

3.2 WD Accretion of Tidally Disrupted Asteroids

Twenty years prior to the work of Koester & Wilken (2006), a potentially ground breaking paper appeared by Alcock et al. (1986) who applied the Alcock & Illiaronov (1980) calculations of heavy element diffusion in a hydrogen-rich WD, to invoke an accretion event due to cometary impact as the source of the photospheric calcium in the DAZ star G74-7. They extended their model of cometary accretion events to constrain the extent of Oort clouds around the main-sequence progenitors of DAZ WDs. This marked the beginning of the realization that quite possibly, exoplanetary system material that survived the heavy mass loss in the advanced stages of evolution of the main-sequence progenitors of WDs could be the source of metals in DAZ stars. Indeed, for the helium-rich WDs, Sion et al. (1990) compared the helium-dominated DZ star CBS 127 to several other DZ WDs and presented evidence that interstellar hydrogen screening is inefficient and cannot be operating at the lowest DZ temperatures. Instead, they proposed that DZ stars were accreting volatile-depleted asteroidal or dust material or collisionally disintegrated, chemically differentiated planetary material. This was analogous to the cometary accretion episode used to explain the DAZ star, G74-7, by Alcock et al. (1986).

At the same time, Zuckerman & Becklin (1987) in a search for cool companions to WDs, discovered an infrared excess was associated with the cool ZZ Ceti DAZ WD G29-38 (Zuckerman & Becklin 1987) between wavelengths of 2 and 5 microns, that was clearly not emitted by WD itself. They derived the infrared color temperature of the IR excess source to be 1200 ± 200 K. At the distance of G29-38, the corresponding luminosity of the source is $5 \times 10^{-5} L_\odot$. Reach et al. (2005) obtained observations of the WD G29-38 with the cameras (4.5 and 8 mm), photometer (24 mm), and spectrograph (5.5 and 14 mm) of the Spitzer Space Telescope which is shown in Figure 3.2. In Figure 3.2, the spectral energy distribution of the G29-38 and the IR excess radiated by the dust cloud are shown with data obtained with the Spitzer Space Telescope.

Koester et al. (1997) obtained Hubble spectroscopic observations of G29-38 showing accreted metals from which they derived large abundances which revealed a high rate of accretion onto the WD. The accretion rate they derived from the high calcium abundance yielded $\dot{M} = 5 \times 10^{-15} M_\odot$ yr^{-1}. This rate of accretion, if interstellar in origin, is comparable to what would be expected if G29-38 was accreting from a dense interstellar cloud at the Bondi & Hoyle "fluid" rate. While the authors did not rule out interstellar accretion, they did report that circumstellar material (dust)

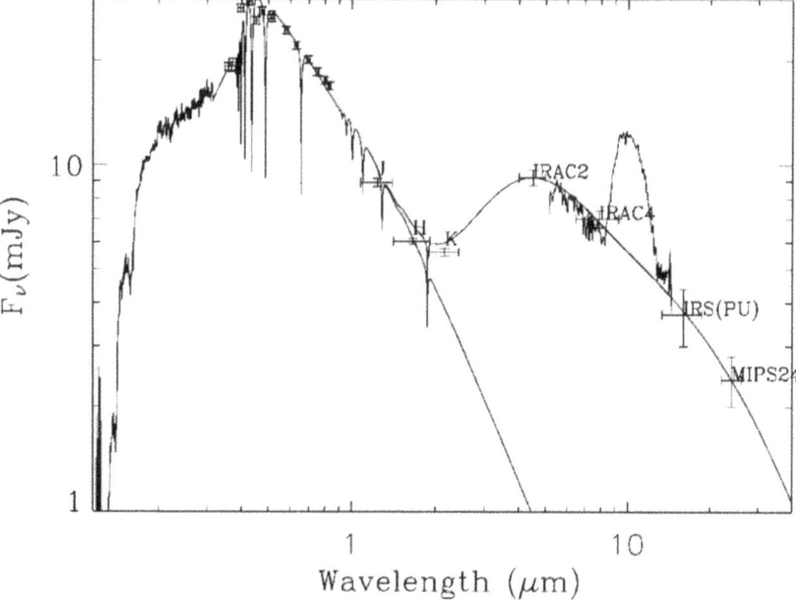

Figure 3.2. Spectral energy distribution of the WD G29-38 from the ultraviolet through infrared. The infrared observations and the ultraviolet through near-infrared observations and models are described in Reach et al. (2005) The solid line through the infrared data is a modified blackbody fit to the continuum. (Reproduced with permission from Reach et al. (2005), their Figure 1. © 2005. The American Astronomical Society. All rights reserved.)

was being heated. The source of the IR excess in G29-38 remained uncertain until a brown dwarf companion was definitively eliminated as the source of the IR excess by Kuchner et al. (1998) using the 10 m Keck telescopes. Instead, Kuchner et al. (1998, and references therein) showed that the IR excess was due to a heated dust cloud surrounding G29-38. The presence of dust near G29-38 has been verified by infrared observations with the Spitzer Space Telescope (Reach et al. 2005).

Although the hypothesis of interstellar accretion as the source of photospheric metals in cool WDs was refuted for the local WDs, an entirely new frontier of accretion of exoplanetary debris by WDs was rapidly emerging as a uniquely promising analytical probe of the remaining bodies of former exoplanetary systems. Finally, if one observes WDs very far from the local bubble in which the Sun resides, in regions of the galactic disk more populated with dense clouds, observations may well reveal evidence of both interstellar accretion and the accretion of exoplanetary debris.

It has become increasingly evident that the origin of metals in the atmospheres of cool WDs ($T_{\mathrm{eff}} < 25,000$) is tidal fragmentation of exoplanetary asteroids and, in some cases, extrasolar comets. This evidence arose from the discovery of circumstellar debris disks around DAZ WDs, beginning with the detection of the circumstellar dust disk around the ZZ Ceti pulsating DAZ star G29-38 (Zuckerman & Becklin 1987; Graham et al. 1990; Jura & Young 2014), and the eventual detection of dusty gaseous disks (Gänsicke et al. 2006) are now commonly interpreted as due to disruption of extrasolar asteroids or planetesimals that currently accreted onto the WD (Debes & Sigurdsson 2002; Jura 2003) thus explaining the source of the strong absorption lines of metals polluting their otherwise pristine monoelemental atmospheres.

To fully appreciate the breakthrough explanation of Debes & Sigurdsson (2002), it is important to place it within the broader context of planet formation and stellar evolution. In an exoplanetary system in the process of formation, even hundreds of planetesimals are present, each growing by accretion via collision coalescence with other planetesimals while some are also being destroyed by collision. The net result of accretion–destruction process is that fewer and fewer of the planetesimals survive while others grow in mass to have enough gravitational influence to dominate their orbital zones. As a system of planets establishes equilibrium in their stable orbits, this dynamical stability should endure for the hundreds of millions to billions of years while a main sequence star undergoes stable hydrogen burning (Duncan & Lissauer 1998). During this time, there are relatively little changes in their orbits. The Hill spheres (Hill 1886) of gravitational influence (the gravitational sphere of influence of a smaller body in the face of perturbations from a more massive body), which depend upon the ratio of a body of smaller mass to the mass of a larger body, should remain unchanged.

The radius of a Hill sphere, r_{H}, is given by

$$r_{\mathrm{H}} \approx a(1 - e)[m/3M]^{1/3} \tag{3.1}$$

where a is the semimajor axis, e is the orbital eccentricity, m is the mass of a small body (e.g., a natural satellite) orbiting a larger body of mass M (a WD, large planet, or the WD's main-sequence progenitor or advanced evolutionary progenitor). However, the ratio of the planet's mass to, for example, the central star's mass will eventually change when a main sequence star becomes a mass losing red giant and eventually an AGB star when the rate of mass loss increases by an order of magnitude or more via thermal pulse superwinds and planetary nebula ejection. This loss of mass by the central star marks the beginning of the potential de-stabilization of planetary orbits. The ratio of planet mass to stellar mass increases and thus the orbits and Hill spheres of exoplanetary bodies orbiting the central star will widen. Hence two orbiting bodies that were marginally stable to close encounters during the main sequence stage, can become unstable during the post-main sequence stage when mass loss of the central star become very substantial. The overall effect is to serve to de-stabilize the orbits of smaller bodies like asteroids or comets close to large planets or the satellite bodies of planets. If these stony, stony-iron, or icy bodies moved into orbits within the Roche limit of the central star, they would be tidally disrupted. The shredding effect of this differential gravitational force would lead to debris from the disintegrating body falling into a WD. As long as the smaller body is within the Hill sphere of gravitational influence of the larger body, its orbit will remain stable and not be perturbed such that it would escape and enter an independent orbit. Such gravitational perturbations could also disturb a putative Oort cloud of comets (an exo-Oort cloud) resulting in enhanced comet accretion (Caiazzo et al. 2017).

Given the breakup of asteroids by either destructive collisions between asteroids of different size scales or rotationally-induced fission of asymmetric bodies (see below), and barring direct impact, the material with angular momentum will orbit the central WD and form a disk. Unlike a mass transferring compact binary where orbital angular momentum loss shrinks the Roche lobe into contact with the mass donor star, and a thermal viscous instability (e.g., magneto-rotational instability) in a disk would drive accretion, what are the physical mechanisms that could drive accretion of this exoplanetary material onto a WD? It remains unclear how a ring or disk of pebbles and dust grains reach the WD surface, traversing a gap between the innermost ring and the WD photosphere. Within the realm of studies of interplanetary matter around the Sun, there are at least three potential candidates of interest. They are, in the order of potential importance, the Poynting–Robertson effect, the Yarkovsky effect, and the YORP effect.

3.2.1 Poynting–Robertson Effect

The Poynting–Robertson effect, also known as Poynting–Robertson drag (hereafter PR drag), named after John Henry Poynting and Howard P. Robertson, is a process by which solar radiation causes a dust grain orbiting a star to lose angular momentum relative to its orbit around the star. Solar photons are incident preferentially on the front or leading side of a dust particle that is orbiting at a right angle to the incoming solar radiation. The effect of PR drag is always to cause the particle to spiral inward toward the star. In other words, with respect to the star, the dust grain absorbs sunlight entirely in a radial direction, thus not affecting its orbit. However, the re-emission of photons, which is isotropic with respect to the grain, is anisotropic with respect to the star. This anisotropic emission causes emitted photons to carry away orbital angular momentum from the dust grain. The basis of PR drag is displayed in Figure 3.3.

In the case of the solar system, this affects dust grains with diameters between 1 μm to 1 mm in diameter. Farihi et al. (2008) showed that since the luminosities of the WDs manifesting dusty debris disks lie within the range of $10^{-3}L_\odot$ to $10^{-1.5}L_\odot$, then radiation pressure is not strong enough to remove grains of any size. Hence, there is no lower limit to the size of dust grains in the debris disks of WDs. They point out that the Poynting–Robertson effect will deplete the optically thin inner-most regions of the dusty debris disk of small dust grains on timescales of $\sim 10^5$ years or less (e.g., Hansen et al. 2006).

The timescale for particles of radius a and density rho to spiral inward from orbital radius r toward a star of luminosity L is (Burns et al. 1979)

$$t_{\mathrm{pr}} = \frac{4\pi}{3}\left[(c^2 r^2 \rho a)/LQ_{\mathrm{pr}}\right] \tag{3.2}$$

where ρ is the density, r is the orbital radius, a is the particle size, L is the WD luminosity and Q_{pr} is the coupling coefficient for radiation pressure and is of order unity for the case of geometric optics (appropriate for micron-sized and larger particles; Artymowicz 1988).

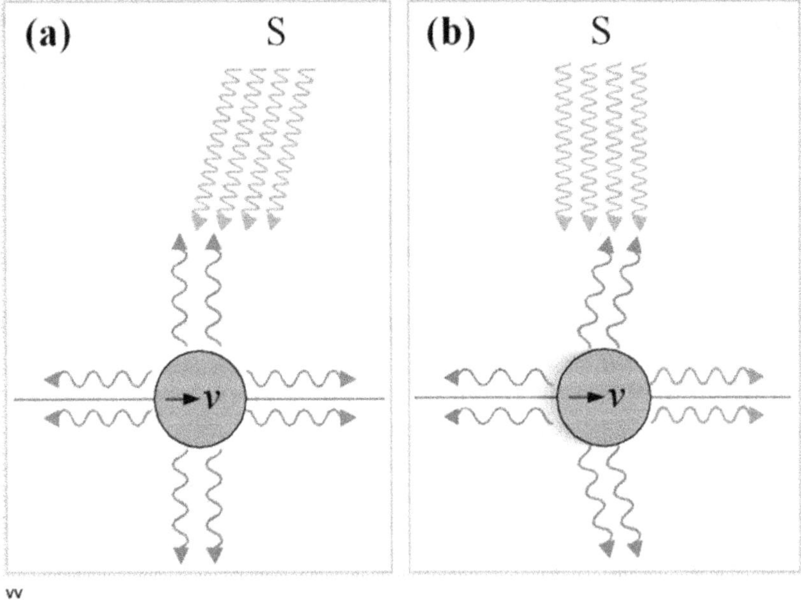

Figure 3.3. A particle's orbital momentum and thus its orbital speed is reduced due to continual PR drag on the particle due to the apparent displacement of photons from the direction of the Sun. This leads to a gradual spiral in toward the Sun. PR drag occurs on small particles (mm to a few cm range). (Courtesy of Observatoire Solaire. Michael Schmid, Wikimedia. CC BY-SA 2.5.)

3.2.2 Yarkovsky Effect

Yarkovsky effect, Polish investigator Ivan Yarkovsky noted that the daily heating of a rotating object in space would cause it to experience a force that, while tiny, could lead to large long-term effects in the orbits of small bodies, especially meteoroids and small asteroids (Yarkovsky circa 1900; see Opik 1951). When solar radiation during the "day" heats a rotating body and subsequent cooling occurs at "night", there is a delay between the absorption of radiation from the Sun (or central star), and the emitted photons as heat. This causes the warmest part of the rotating body to occur around the "early afternoon" region on its surface. Therefore, there is a difference between the direction of absorption and the direction of cooling re-emission of radiation. Due to this difference in direction, a net force (like a mini-rocket exhaust) is exerted along the direction of motion of the orbit. If the rotation and the orbital revolution of the small body are both prograde, then the body will be thrust into an expansion of its orbit, moving it away from the central star (Hartmann 2005 and references therein). However, if either the rotation or the orbital motion is retrograde, then the body will gradually spiral inward, and thus eventually accrete onto the central star. However, since either the revolution or rotation need to be retrograde in order to have in-spiral, it is more likely that most objects will spiral outward and thus the Yarkovsky effect is more likely, for the most part, to be ineffective for accretion onto WDs. The Yarkovsky Effect is illustrated in Figure 3.4.

Over astronomical timescales, a small body like an asteroid or a comet could suffer sufficient perturbations to carry it from a great distance into orbits much closer to the WD. This amplifies the probability that the small body will eventually be perturbed into crossing the WD tidal radius (its Roche limit) and hence eventually accrete onto the central WD. This is the dominant mechanism for bodies with diameters greater than ~100 m. This day to night (diurnal) variation in the direction of incoming radiation versus the direction of cooling re-emission is the dominant mechanism underlying the Yarkovsky effect but there is also a seasonal Yarkovsky effect stemming from high obliquity that would serve to cause small bodies to spiral inward. Rafikov (2011a, 2011b) noted that observations do not reveal WDs devoid of accreted photospheric metals yet having detected debris disks around them, which one might expect if outward migration of particles was a significant effect in the evolution of a debris disk. Of course, if the rotation and orbit of these larger bodies (~100 m) was retrograde, then they would migrate inward along with the flow of smaller particles due to PR drag.

3.2.3 YORP Effect

The YORP Effect is relevant to the accretion of exoplanetary material by WDs if an asteroidal body has protrusions that could act as "fins" projecting outward from an otherwise spherical former body. If say, one re-emits absorbed radiation from the central star from one fin, then the re-radiated light from the other fin is in the same amount but is not parallel to the incoming starlight. This introduces a torque which can affect the spin rate and the obliquity of its spin axis. This is one way in which

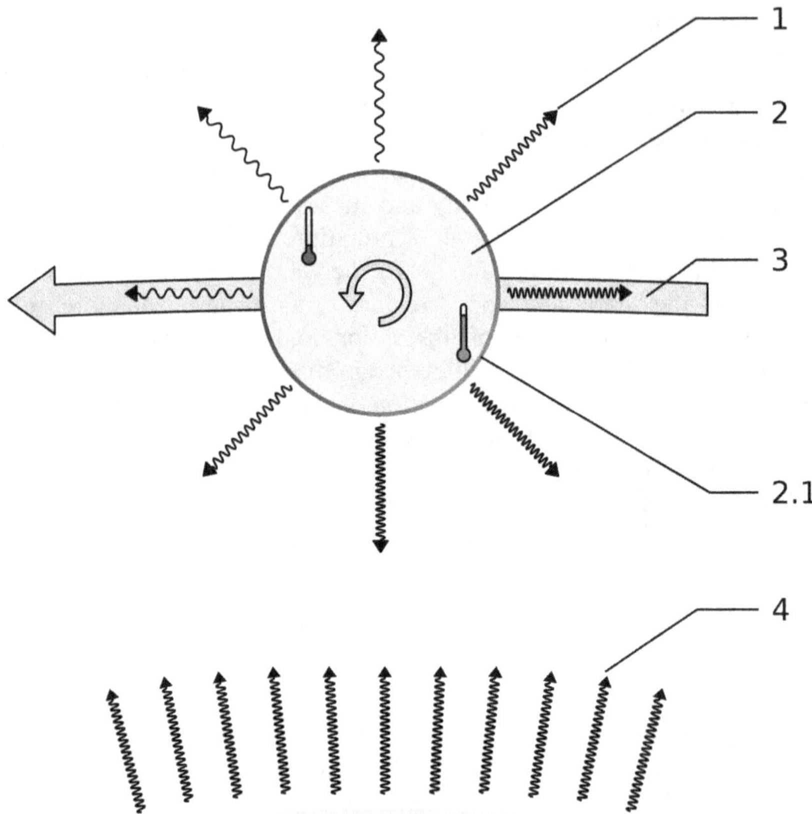

Figure 3.4. The re-emission of energy at infrared wavelengths causes the Yarkovsky effect. The scale of the Yarkovsky effect is dependent upon diurnal temperature differential and the mass, thermal conductivity, morphology, rotation rate and obliquity of the asteroid. (Courtesy of Observatoire Solaire. Graevemoore, Wikimedia. CC BY-SA 3.0.)

tumbling (rotating) and binary asteroids are produced. Most asteroids have orbital periods that are much longer than their spin periods. Thus, for most asteroids, the YORP effect is the secular change in the rotation state of the asteroid after averaging the solar radiation torques over first the rotational period and then the orbital period.

Veras et al. (2014a) have pointed out that many asteroids will not survive the giant branch stages of stellar evolution intact due to radiation-induced rotational fission. As a result, the available reservoir of debris to form disks around WDs would consist primarily of smaller fragments than expected. If the tensile strength of extrasolar asteroids is like a loosely bound "rubble pile" as for many asteroids in our planetary system, then it has been shown by Harris (1994) that those with radii greater than ~250 m disintegrate by rotational fission breakup when their spin periods becomes shorter than 2.33 h. The size of the asteroidal body as well as shape determines the amount of the effect and the effect accelerates with time. Smaller objects will spin up or down much more quickly. Similarly, the YORP effect intensifies for objects closer to the Sun. As the spin period gets shorter, the body would reach breakup at a spin period of ~2 hours (Veras 2016).

3.3 Comet Accretion onto a WD

The vast reservoir of comets in the solar system is virtually unlimited as first hypothesized by Oort (1950). Therefore, in the context of accretion events that contaminate a pristine WD atmosphere, comets and asteroids are both regarded as the strongest candidates. While their compositions and lower masses, compared to asteroids, have ruled them out in most cases of DA WD pollution (e.g., Zuckerman et al. 2003), it is inevitable that at least a few or more WD accretion events involve comet accretion. Moreover, while a number of authors have dismissed the role of comets in delivering metals to WDs, they are still regarded as being a viable source for the delivery of hydrogen onto cool WDs. Indeed, the mass of hydrogen appears to increase with the cooling age of accreting DA WDs (Dufour et al. 2007; Veras et al. 2014b). Thus, it is important to understand the physics of comet accretion and to recognize their spectroscopic signature.

The reservoir of comets associated with the work of Oort (1950) is now seen as consisting of three different components (Parriot & Alcock 1998): The innermost component, having a disk-like structure, is known as the Kuiper Belt. It lies between a distance of 40 AU to 1000 AU and is estimated to contain $\geqslant 10^{10}$ comets. There is a middle component first recognized by Hills (1981), consisting of a larger disk-like structure, lying between 1000 AU and 10^4 AU and containing between 10^{12} and 10^{13} comets known as the Hills Cloud. The outermost component of the Oort Cloud lies between 20,000 AU and 2×10^5 AU with a spherical distribution of roughly 10^{11} to 10^{12} comets. The innermost and middle components of the Oort cloud with their disk-like morphology are thought to have condensed from the protoplanetary disk while for the outermost component, the origin remains unclear.

The pioneering work by Alcock et al. (1986) first cited in Chapter 2 on comets as the possible source of metals in DAZ stars still represents a potentially viable, at some level, likely, source of contamination of the pure H atmospheres of DA stars. As such, it is useful to include a discussion of the stages in the process of accretion of extrasolar cometary debris first outlined in Alcock et al. (1986). The following discussion is a condensed version of what is presented in Alcock et al. (1986). The accreted mass of a typical comet is orders of magnitude smaller than the combined mass of the atmosphere and convection zone masses of DA stars. The masses of comets range from 5×10^{15} g to an average mass of 10^{17} g. The composition can be assumed to have the cosmic abundance. The metal abundances in a comet (Delsemme 1982) have a mass fraction of Ca of 0.004. Thus, a comet of mass 10^{17} g contains 6×10^{36} Ca atoms. This translates into a Ca/H abundance of 10^{-9} to 10^{-10} (Alcock et al. 1986).

From the most realistic WD envelope models (e.g., Fontaine & Van Horn 1976), the mass of the atmosphere layers plus the subsurface convection zone is always below 10^{23} g. However, typical accreted comet masses are $\leqslant 10^{18}$ g (Delsemme 1982).

For DA stars with their shallower convection zones than non-DAs at the same temperature, the diffusion timescales for DAs hotter than 15,000 K can be as short as tens to hundreds of days whereas for cool ($T_{\text{eff}} < 10,000$ K) DAs the diffusion timescales range between 1000 to 100,000 years. The spectroscopic signature of accreted calcium should be detectable for between $\sim$5000 and 10,000 years. For a WD with an Oort-type cloud, the estimated cometary accretion rate onto a WD is roughly 10^{-4} per year, when scaling down by 10^{-2} of the solar radius compared with the infall rate into the Sun (Joss 1974). In this shortened version of the processes involved in the accretion of a comet by a WD, five stages were identified by Alcock et al. (1986) as infall, impact, explosion, spreading phase, and diffusive separation.

The **infall** phase commences when the comet's orbit is perturbed into a trajectory that carries it within the Roche lobe of the WD, the tidal stress, Y, felt by the center of the comet, assuming a spherical nucleus, is

$$Y = GM_{\text{wd}}\rho b^2/2r^2 \tag{3.3}$$

where M_{wd} is the mass of the WD, ρ is the density of the ice-rich cometary nucleus (1.0 g cm^{-3}), b is the radius of an assumed spherical nucleus, r the distance from the center of the WD to the comet, and G the constant of gravitation. When $Y > Y_{\text{c}}$, where Y_{c} is the critical fracture strength of cometary material, the comet will fragment. The Shoemaker–Levy accretion event into Jupiter provides a possible example of such fragmentation albeit on a vastly smaller physical scale than comet accretion into a WD. Assuming that the comet is solid ice, $Y_{\text{c}} = 10^7$ dynes cm^{-2} (Hobbs 1974). With the above values, Alcock et al. (1986) found that the nucleus is large enough that fragmentation should occur before the comet impacts the WD surface layers. It should fragment before reaching the surface of the WD star if $b > 100$ m which corresponds to a mass of $\sim 10^{13}$ g. Fragments will be this size or smaller as they accrete onto the WD.

With the increasing tidal stress and heating, the comet will be ablated as it undergoes sublimation of its ice down to some depth in the nucleus and will be expected to undergo fragmentation. For a comet falling into the Sun, the sublimation of surface ices was estimated by Weissman (1980) to be only 5–15 meters lost prior to impact. However, the higher tidal stress of comet accretion onto a WD should lead to more extreme sublimation.

How deep will the cometary fragments penetrate upon impact before stopping and what is the mass of material above that depth of penetration? How much of the comet's mass ends being deposited in the convection zone? The impact velocity will typically be $\sim$5000 km s^{-1}. Using Eugene Shoemaker's (1963) estimate of the penetration depth of the iron meteorite that blasted out the Barringer crater (meteor crater) in northern Arizona, Alcock et al. (1986) estimate that the column mass above the depth, D, into the WD that the comet penetrates can be estimated by considering a high velocity jet of material with density ρ_j onto a target surface of density ρ_t by

$$D = L\left(\frac{\rho_j}{\rho_i}\right)^{1/2} \tag{3.4}$$

where L is the length of the jet. Alcock et al. (1986) use a more physical parameter for comet accretion, which is the amount of mass above the depth at which the comet comes to rest. For a WD,

$$M_{\mathrm{p}} \approx 4\pi R_{\mathrm{fwd}}^{2} L (\rho_j \rho_t)^{1/2}. \tag{3.5}$$

For $L = 1$ km and $\rho_t = 1$ g cm^{-3}, then M_{p} is $\cong 10^{24}$ g. This value of M_{p} is greater than the depths of convection zones of all WDs of any mass. Thus the comet should actually stop below the bottom of the convection zone in any cool WD (Alcock et al. 1986). Internal shocks and rarefaction waves heat and bring the comet to a halt. Buoyancy forces from the tremendous shock heating would cause the high entropy material, including the heated gas surrounding the remnant of the comet, to expand to reach pressure balance with the ambient, unshocked gas. This buoyancy causes the cometary material to float upward to the surface layers, with or without an explosion, and hence would be detected spectroscopically. This circulation will occur more rapidly than diffusion can remove the accreted material. There would be complete mixing of the material over the entire stellar envelope.

In the **spreading** phase, whether or not the explosion widely disperses the cometary matter over the surface of the WD, convection will efficiently spread the material over the entire photosphere of the WD. Even if the material is added over a small area initially, the expected differential rotation of the WD would spread the accreted metals along a latitude circle after which the material would start to spread latitudinally toward the poles. This circulation will occur more rapidly than diffusion can remove the accreted material. Hence, there would still be the complete mixing of the material over the entire stellar envelope. The spreading with time should lead to increases in the equivalent widths of the absorption lines of metals. If the material spread less rapidly because the WD is rotating slowly, then there should be marked variations in the absorption line profiles as a function of time. Hence, variations in the strength of line absorption should be detectable from time-tagged ground-based observations.

Pertinent to the above prediction in the spreading phase that marked variations in the absorption line profiles should be detectable, Farihi et al. (2018) undertook a long-term time series study of the DAZ star, GD56, using 3–5 micron infrared observations and optical spectroscopy to look for flux variations that might be manifested in both wavelength ranges. While the absorption lines remained constant, infrared flux variations were detected that were not manifested by a variation in the absorption line profiles of the accreted metals. Surprisingly, despite the observed IR flux variations as a function of time, the absorption line profiles remained unchanged. On the other hand, this may indicate that the spreading is considerably more rapid than is thought to be the case for a slowly rotating WD.

In the **diffusive separation** phase, the sinking of the accreted material by diffusion will lead to the exponential decay of the metal abundances as the accreted metals sink downward from the bottom of the convection zone. The absorption lines will weaken as this occurs until, in the absence of a new accretion event, the atmosphere of the DAZ star should revert relatively quickly to a pure DA spectrum.

If stars evolve through their advanced stages of evolution without losing their putative ex-Oort clouds of cold comet nuclei, then this spectroscopic signature of comet accretion will persist until diffusive separation has removed the heavy elements from the WD atmosphere; if the mean time between accretion events from an exo-Oort reservoir of comets (Hills 1981) is comparable to the diffusion timescales for cool DA stars, then this spectroscopic signature of accreted metals should persist if cometary systems within Oort clouds are common. In addition, the resulting contamination will be even greater if the comet is more massive than the average mass of 10^{17} g, or if the interval between accretion events due to comet influx (comet showers) is short compared to the diffusion time.

3.4 Does Accretion by ISM or Detached DA + dM Binaries Still Have a Role?

In one of the landmark papers in the field, Zuckerman et al. (2003), observed 120 WDs with the Keck telescope HIRES echelle spectrograph to assess the prevalence of atmospheric pollution by exoplanetary debris while also looking any for evidence for interstellar accretion, especially from the warm component of the ISM or accretion of wind or flare material from the M dwarf components of detached, moderately close DA + dM binaries. Long range, wind, flare, and CME accretion in such systems could still be a viable process for spectroscopic signatures of accretion into the WD. In 17 such binaries, they detected the Ca ɪɪ K-line in only 6 out of 10 showing possible contamination of the DA stars by wind and/or flare accretion. Of their sample of 100 DA stars without dM companions, 25% revealed Ca ɪɪ K absorption (Zuckerman et al. 2003).

An ISM origin for the Ca, even from the warm component of the ISM, with its lower particle densities than the cold component, is questionable unless the density is sufficient for Bondi–Hoyle accretion. No DAZ star showed interstellar Ca absorption. They did find three DA stars exhibiting ISM calcium and sodium D lines but they have no physical association with the WD. Zuckerman et al. (2003) also considered comet accretion from an exo-Oort cloud but found that the cometary model disagreed with the observed distribution of [Ca/H] unless the diffusion times of metals were much longer than diffusion theory predicted for a DA star (Koester 2009). This could happen if the DA star had unseen helium in the atmosphere but none was evident in their sample. On the other hand, they also found that asteroidal dust from a tidally disrupted asteroid as the source of accreted matter was not strongly supported by mid-infrared photometry of DAZ stars. However, the existing IR observations were not sensitive enough to reveal IR excesses at the WDs. in the Zuckerman et al. (2003) survey. They anticipated that the existing IR observations were not sensitive enough to reveal IR excesses at the WDs in their survey and pointed out that more sensitive observations with the Spitzer Space Telescope could change their conclusion about asteroidal dust as the source of the accreted matter. Indeed, this is exactly what happened as seen below.

The above conclusions about ISM accretion and WD pollution were conclusively strengthened in a further study by Farihi et al. (2010). Their study involved 167 DZ and DZA stars from the SDSS. They showed that while Eddington-type accretion rates (see Chapter 2) could explain the observed hydrogen abundances, it was clear that the Ca/H abundances being largely supra-solar in the DZA stars, could not be explained by ISM accretion. On the other hand, they presented evidence that the polluted cool WDs typically contain $10^{20} \pm 2$ grams of Ca, which is roughly the amount of calcium one finds in large asteroids. They did find however, based upon the hydrogen abundances in 37 DZA stars in their sample, that the amount of hydrogen increased with increasing cooling age. They also examined the space distribution and kinematics of the 146 SDSS DZ WDs from the Sloan Digital Sky Survey and found no correlations between their accreted calcium abundances and spatial kinematical distributions compared with known interstellar material. Farihi et al. (2010) also noted that two thirds of their sample are located above the Galactic gas and dust in a region where there is little evidence of any interstellar matter.

3.5 Disk Structure and Accretion Rates onto WDs

The structure and morphology of debris disks that form from tidally disrupted asteroid and comets accreting onto a WD was first firmly established with the breakthrough detection of emission lines revealing gas motions around accreting WDs (Gänsicke et al. 2006). Prior to this discovery, only rudimentary disk models were constructed and found to successfully fit the observed SEDs from optical to infrared (Jura et al. 2007). However, these early models fell short of providing the detailed disk structure on a solid footing. Jura et al. (2007) were among the first to fairly successfully model the observations with simple disk models. These insights were only possible from studies of the gas motions and not from the presence of an IR excess or models of dust emission. On the other hand, WDs that do not show IR excesses do not rule out the presence of disks. In Figure 3.5, the sizes of debris disks around the central WD are displayed to scale with respect to relative sizes.

Early descriptions of debris disk morphology were presented by Jura et al. (2003) who first showed that the explanation of the observed properties of the pulsating DAZ star G29-38 was due to a tidally-destroyed asteroid. Innumerable collisions among the debris of the asteroid grinds the fragments down to boulders, stones and dust. Over time, dynamical relaxation would lead to a flat ring over multiple orbital timescales. This opaque, flat disk/ring of dust would most closely resemble the extreme vertical thinness one sees in the rings of Saturn. The disk dust grains gradually accrete onto the photosphere, turning a pure DA atmosphere into one laden with accreted metals. Recent models have indicated that not all disks are as flat as the Jura (2003) models (cf. Kenyon & Bromley 2017).

Models of geometrically thin, optically thick dust disks with temperatures between 600 K and 1200 K gives well-fitted SEDs but a strong 10 micron silicate emission peak requires more modeling. This additional modeling needed to fit the broad 10 micron emission feature was provided by Jura et al. (2007b) and is displayed in Figure 3.6.

While only a small fraction of all metal polluted WDs show IR excess stars, the much larger sample of polluted WDs showing no IR excess, reveal spectroscopic evidence of having metal pollution from a tidally disrupted parent body. Of those WDs

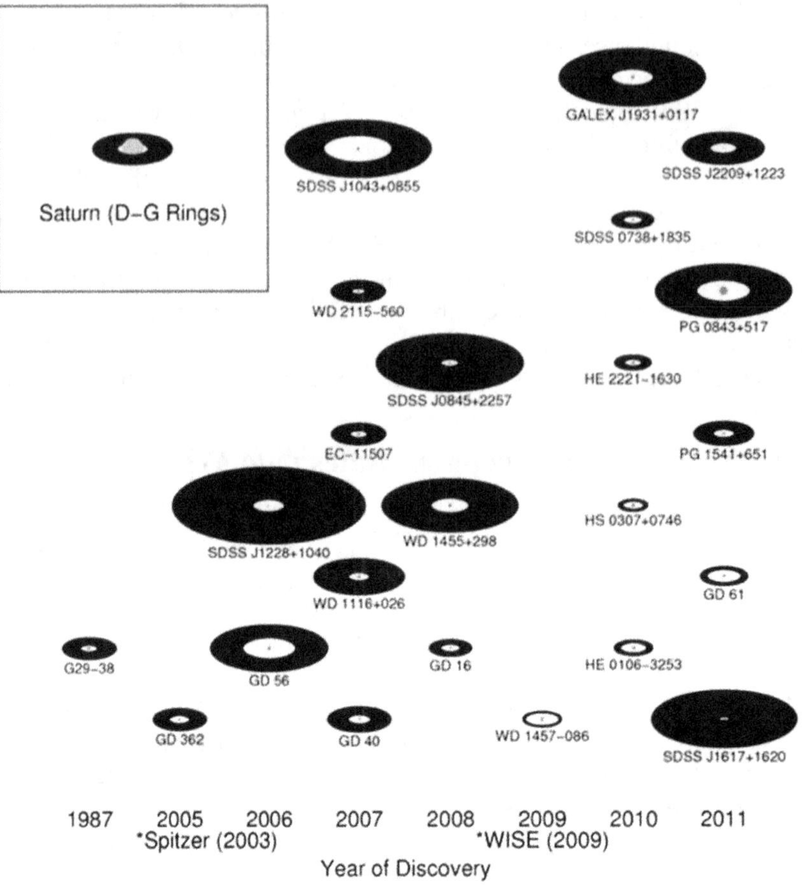

Figure 3.5. Relatives sizes of dusty debris disks from tidally disrupted asteroids and comets around WDs detected by Spitzer Space Telescope and WISE with the name of the WD and the year of discovery of the debris disks. The relative size of Saturn's rings is shown by comparison. (Reproduced with permission from Dr. John Debes, STScI.)

that exhibit the highest degree of metal pollution, only about 50% show a dust debris disk. It can be stated with confidence that metal pollution is strongly connected to the presence of a debris disk. If however, there is no IR excess, then the heating of the orbiting dust by the stellar radiation is insufficient to produce IR emission.

In this geometry, the disk height to width ratio is negligible, and only a tiny fraction of particles are subject to the full stellar radiation and drag force. For stellar radius R and effective temperature T_{eff}, the dust temperature T at disk radius r is given by (Chiang & Goldreich 1997)

$$T = \left(\frac{2}{3\pi}\right)^{1/4}\left(\frac{R}{r}\right)^{3/4} T_{\mathrm{eff}}.$$

(3.6)

Like accretion disks in cataclysmic variables, this disk model has three adjustable free parameters, the inner disk radius r_{in} and the outer disk radius r_{out}

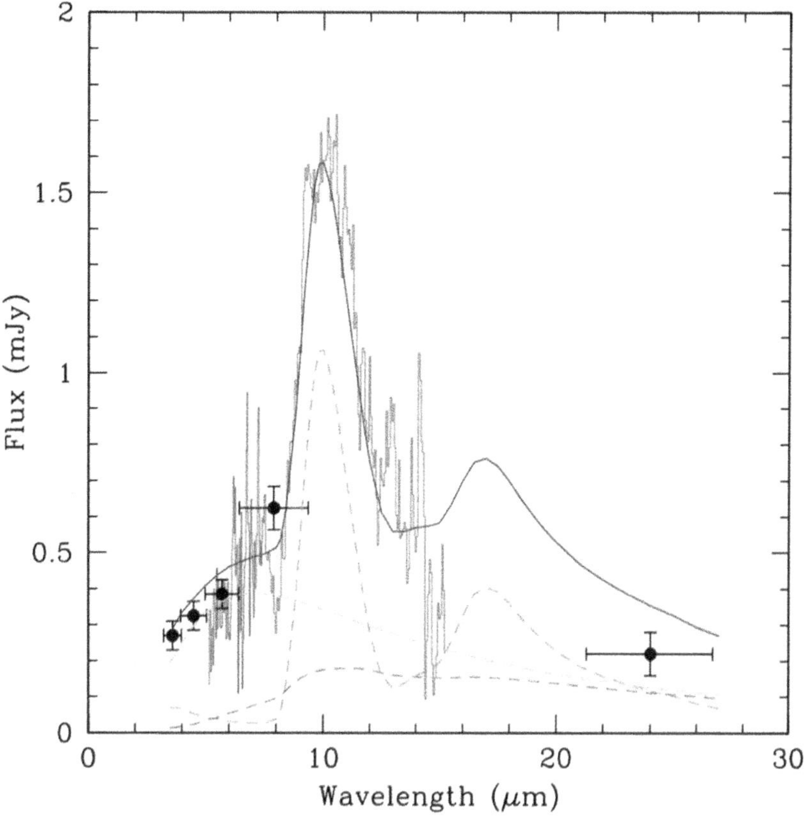

Figure 3.6. A complete disk model for the infrared continuum and the strong dust silicate emission features in this case for GD 362 from Jura et al. (2007b). Data points with error bars are Spitzer photometry with IRAC and MIPS, while the magenta line is the IRS low-resolution spectrum. The solid blue line is the total model dust spectrum, which is the sum of emission from three disk components: (1) cyan is an innermost, opaque disk; (2) red is a radially intermediate zone with a vertical temperature gradient; (3) green is an outermost, optically thin region (Jura et al. 2007b). The only dust species used in the model are olivines. (Reproduced with permission from Jura et al. (2007b). © 2007. The American Astronomical Society. All rights reserved.)

and the disk inclination i. Following Jura et al. (2003), the flux radiated by the ring at distance d is given by

$$F = \frac{2\pi \cos(i)}{d^2} \int_{r_{\rm in}}^{r_{\rm out}} B_\nu(T) r\, dr \qquad (3.7)$$

where $B_\nu(T)$ is the Planck function. Defining $x = h\nu/kT$ with the dust temperature in Equation (3.7), the flux radiated by the flat debris disk is expressed by

$$F \sim \frac{12\pi^{1/3} R^2 \cos(i)}{d^2 (2kT_{\rm eff}/3h\nu)^{8/3}\left(\dfrac{h\nu^3}{c^2}\right)} \int_{x_{\rm in}}^{x_{\rm out}} \left(\frac{x^{5/3}}{e^x - 1}\right) dx. \qquad (3.8)$$

From this expression, synthetic spectra can be generated for comparison with the infrared observations from Spitzer and other IR missions. The success of this theoretical formulation for the flux and remaining uncertainties have been discussed in detail by Farihi (2016).

3.6 Gaseous Disks

For gaseous disks, Hartmann et al. (2011, 2014) constructed non-LTE, axially symmetric models of gas using a standard Alpha-disk prescription, where the gas emission is driven by viscous heating. The full disk components are built up from concentric rings, where each individual ring is fully solved for radiative transfer and structure, comprising hydrostatic and radiative equilibrium, population of atomic levels, metal-line blanketing, and irradiation by the central object. These models are relatively successful at reproducing the broad morphologies of the Ca II triplet line profiles, including asymmetries due to eccentric rings. If accurate, these models may constrain the chemical composition of the debris; the lack of observed C II emission just redward of the Ca II triplet is inconsistent with a chondritic mass fraction of carbon (Hartmann et al. 2011), but consistent with the terrestrial value (Hartmann et al. 2014). The models also yield a strong upper limit to the H mass fraction of 1%, consistent with volatile-depleted parent bodies, and finds innermost radii for the gas disk at modestly wider orbits than those found by previous studies (Gänsicke et al. 2008; Melis et al. 2010), supporting previous conclusions that the gas and dust coexist spatially. However, caution is warranted in the above interpretations as a major weakness of the accretion disk model are strongly over-predicted emission lines of Ca II H & K, and additional metal lines (e.g., Mg II) that are not observed.

Until very recently, only a handful of gaseous disks were known to exist. Manser et al. (2020) found only five out of 7702 WDs show evidence of a gaseous disk component based on the Ca II emission line triplet at 8600Å. Their finding implies that only ~4% of dusty debris disks have gaseous components. However, Melis et al. (2020) have reported the optical, low resolution, spectroscopic discovery of nine more accreting WDs with gaseous debris disks based upon the emission line signatures in their spectra and point out that the present searches have concentrated in the northern hemisphere and suggest that more will be detected among southern hemisphere degenerates.

3.7 Accretion Rates

One of the earliest estimates of the expected accretion rate of disrupted asteroidal dust onto a WD was presented by Graham et al. (1990). From Poynting–Robertson drag alone, their rate was of order 10^{15} g yr^{-1}, which is within the range of subsequent estimates given below from studies of much larger samples of accreting cool WDs. Their estimate assumed a steady state between dust accretion and gravitational diffusion.

Accretion at a rate this small ($\sim 1.6 \times 10^{-18} M_\odot$ yr^{-1}) is nine orders of magnitude smaller than the average accretion rate seen in cataclysmic variables (Dubus et al. 2018). While this accretion rate is very small, it is roughly what is expected for an average mass DA WD moving through the intercloud medium of the ISM at a thin disk space velocity accreting at the Eddington rate, and is sufficient to be manifested in the WD photosphere as absorption lines of metals. Indeed, the observed abundances of accreted heavy elements in cool WDs pointed to accretion rates of metals between 10^6 g s^{-1} and 10^{10} g s^{-1} (Farihi et al. 2009, 2010). Poynting–Robertson drag on a dusty debris disk for grains of any size yields accretion rates $\sim 10^8$ g s^{-1} or more. On the other hand, radiation pressure for the luminosity range of cool polluted WDs (10^{-3} to $10^{-1.5} L_\odot$) was insufficient to remove grains of any size (Farihi et al. 2008).

For non-DA WDs, for example DBZ stars, the average accretion rate is found to be 1.4×10^8 g s^{-1} (Gänsicke et al. 2012; Jura & Xu 2012). However, as pointed out by Rafikov (2011a), some polluted, cool non-DA stars require much higher accretion rates ($>10^{10}$ g s^{-1}) derived from their observed abundances, due in large measure to their deeper, more massive convection zones compared with DA stars in the same temperature range. To achieve these higher accretion rates, Rafikov (2011a) formulated a scenario whereby rapid accretion of metals by a WD could occur if there is interaction between the dust disk particles, which spatially coexist with a metallic gas disk component. As dust particles from the dusty component evaporate at the sublimation radius some distance from the WD, they intersperse with the metallic gas which leads to a difference in the orbital velocities of the solid and the gaseous components, with the latter having a slower orbital velocity. According to Rafikov (2011a), this would lead to an aerodynamic drag on the solid component of the debris disk and cause solid particles to migrate inward toward the central WD. Since at the sublimation radius, the solid particles would evaporate, thus adding to the gas, they would increase the gas density of the disk and a consequent increase of the drag on the solid particles, which in turn would increase the sublimation rate. Depending on the viscosity of the gas disk (it must be low) and the degree of aerodynamic coupling between the dust disk component and the gas disk component (it must be efficient), this positive feedback could lead to runaway accretion thus providing the higher accretion rates seen in some polluted non-DA WDs.

Using Spitzer Space Telescope IRAC observations, Farihi et al. (2016) derived accretion rates for 108 polluted WD of which 49 were DAZ stars and 59 were DBZ stars. These objects have known or estimated calcium abundances and are displayed in Figure 3.7 plotted versus the log of the accretion rate in g s^{-1}. The ordinate of this plot clearly displays the range of known accretion rates onto metal-polluted DA stars and metal-polluted non-DA stars. The highest accretion rate for DAZ stars is 2×10^9 g s^{-1} while some DBZ stars have accretion rates exceeding this threshold for DAZ stars.

This subset of DBZ stars have accretion rates one to two orders larger than the DAZ stars. This can almost certainly be attributed to the deeper, more massive convection zones in DB stars and hence the much longer diffusion timescales. If one assumes a constant accretion rate over a given DBZ WD cooling lifetime, then estimates of the maximum accreted mass can be obtained. For example, if one takes the typical average accretion rate of 1.4×10^8 g s^{-1} for DBZ stars (Gänsicke et al. 2012; Jura & Xu 2012) and a cooling age of 2×10^8 years, then an average accreted mass of $\sim 10^{24}$ g lands on a typical DBZ star. This is roughly the mass of the dwarf planet, Ceres, the largest asteroid in our solar system. However, if one takes typical dusty disk lifetimes to be

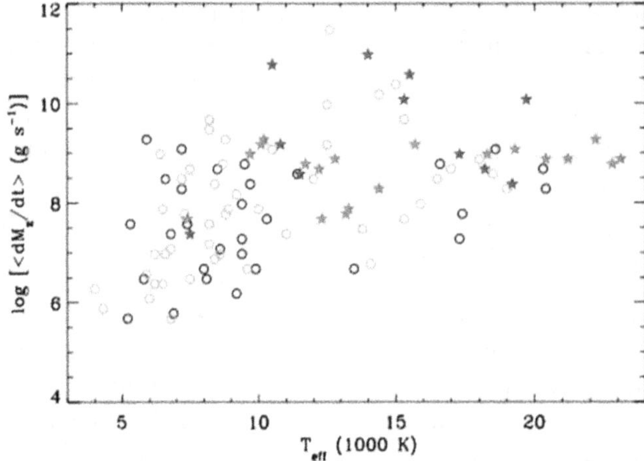

Figure 3.7. Total heavy element accretion rate as a function of effective temperature for 108 WDs with known (or estimated) Ca abundance that have been observed with Spitzer IRAC 49 hydrogen-rich WDs exhibiting infrared excesses are displayed as red stars, but with black circles no IR excess was detected. Similarly, helium atmosphere WDs ($n = 59$) with infrared excess are plotted as blue stars, and as gray circles if no IR excess. (Reproduced with permission from Farihi (2016); his Figure 10. Copyright © 2016, with permission from Elsevier.)

10^5 to 10^6 years, then the total accreted masses are $M_{\odot z}\tau_{\mathrm{disk}} \sim 10^{21}$ g, three orders of magnitude lower than the average accretion rate for the DBZ stars cited above.

The most typical dusty disk lifetimes are 10^5 to 10^6 years implying total accreted masses are $M_{\odot z}\tau_{\mathrm{disk}} \sim 10^{21}$, three orders of magnitude lower than the value for the average for the DBZ stars cited above. Stepping back with an overall examination of debris accretion onto WDs, there specifically appears to be a tension between the estimated overall percentage of debris accretion onto H-rich and He-rich degenerates which amount to as much $\sim 10^{22}$–10^{23} g for the highest mass disrupted bodies residing within the reservoir of the convective zone) and the dynamical rarity of the largest asteroids like Ceres and Vesta or larger, tidally disrupting for $\sim 1\%$ of systems at any instant of time. The sheer large fraction of WDs that show any accretion: 25%–50% based on spectroscopic estimates, implies that the process is very robust. Thus, it is possble that resonances can play as well (Debes et al. 2012) because resonances can drive a slowly declining steady flux of smaller bodies which may help to account for the full population of WD accretion.

Recently, Bauer & Bildsten (2019 and references therein), found ongoing accretion rates up to 10^{13} g s^{-1} are possible if hydrogen-rich WDs suffer the thermohaline instability in their mixing layers where metals are present. This type of convective transport results from having two different density gradients in a gas whose fluid components, say metals and hydrogen or helium, have different rates of diffusion. Since the accreted material's mean molecular weight is larger than a DA WD's atmosphere, an inverted mean molecular weight gradient can lead to additional mixing due to thermohaline instability even if the gas is convectively stable. This process is also referred to as double diffusive convection. While some doubts exist that a thermohaline instability can operate in an accreting WD (see for example Koester 2013), this possibility merits further exploration because it would lead to even higher required accretion rates to be consistent with the absorption line abundances derived with ordinary mixing length convection (e.g., Wachlin et al. 2017; Bauer & Bildsten 2019). The magnitudes of the accretion rates that are documented above are based upon the expectations of diffusion theory, but thermohaline mixing, if feasible in an accreting WD, would require higher rates of accretion by an order of magnitude or more, in order to be consistent with the derived metal abundances of accreted metals.

If the Bauer & Bildsten (2018) maximum accretion rates derived for H-rich degenerates are correct, then one must explain why DAZ stars would have higher accretion rates than DBZ degenerates. The difference between the maximum accretion rates could be resolved if the thermohaline instability is present but not as strongly as calculated by Bauer & Bildsten (2018). At the time of this writing, the question of the maximum accretion rates for DAZ versus DBZ stars remains unresolved.

An important new finding on accretion rates was presented by Cunningham et al. (2020) who reported a 4.4σ detection of X-rays from the exoplanetary debris-polluted WD, G29-38. From the measured X-ray luminosity, they obtained an instantaneous accretion rate of $\dot{M} = 1.63 - 0.40 + 1.29 \times 10^{9}\,\mathrm{gs^{-1}}$. This accretion rate is completely independent of all other debris accretion rates, which are model-derived. The authors point out that this accretion rate is higher than estimates from past studies of the photospheric abundances of G29-38, which suggests the possibility that convective overshoot may be an important factor when modeling the spectra of debris-accreting WDs.

3.8 Diffusion in WDs Accreting Exoplanetary Matter

If exoplanetary debris is accreting at the same rate that it is diffusing, then accretion–diffusion equilibrium has been established. Thus, the metals abundances corresponding to this equilibrium will follow from absorption line profile fitting.

Koester (2009) determined diffusion velocities and diffusion timescales for C, Na, Mg, Si, Ca, and Fe. Following Dupuis et al. (1992) and Koester (2009) but with slightly different notation, we have the time variation of the mass fraction of a given element to be

$$M_{\mathrm{cvz}}\frac{dX}{dt} = X_{\mathrm{acc}}\dot{M} - 4\pi^2 X\rho v$$

$$= X_{\mathrm{acc}}\dot{M} - \frac{XM_{\mathrm{cvz}}}{\tau_{\mathrm{D}}} \tag{3.9}$$

where the first term on the right-hand side is the accretion rate of that element, and the second term is the rate of gravitational diffusion (where we use a positive velocity to represent elements diffusing downward). Taking the diffusion timescale, accretion rate and diffusion velocity (always positive for downward diffusion) to be constant, then the solution to this equation is

$$X(t) = X(0)e^{-t/\tau_D} + \left(\frac{\tau_D X_{acc}\dot{M}}{M_{cvz}}\right)[1 - e^{-t/\tau_D}]. \tag{3.10}$$

At the very onset of accretion, when $t \ll \tau_D$, diffusion will have little effect and therefore the chemical abundances will build up linearly as a function of time with the slope given by the rate of accretion. The abundances will manifest the composition of the original parent object. However, after several diffusion time-scales, the rate of accretion and rate of diffusion of a given element should reach an equilibrium or steady state which will sustain the abundances at an unchanging or steady-state level, X_{ss}, given by

$$X_{ss} = \frac{\tau_D X_{acc}\dot{M}}{M_{cvz}}. \tag{3.11}$$

In that case, the abundances ratios of any two diffusing elements will be in the same ratio as their diffusion timescales. Thus, in terms of the abundance ratio of the two elements in the parent body

$$\frac{X_1}{X_2} = \frac{\tau_{D1}X_{acc1}}{\tau_{D2}X_{acc2}}. \tag{3.12}$$

Once the accretion rate has started to decline because the source of disk material is being consumed without being replaced by new material entering the disk, the photospheric metal abundances will start to undergo an exponential decline proportional to each element's respective diffusion timescale. It follows that the abundance ratios will no longer be those of the parent disrupted body.

3.9 Spectroscopic Abundances and the Chemical Composition of Exoplanetary Matter Accreted by WDs

The extraction of chemical abundances from the spectra of WDs that are accreting exoplanetary debris proceeds with essentially the same standard techniques of stellar spectroscopy. WD synthetic spectral codes are employed for helium-rich or hydrogen-rich atmospheres to secure robust determinations of the surface gravity and effective temperature of the polluted WDs. The most widely used codes having the most advanced input parameters (line broadening, damping constants, etc.) for cool H-rich (DA) stars and He-rich (non-DA stars) are from D. Koester (2009) in Kiel, Germany and Blouin et al. (2019 and references therein) at the University of Montreal, Canada. The detected lines for each chemical element are compared to theoretical line profiles and chemical abundances are determined for each element by curve of growth analysis. The observed absorption line profiles and their equivalent widths are fitted to those line profiles calculated from the synthetic spectra and the abundances are varied until there is a match between the theoretical equivalent widths and the observed equivalent widths for several lines of a given element.

On the basis of chemical composition of the accreted matter onto WDs, the infrared spectroscopy of Reach et al. (2005) provided the first key compositional clue that the accreted matter could not be interstellar because the broad 10 micron silicate emission feature was in strong disagreement with interstellar silicates (see Figure 3.5 and accompanying discussion). They also identified glassy, silicate dust grains of olivine minerals commonly found in the mantles of the terrestrial planets, meteorites and asteroids. Furthermore, the absence of the carbon-rich polycyclic aromatic hydrocarbons (PAHs) in the IR spectra coupled with the known rocky debris disks, underscored a carbon deficiency in the accreted matter. Indeed, the range of metal abundance ratios of the rock-forming elements demonstrates that the impacting bodies were differentiated with metal-rich cores, rocky mantles, and crusts that are very similar to the cores, mantles, and basaltic crusts in our own solar system. They are markedly volatile-depleted, notably carbon deficient and dry like rocks on Earth, containing virtually no water (Gänsicke et al. 2012; Jura 2006, 2008; Jura et al. 2012; Xu et al. 2014). In Figure 3.8, an important example of the use of the range of abundances ratios derived from polluted WDs with the abundances ratio variations of solar system objects like meteorites, asteroids.

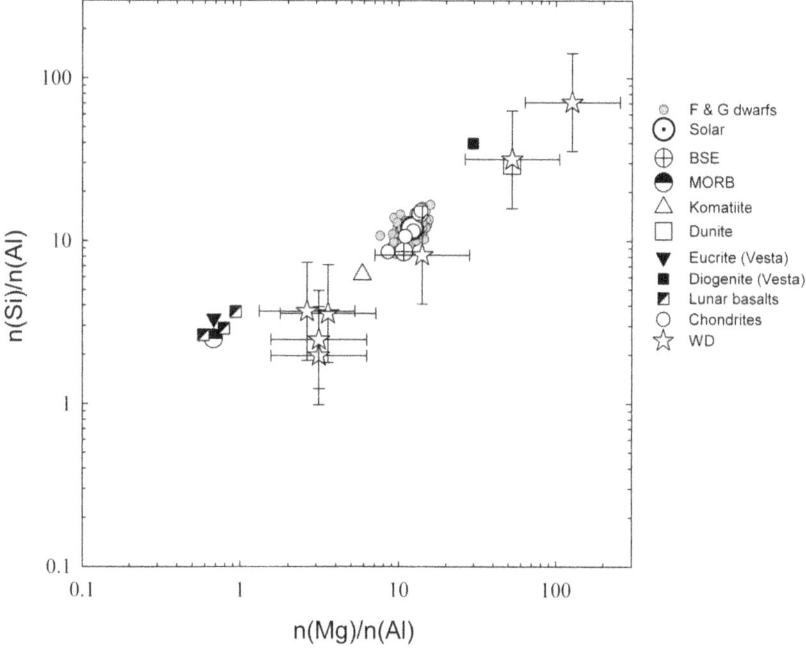

Figure 3.8. Si/Al versus Mg/Al abundance ratios by number for debris-accreting WDs (yellow stars), F and G main-sequence star progenitors of WDs (green circles), and various solar system objects and materials for reference. For example, the bulk Earth (BSE), the Sun, and CI chondrite meteorites (CI). The F and G star abundance ratios are similar to the Sun (the solar symbol) and chondritic meteorites. Olivine-rich HED meteorites associated with the volcanic asteroid 4 Vesta. Volcanic basaltic lava proliferates on the surfaces of planets from the solidification of lava, including terrestrial mid-oceanic rift zone basalts (MORB) including Komatiite, a primitive type igneous rock. The spread in WD data is similar to the trends defined by igneous differentiation of mantle and crust. Data are from Jura & Young (2014) and references therein. (Reproduced with permission from Zuckerman & Young (2018), their Figure 3.3. Copyright © 2018, Springer International Publishing AG, part of Springer Nature. With permission of Springer.)

The igneous differentiation of mantle and crust referred to in the caption to Figure 3.8 is brought about by the melting of a parent body due to rapid heating by short-lived radionuclides. An important discovery regarding these radionuclides by Jura et al. (2012) reveals that extrasolar asteroids show a factor of 100 variation in the iron to aluminum ratio. They attribute this variation to igneous differentiation resulting from intense heating by the radioactive decay of ^{26}Al, that is comparable to that in the solar system's protoplanetary disk (i.e. the solar nebula). They point out that conventional view of the formation of the solar system being associated with an unusually high amount of ^{26}Al is wrong. Overall, the chemical makeup of the majority of disrupted extrasolar asteroids supports the conclusion that the exoplanetary physical processes and evolution that they underwent are not unlike the physical processes that produced rocky, metallic, differentiated parent bodies in our planetary system. This implies that both collisions and differentiation are key processes in exoplanetary systems.

There is evidence of accreted debris onto WDs suggesting that only the lithosphere of a differentiated parent body accreted but not surprisingly, other accreted debris that suggests the lower mantle and core accreted onto a WD. The analysis of the polluted WD NLTT43806 by Zuckerman et al. (2011) reveals that the aluminum abundance is enhanced as well as calcium. Both of these elements are refractory (i.e., condense at the highest temperatures in the innermost regions of a protoplanetary disk). The enhanced aluminum may point to an accreted parent body in which only the outermost layers of a differentiated rocky parent body was accreted. This could happen if the parent body suffered a collision with another rocky body resulting in much of its lithosphere stripped off before the trajectory of the fragments carried them to within the tidal radius of the WD. On the other hand, they point to other cases, for example, GD40 where the photospheric abundance ratios indicate accretion of the lower mantle and core of a shattered rocky parent body (Klein et al. 2010).

The case of NLTT43806 is illustrative of the methodology that is widely employed in the field to ascertain the nature of the disrupted parent body. In view of the enhanced abundance of aluminum in the atmosphere of NLTT43806 and given that Al is the third most abundant element in Earth's crust, could only the disrupted parent body's crust have been accreted? The standard analytical tool used to identify the nature of the parent body is to compare the abundances ratios from the WD model atmosphere absorption line fits with the composition of various types of meteorites both differentiated (e.g., irons, stony irons, brecciated, achondrites) and undifferentiated (e.g., Carbonaceous chondrites, ordinary chondrites). To relate photospheric element abundances to relative abundances in a rocky parent body or bodies requires knowledge of whether a given polluted WD is in the "building-up," "steady-state," or "declining" phase of accretion (these phases are discussed in Section 3.8). These terms are most clearly defined and visualized in a situation where only a single parent body is or has been accreted onto a star.

The work by Zuckerman et al. (2011) provides a superb analytical example of the intercomparison of abundance ratios of the accreted metals detected in the photosphere of WDs with the abundance ratios of various solar system objects. in Table 3.1 below,

Table 3.1. Element Abundances by Number in NLTT 43 806, G149-28, and the Solar System

Z/Al	NLTT 43 806 Photosphere	NLTT 43 806 Steady-state Accretion	G149-28 Photosphere	Solar	Bulk Earth	Earths Continental Crust	30% Crust 70% Upper Mantle (by Weight)	Lunar Mare Basalt	CI	Euc	How
Ca/Al	0.5	0.61	1.35	0.76	0.72	0.425	0.57	1	0.73	0.76	0.70
Mg/Al	3.16	2.82	8.5	12.3	11.8	0.428	3.11	0.77	12.6	0.74	1.54
Fe/Al	0.63	1.07	5.75	10.2	9.1	0.405	0.61	1.31	10.5	1.1	1.26
Si/Al	2.5	2.68	<9.3	12.0	11.0	3.07	4.48	3.65	12.0	3.3	4.1
Na/Al	0.32	0.27	<4.7	0.69	0.19	0.32	0.26	0.043	0.69	0.038	0.052
Ti/Al	0.011	0.0166	0.05	0.029	0.0265	0.036	0.0343	0.236	0.029	0.037	0.031
Cr/Al	0.011	0.0178	<1	0.155	0.148	0.001	0.0147		0.158	0.019	
Ni/Al	0.032	0.0575	0.69	0.57	0.49	0.000 57	0.0105		0.575		

Notes. The 30/70 Earth crust/mantle division is apportioned by weight. However, all abundance ratio entries in the body of the table are by number of atoms, including those in the column headed 30% crust 70% upper mantle. Abundance ratios for G149-28 in the steady-state phase are approximately the same as the photospheric ratios because the various settling times are so similar (see Table 2 in Zuckerman et al. (2011)). See Sections 4.1 and 4.2 for explanation of the steady-state accretion and crust/mantle columns. Solar abundances from Lodders (2003); bulk Earth from Allegre et al. (1995); Earth crust and upper mantle from Anderson (2007); CI meteorites from Lodders (2003) and Jarosewich (1990); Eucrite and Howardite meteorites from Jarosewich (1990) and Kitts & Lodders (1998). The Ni abundance in Eucrites is quite variable (Kitts & Lodders 1998). See also Hawkesworth & Kemp (2006) for slightly different average Earth crustal abundances.

adapted by permission of Zuckerman et al. (2011), it is seen how closely or dissimilar the derived abundance ratios are, compared with the abundance ratios of different solar system objects. In columns in Table 3.1 are, (1) the abundance ratio of each detected element with respect to Al in the photospheres of the two DAZ WDs NLTT43806 and G149-28, (2) the abundance ratio in the photosphere of LTT43806, (3) the abundance ratio for NLTT43806 by assuming a steady-state equilibrium between accretion and diffusion, (4) the photospheric abundance ratio of G149-28, (5) the abundance ratio in the Sun (from Lodders 2003), (6) the abundance ratio of bulk Earth (from Allegre et al. 1995), (7) the abundance ratio in the Earth's continental crust (from Anderson 2007), (8) a compositional composite of 30% crust and 70% mantle (Anderson 2007), (9) lunar mare basaltic, (10) the CI carbonaceous chondrite meteorite (from Lodders 2003 and Jarosewich 1990), (11) the Eucritic meteorites and (12) Howardite meteorites from Jarosewich (1990) and Kitts & Lodders (1998).

When a debris accreting WD has left the steady state accretion phase and the accreting material is not being replenished, the WD enters the declining phase where in the convection zone, the heavier elements like Ni and Fe become less abundant relative to the lighter lithophile elements. This is seen in the photosphere of NLTT 43806 and is consistent with the non-detection of a dusty debris disk which is expected with NLTT 43806 being in the declining accretion phase. If one assumes a model in which during the steady state phase, the ratio by number of Fe/Al would be $\sim$10.5, which is the same ratio of Fe to Al in CI chondritic meteorites. Using the relative abundances of Fe and Al as well as the tabulated diffusion timescales, Zuckerman et al. (2011) deduced that 6.6×10^4 years ago, the [Fe/Al] ratio would have been approximately 10 which is what is expected for the steady state accretion phase. They showed that at that time 66,000 years ago, the mass accretion phase of Fe would have been 735 larger than in the declining accretion phase and the mass accretion rate of Al would have been approximately 50 times than Al. The increase of the accretion rates for the seven additional detected elements would have ranged from a factor of 25 for Na to 1180 for Ni.

While Zuckerman et al. (2011) did raise the possibility that NLTT43806 could be in the declining accretion phase and could not entirely rule it out, it would require extreme conditions and low probability events. To decide definitively between their lithospheric model and the declining accretion phase, more abundance data is needed. Specifically, the detection or non-detection of relatively heavy, yet relatively volatile elements like Mn and K would be substantially more abundant in the lithospheric model.

Both the Eucritic and Howardite meteorites are achondritic meteorites associated with formation on the surface of the second largest asteroid, 4 Vesta and thus members of the Howardite–Eucrite–Diogenite (HED) meteorite group. These achondrites are thought to have crystallized within the range of ages between 4.43 and 4.55 billion years ago. Our knowledge of this association with 4 Vesta was solidified by NASA's Dawn spacecraft mission to 4 Vesta during which Dawn remained in orbit for a year carrying a visible camera, a visible and infrared mapping spectrometer, and a gamma-ray and neutron spectrometer to analyze 4 Vesta's chemistry and geology.

Numerous observed examples linking the accreted chemical abundances to the different types of parent bodies that were tidally disrupted are documented in the excellent reviews by Farihi (2016) and Zuckerman & Young (2018). One of the most remarkable cases is the abundance measurements of the polluted DAZ WD GD362 which reveals 17 separate elemental identifications in the accreted material (Zuckerman et al. 2007). Xu et al. (2013a, 2014) found that the composition of the accreted material suggests that the parent body is similar to a class of stony-iron meteorites known as mesosiderites which appear to originate from the mantle-core interface of a differentiated asteroid. These meteorites are rare in our solar system, comprising only ~1% of falls yet the fortuitous detection of a mesosiderite-like accreting body provides confirmation that collisional destruction of differentiated asteroids occurs in exoplanetary systems as well. However, of equally great interest are polluted WDs in which the disrupted, accreted parent body was volatile-rich and undifferentiated. Do they exist? The answer is a resounding yes but they are more rare, as discussed below.

In the solar system, the inner asteroid belt and much of the mid-belt is dominated by differentiated bodies like the refractory-rich S-type and M-type asteroids. However, among the most important discoveries in the entire field of polluted WDs is the detection of disrupted planetesimals that are essentially undifferentiated, with a primitive composition suggesting these objects formed from direct nebular condensation during the collapse of the protostellar cloud. Such objects are implicated in the delivery of water, hydrocarbons and other elements of considerable interest to geochemistry, astrobiology and the origin of life.

Is there evidence of the accretion of tidally-disrupted, volatile-rich, undifferentiated (primitive) bodies, some of which might have formed primordially by direct nebular condensation during the collapse of the original protostellar cloud without undergoing any subsequent differentiation? The DBZA WD GD40 is a very interesting example. While most disrupted asteroids accreted by WD stars appear to be dry; that is, if all of the measured oxygen content from spectroscopy can be accounted for by accretion of oxides of the abundant elements Fe, Mg, Si, and Ca (e.g., CaO FeO, MgO SiO_2, etc.), then there is no indication of a water component. However, if the given percentage of oxygen required by charge balance in rock-forming metal oxides is less than the total oxygen content inferred by spectroscopy, then that difference is assumed to be tied to the presence of water (cf. Melis et al. 2011 and references therein). Using this method in GD40, the tidally disrupted impacting body contained 10% water and its mass was roughly the size of 4 Vesta (Doyle et al. 2019).

Jura et al. (2012) have uncovered a DBZ WD, G241-6, which revealed 12 polluting elements, compared to 13 for GD40. Combining optical and ultraviolet spectra of these stars with He-dominated atmospheres, 13 and 12 polluting elements are confidently detected in GD 40 and G241-6, respectively.

Both GD40 and G241-6 reveal a similar abundance pattern with the volatile elements C, and in GD40, S, being deficient by more than a factor of 10, and N by at least a factor of 5 compared to their mass fractions in primitive CI chondrites, close to what is inferred for the bulk Earth. The only appreciable difference between the two DBZ spectroscopic abundances is that in G241-6, sulfur is undepleted. Jura et al. (2012) also newly detected (or placed meaningful upper limits on) the abundances of Cl, Al, P, Ni, and Cu in the accreted matter.

Another extraordinary example of the accretion of a water-bearing impacting object by a WD is the DBZ star GD61. According to Farihi et al. (2013), the original disrupted accreting body contained an extraordinary content of oxygen. The overabundance of oxygen relative to the rock-forming oxides reveals that the parent body had a composition of 26% water by mass. Both GD 40 and another DBZ star, G241-6, manifest an abundance of carbon an order of magnitude lower than in a carbonaceous chondrite implying that the most likely parent body was like an ordinary chondrite, having less carbon, less water than a carbonaceous chondrite but with chondrules (Jura et al. 2012).

One of the most exotic cases of an undifferentiated parent body was reported by Xu et al. (2017). They used optical and ultraviolet spectroscopic observations of WD1425+540 which revealed a volatile-rich composition of water, carbon, nitrogen in the accreted parent body. The parent body of this object must have formed far from its central star. The large abundance of nitrogen strongly suggests that the parent body of the accreted material had to have formed in the cold outer reaches of an exoplanetary disk far from the WD's main sequence progenitor. The heavy elements they detected were C, N, O, Mg, Si, S, Ca, Fe, and Ni. This work was the first reported detection of the element nitrogen in any exoplanetary debris. The authors measured the mass fraction of N to be ~2%, which they point out is similar to the abundance of N in Halley's comet, and higher than in any other known solar system object (Xu et al. 2017). This would be analogous to a Kuiper Belt object in our planetary system, the frigid source region of the short-period comets.

Hoskin et al. (2020) reported the discovery of a polluted WD, WD J20413.76-12508.9, that accreted the debris of an essentially undifferentiated parent body. These authors used the COS spectrograph on Hubble and the ground-based Very Large Telescope (VLT) to obtain spectra revealing photospheric absorption lines due the heavy elements C, O, Mg, Si, P, S, Ca, Fe, and Ni. The abundances and abundance ratios reveal that the accreting material was rich in the volatile elements C and O, with respect to the composition of the bulk Earth. They suggest that the parent body was similar in composition to the primitive carbonaceous chondrites in our solar system. Under the assumption of steady-state accretion, they determined that the parent body had a mass of at least 1.6×10^{20} g, which is approximately the mass of a small asteroid with a diameter of about 26 miles.

Exotic discoveries, perhaps in the past thought to be unimaginable, continue to accumulate as more and more debris-accreting white dwarfs are identified. Gäansicke et al. (2019) have detected a hydrogen-rich gaseous disk surrounding the hot WD, WD J091405.30 + 191412.25, in which hydrogen, oxygen, and sulfur are accreting onto the WD at $\sim 3.3 \times 10^9$ g/s. This is the first spectroscopic evidence of a polluted WD that has likely accreted the atmosphere of a giant planet (Gänsicke et al. 2019). Optical spectroscopy by Gänsicke et al. (2019) has revealed that the WD has accreted hydrogen, oxygen, and sulfur from a debris disk that is unique among all others known at the time of this writing. This chemical composition most closely matched the expected composition of the deeper atmospheric layers of an ice giant planet, perhaps like Neptune. It is known that the deeper cloud decks of the Jovian planets contain layers of water, ammonium hydrosulfide, methane and ammonia and hydrocarbons, each with different condensation temperatures. It is very likely that a Jovian-like exoplanet is being photo-evaporated by the WD whose temperature is $\sim$27,750 K and it seems likely that photoevaporation of other cold Jovian-like planets hot WDs occurs (cf. Schreiber et al. 2019). For WD surface temperatures higher than 20,000 K, radiative forces levitation cannot be ruled out, which may complicate the interpretation of the photospheric abundances but the mix of elements and the abundance ratios support the accretion of an icy giant planet. Further observations, especially in the ultraviolet are needed to further confirm this exciting discovery.

Still another surprising discovery is the first detection of the element beryllium in two helium-atmosphere WDs, GALEX J2339-024 and GD378 (Klein et al. 2021). Further, the authors found that the beryllium is overabundant by roughly two orders of magnitude relative to magnesium, silicon, and iron. The tidally disrupted accreted bodies that polluted these two WDs almost certainly had to have been in an intensely high energy environment where O and/or C, N, nuclei and protons were subjected to spallation reactions brought about by high energy collisions. This is one of several possibilities for how the production rate of beryllium could be so high, as implied by the large overabundance. In a related paper by Doyle et al. (2021), the authors argue instead that the Be was produced by energetic proton irradiation of ice mixed with rock. They showed that the MeV proton bombardment needed to form the high Be/O ratio in the accreted parent body is consistent with irradiation of ice in the rings of a giant planet like Saturn, within its radiation belt, followed by accretion of the ices to form a beryllium-enriched moon that is eventually tidally disrupted and accreted by the WD. Could the polluted parent body have been an icy exomoon formed in the ring system of a giant planet?

Finally, Vanderburg et al. (2021) report the observation of a giant planet candidate transiting the white dwarf WD1856+534 (TIC 267574918) with an orbital period of 1.4 days. The authors modelled the observed periodic dimming of the white dwarf caused by the transit of the planet candidate across the photosphere of the star. They found that the planet candidate is roughly the same size as Jupiter and estimate an upper limit mass of roughly 14 Jupiter masses. The most important implication of this finding is that very likely the planet plus WD pair is not the product of common envelope evolution. The authors argue that the planet candidate's small mass and its long orbital period (relative to other WD plus brown dwarf binaries) rule out the pair being the product of a common envelope stage as is the case with more compact pairs. The implication would be that Jovian planets in close orbits with the parent star could have stable orbits and thus avoid tidal disruption. It will be interesting to see if this important discovery and its implication is confirmed with further observations.

3.10 Summary

The differentiation of a molten body, as happens in small bodies like the parent meteorite planets (asteroids) of sufficient mass in the solar system, would be expected to occur in all exoplanetary systems around WDs and other stars with compositions similar to the Sun and its neighbors. Early on, H. C. Urey (1955, 1956) recognized that the naturally occurring long-lived radionuclides such as 40 K, ^{238}U, ^{235}U, and ^{232}Th would not have generated sufficient heat so early and quickly in the solar system to have melted small bodies. Now this very process has indeed been shown to occur in exoplanetary systems orbiting WDs. Among the rare and now extinct radionuclides (i.e., extremely short half-lives) in the solar system capable of melting growing planetesimals are Al^{26}, Fe^{60}, I^{129}, Be^{10}, Cl^{36}, Np^{237}, Pu^{244}, and Cm^{247} and probably others (Wood 1968), all of which originate in differing degrees from novae, supernovae, and Wolf–Rayet winds. The extinct radionuclide I^{129} has provided the most reliable information regarding the timescale leading up to the formation of solid planetesimals in our solar system (Reynolds 1960). It is likely that such short-lived radionuclides play the same role in exoplanetary systems.

The exo-geochemical melt in these intensely-heated small bodies, initially having compositions essentially the same as their parent main sequence star, would lead to differentiation with a resulting core, mantle, crust layered structure and grouping of elements and minerals into families stemming from chemical affinities similar to the lithophile, chalcophile and siderophile groups seen in meteorites and the terrestrial planets in our solar system. Hence, the vaporized fragments of tidally disrupted, differentiated rocky metallic exo-asteroids have been spectroscopically confirmed in the atmospheres of cool WDs through the mix of accreted elemental species and their respective abundances and abundance ratios (Jura 2003; Zuckerman et al. 2003, 2010; Koester et al. 2014). Likewise, the vaporized fragments of undifferentiated (primitive) bodies have also been spectroscopically confirmed in the atmospheres of cool WDs (Xu et al. 2013a, 2013b, 2017). Compositions of the parent bodies have been inferred through detailed comparisons of the observed chemical abundance ratios with the Sun, comets, C1 chondrites (the most primitive, water-rich meteorites), bulk Earth composition with 85% by mass composed of oxygen, magnesium, silicon, and iron (Xu & Jura 2015), volatile elemental species such as H, He, C, S and the rock-forming elements, Mg, Si, Fe, Ca (see Zuckerman & Young 2017 and references therein).

One can only imagine that new surprises will emerge in this fertile field of investigation as increasingly deeper optical surveys identify hundreds of thousands of newly discovered accreting WDs with exoplanetary bodies orbiting them. Probing for these objects out to increasingly greater distances should reveal new, unexpected discoveries and insights into regions of space well outside the solar neighborhood. Extending these studies to the realm of hot WDs will present new phenomena and challenges where the radiation fields of the accreting WDs are far more intense. Higher resolution spectroscopy will enable the separation of line components arising from the interstellar medium, circumstellar matter, debris disks and the gravitationally redshifted features in the WD photosphere. Imminent missions like the James

Webb Space Telescope (JWST) will extend searches for exoplanetary accretion down to much fainter magnitudes along with mid-infrared spectroscopy, imaging and ever larger ground-based telescope apertures. The prospects for enlarging the scope and depth of investigations into exoplanetary accretion onto WDs are promising, to say the least.

References

Alcock, C., Fristrom, C., & Siegelman, R. 1986, ApJ, 302, 462

Alcock, C., & Illarionov, A. 1980, ApJ, 235, 541

Allegre, C J, et al. 1995, E&PSL, 134, 515

Anderson, D. 2007, New Theory of the Earth (Cambridge: Cambridge Univ. Press)

Artymowicz, P. 1988, ApJ, 335, L79

Bauer, E, & Bildsten, L 2018, ApJ, 859, L19

Bauer, E. B., & Bildsten, L. 2019, ApJ, 872, 96

Blouin, S., Dufour, P., Thibeault, C., & Allard, N. F. 2019, ApJ, 878, 63

Burns, J. A., Lamy, P. L., & Soter, S. 1979, Icar, 40, 1

Caiazzo, I., & Heyl, J. S. 2017, MNRAS, 469, 2750

Chiang, E. I., & Goldreich, P. 1997, ApJ, 490, 368

Cunningham, T, Tremblay, P-E l, Gentile, F, et al. 2020, MNRAS, 492, 3540

Cunningham, T., Wheatley, P. J., Tremblay, P.-E., et al. 2022, Natur, 602, 219 (2022?)

Debes, J., & Sigurdsson, S. 2002, ApJ, 572, 556

Debes, J, Walsh, K J, & Stark, C 2012, ApJ, 747, 148

Delsemme, A. H. 1982, in Comets, ed. L. L. Wilkening (Tuscon, AZ: Univ. Arizona Press), 85

Doyle, A. E., Desch, S. J., & Young, E. D. 2021, ApJ, 907, L35

Doyle, A., Young, E. D., Klein, B., Zuckerman, B., & Schlichting, H. E. 2019, Sci, 366, 356

Dubus, G., Otulakowska-Hypka, M., & Lasota, J.-P. 2018, A&A, 617, A26

Dufour, P., Bergeron, P., Liebert, J., et al. 2007, ApJ, 663, 1291

Duncan, M. J., & Lissauer, J. 1998, Icar, 134, 303

Dupuis, J., Fontaine, G., Pelletier, C., & Wesemael, F. 1992, ApJS, 82, 505

Farihi, J. 2016, NewAR, 71, 9

Farihi, J., Barstow, M., Redfield, S., et al. 2010, MNRAS, 404, 2123

Farihi, J., Becklin, E., & Zuckerman, B. 2008, ApJ, 681, 1470

Farihi, J., Gänsicke, B., & Koester, D. 2013, Sci, 342, 218

Farihi, J., Jura, M., & Zuckerman, B. 2009, ApJ, 694, 805

Farihi, J., van Lieshout, R., Cauley, P. W., et al. 2018, MNRAS, 481, 2601

Fontaine, G., & Van Horn, H. 1976, ApJS, 31, 467

Gänsicke, B., Marsh, T., Southworth, J., & Rebassa-Mansergas, A. 2006, Sci., 314, 1908

Gänsicke, B. T., Koester, D., Marsh, T. R., Rebassa-Mansergas, A., & Southworth, J. 2008, MNRAS, 391, L103

Gänsicke, B., Koester, D., Farihi, J., et al. 2012, MNRAS, 424, 333

Gänsicke, B., Schreiber, M. R., Toloza, O., et al. 2019, Natur, 57, 61

Graham, J., Matthews, K., Neugebauer, G., & Soifer, B. T. 1990, ApJ, 357, 216

Hansen, B. M. S., Kulkarni, S., & Wiktorowicz, S. 2006, AJ, 131, 1106

Harris, A. W. 1994, Icar, 107, 209

Hartmann, W. K. 2005, Moons & Planets (5th ed.; New York: Thompson Brooks/Cole)

Hartmann, S., Nagel, T., Rauch, T., & Werner, K. 2011, A&A, 530, 7

Hartmann, S., Nagel, T., Rauch, T., & Werner, K. 2014, A&A, 571, 44

Hawkesworth, C. J., & Kemp, A. I. S. 2006, Natur, 443, 811

Hill, G. W. 1886, AcMa, 8, 1

Hills, J. G. 1981, AJ, 86, 1730

Hobbs, P. 1974, Ice Physics (Oxford: Clarendon)

Hoskin, M. J., Toloza, O., Gänsicke, B. T., et al. 2020, MNRAS, 499, 171

Jarosewich, E. 1990, Metic, 25, 323

Joss, P. 1974, ApJ, 191, 771

Jura, M. 2003, ApJ, 584, L91

Jura, M. 2006, ApJ, 653, 613

Jura, M., Farihi, J., & Zuckerman, B. 2007a, ApJ, 663, 1285

Jura, M., Farihi, J., Zuckerman, B., & Becklin, E. E. 2007b, AJ, 133, 1927

Jura, M. 2008, AJ, 135, 1785

Jura, M., & Xu, S. 2012, AJ, 143, 6

Jura, M., Xu, S., Klein, B., Koester, D., & Zuckerman, B. 2012, ApJ, 750, 69

Jura, M., & Young, E. 2014, AREPS, 42, 45

Kenyon, S. J., & Bromley, B. C. 2017, ApJ, 844, 116

Kitts, K., & Lodders, K. 1998, M&PSA, 33, 197

Klein, B., Doyle, A., Zuckerman, B., et al. 2021, ApJ, 914, 61

Klein, B., Jura, M., Koester, D., Zuckerman, B., & Melis, C. 2010, ApJ, 709, 950

Koester, D. 2013, in Planets, Stars and Galaxies, Vol. 4, Stellar Structure and Evolution,
    ed. T. Oswalt, & M. Barstow (Dordrecht: Springer), 559

Koester, D. 2009, A&A, 498, 517

Koester, D., Gänsicke, B., & Farihi, J. 2014, A&A, 566, A34

Koester, D., Provencal, J., & Shipman, H. L. 1997, A&A, 320, L57

Koester, D., & Wilken, D. 2006, A&A, 453, 1051

Kuchner, M. J., Koresko, C. D., & Brown, M. E. 1998, ApJ, 508, 81

Lodders, K. 2003, ApJ, 591, 1220

Manser, C., Gänsicke, Boris, T., & Fusillo, G. 2020, MNRAS, 493, 2127

Melis, C., & Dufour, P. 2017, ApJ, 834, 1

Melis, C., Farihi, J., Dufour, P., et al. 2011, ApJ, 732, 90

Melis, C., Klein, B., Doyle, A. E., et al. 2020, ApJ, 905, 56

Oort, J. 1950, BAN, 11, 91

Opik, E. J. 1951, PRIAA, 54A, 16

Parriot, J., & Alcock, C. 1998, ApJ, 501, 357

Rafikov, R. 2011a, ApJ, 732, L3

Rafikov, R. 2011b, MNRAS, 416, L55

Reach, W. T., Kuchner, M. J., von Hippel, T., et al. 2005, ApJ, 635, L161

Reynolds, J. H. 1960, JGR, 65, 3843

Schreiber, M. R., Gänsicke, B. T., Toloza, O., Hernandez, M. S., & Lagos, F. 2019, ApJ, 887, L4

Shoemaker, E. M. 1963, The Moon Meteorites and Comets, ed. G. P. Kuiper, & B. Middlehurts
    (Chicago, IL: Univ. Chicago Press), 301

Sion, E. M., Hammond, G. L., Wagner, R. M., Starrfield, S. G., & Liebert, J. 1990, ApJ, 362, 691

Stephan, A., Naoz, S., & Zuckerman, B. 2017, ApJ, 844, L16

Urey, H. C. 1955, PNAS, 41, 127

Urey, H. C. 1956, PNAS, 42, 889

Vanderburg, A., Rappaport, S. A., Xu, S., et al. 2021, Natur, 585, 363

Veras, D. 2016, RSOS, 3, 150

Veras, D., Jacobson, S., & Gänsicke, B. 2014a, MNRAS, 445, 2794

Veras, D., Shannon, A., & Gaensicke, B. 2014b, MNRAS, 445, 4175

Wachlin, F., Vauclair, G., Vauclair, S., & Althaus, L. 2017, A&A, 601, 13

Weissman, P. 1980, A&A, 85, 191

Wood, J. A. 1968, Meteorites and the Origin of the Planets (New York: McGraw-Hill)

Xu, S., Jura, M., Klein, B., Koester, D., & Zuckerman, B. 2013a, ApJ, 766, 132

Xu, S., Jura, M., Koester, D., Klein, B., & Zuckerman, B. 2013b, ApJL, 766, L18

Xu, S., Jura, M., Koester, D., Klein, B., & Zuckerman, B. 2014, ApJ, 783, 79

Xu, S., & Jura, M. 2015, in Proc. Conf. Pathways Towards Habitable Planets

Xu, S., Zuckerman, B., Dufour, P., et al. 2017, ApL, 836, L7

Young, E. D. 2014, E&PSL, 392, 16

Young, E. D. 2016, ApJ, 826, 129

Zuckerman, B., & Becklin, E. 1987, Natur, 330, 138

Zuckerman, B., Koester, D., Dufour, P., et al. 2011, ApJ, 739, 101

Zuckerman, B., Koester, D., Melis, C., Hansen, B. M., & Jura, M. 2007, ApJ, 671, 872

Zuckerman, B., Koester, D., Reid, I. N., & Hünsch, M. 2003, ApJ, 596, 477

Zuckerman, B., Melis, C., Klein, B., Koester, D., & Jura, M. 2010, ApJ, 722, 725

Zuckerman, B., & Young, E. D. 2018, in Handbook of Exoplanets, ed. H. Deeg, & J. Belmonte (Berlin: Springer), 1545

Accreting White Dwarfs
From exoplanetary probes to classical novae and Type 1a supernovae
Edward M Sion

Chapter 4

Accreting WDs in Roche Lobe-detached Post-common Envelope Main Sequence–White Dwarf Binaries

The aim of this chapter is to understand the effect of accretion on white dwarfs (WDs) in binaries for which their companion stars are main sequence dwarfs detached from their Roche lobes. These types of systems are particularly important because, as in the case of isolated WDs accreting from exoplanetary debris or the interstellar medium (ISM), the accreting WD photosphere can be directly observed. Thus, we can view the accreting WD photosphere spectroscopically, determine the chemical abundances and the mix of ion species in the accreted material, test and extend diffusion theory and gain insights into the long-term consequences of accretion on the WD's evolution. Unfortunately, while advances in this area are a lofty goal, the detection of metals in their spectra may owe their presence to multiple physical processes, many aspects of which remain poorly understood; e.g., diffusion theory (e.g., Heinenen et al. 2020), thermohaline convection (e.g., Bauer & Bildsten 2019), radiative forces levitation and weak WD wind outflows (e.g., Chayer et al. 1987; Chayer 1995) and even to unanticipated sources of the photospheric metals in hot WDs; e.g., exoplanetary debris (Barstow et al. 2020). The interpretation of their spectra over a broad range of WD surface temperatures, differing orbital separations, levels of magnetic activity in the non-degenerate companion and WD masses is challenging to say the least. Nevertheless, considerable progress has been made and this chapter will attempt to elucidate what we know and what remains unclear about the photospheric spectra of WDs in detached compact binaries, revealed from space observations as well as optical spectroscopy.

There are many examples in stellar astrophysics of galactic binaries consisting of WDs in close orbit ($P_{\mathrm{orb}} \leqslant 4$ days) with main sequence dwarfs detached from their Roche lobes. Such systems are generally known as post-common envelope binaries (PCEBs), the reason being that the two stars could never have formed so close together. The orbits of PCEBs are so compact, that they would easily fit inside the Sun yet the WDs had to form inside of red giants or supergiants thus requiring that the initial binaries had to be very widely separated. Paczyński (1976) first demonstrated that the original wide binary orbit shrank due to a process known as common envelope evolution (hereafter CEE). The originally more massive stellar component evolved into a red giant structure first. At this point, the main sequence dwarf began orbiting inside of the extended, non-co-rotating envelope of the red giant where the ensuing frictional drag drained orbital angular momentum and energy, thus shrinking the orbital separation. The energy extracted from the orbit might be used to expel the envelope before the components merge. The importance of establishing the efficiency of deposition of orbital energy into the envelope became evident. The critical parameter that determines the outcome of a CEE interaction is the efficiency with which orbital energy is converted into the ejection of matter from the system. This efficiency, denoted α_{CE}, is defined (see Tutukov & Yungelson 1979; Iben & Livio 1993) by $\alpha_{CE} = \delta E_{bind}/\delta E_{orb}$.

Here δE_{bind} is the binding energy (gravitational minus thermal) of the ejected material and δE_{orb} is the difference in the orbital energy of the binary, between the beginning and the end of the spiraling-in process

$$[(M_1 + M_2)(M_i - M_{1R})]/2a_0 = \alpha_{CE} M_{1R} M_2 \left[\frac{1}{2} a_f - \frac{1}{2} a_0 \right] \tag{4.1}$$

where M_{1R} is the mass of the remnant of the primary after envelope ejection and a_f is the final separation.

If α_{CE} is large, of order unity, then the envelope ejection occurs sooner in the CE interaction. Thus, the majority of the emerging binaries will have relatively long orbital periods (typically 10–100 days; Yungelson et al. 1993). But if α_{CE} is much smaller than unity, then compact binary CE progeny as well as stellar mergers will result. Nebot Gomez-Morani et al. (2011) found that 21% to 24% of all SDSS WD + main sequence binaries, including through data release XII, have undergone CEE which they reported is consistent with published binary population models and Hubble Space Telescope (HST) high-resolution imaging of all such binaries unresolved from the ground. The resulting PCEB objects include planetary nebulae with compact binary planetary nuclei, systems comprised of a Roche lobe-detached red dwarf and a WD, cataclysmic variables (CVs) and, for emergent systems with near Chandrasekhar mass WDs, eventually Type Ia supernovae.

Despite the fundamental importance of the parameter α_{CE}, the basic physics of CEE remains poorly known. Given the broad range of critical astrophysical systems spawned by CEE, it is imperative to fully understand the process. What is the physics that definitively governs the ejection of the CE and whether a merger occurs or whether merger is stopped with the result being a compact binary? Further progress rests in large part upon having increasingly realistic 3D hydrodynamic simulations with ever more sophisticated input physics (see Chamandy et al. 2018). The most obvious and common concern is that of missing input physics regarding: (a) the crucial in-spiral to very small component separations; (b) the role of radiation pressure on the dust grains in the CE; (c) the role played by recombination and equally as important; (d) a deeper understanding of the role of mass outflows and accretion during the CE process (Grichener et al. 2018 and references therein). How do all of these processes affect orbital decay? Since accretion onto the degenerate or semi-degenerate components could play an important role in the CEE process, the focus is on this physical process in this chapter.

4.1 Mass Loss Rates from Late-type Dwarfs in PCEBs

Once a PCEB forms as an isolated binary or as the nucleus of a planetary nebula, it will evolve into a CV once the secondary main sequence dwarf fills its Roche lobe. This waiting period prior to the onset of CV evolution depends upon the rate at which orbital angular momentum is drained from the system thus shrinking the Roche lobe until contact with the secondary's stellar surface. Once this occurs Roche lobe overflow commences due to the thermal dis-equilibrium of the secondary component and the transfer of mass through the inner Lagrangian point, L_1 proceeds toward the primary WD. Until Roche lobe contact occurs, the rate of mass loss by the Roche Lobe-detached secondary star is expected to be orders of magnitude lower than the typical mass transfer rates of 10^{-12} to $10^{-8} M_\odot$ yr^{-1} associated with either disk accreting or magnetic cataclysmic variables. Nevertheless, some mass transfer can still take place via magnetic stellar winds, flares and coronal mass ejections (CMEs). Therefore, it is important to consider the accretion process onto WDs in close orbit with Roche Lobe-detached, K dwarfs or M dwarfs. The most likely accretion mechanisms are the Bondi–Hoyle and Eddington rates that were discussed in Chapter 2 (Equations (2.1) and (2.2) respectively). What is most uncertain are the actual rates of mass loss by K and M dwarfs that are challenging to determine from observations (see below).

To begin with, early work on mass outflow from late-type dwarfs in compact binaries relied upon a value of the solar wind mass loss rate of $\sim 10^{-13} M_\odot$ yr^{-1} and early studies of angular momentum mass loss via magnetic stellar winds adopted this value (e.g., Verbunt & Zwaan 1981). More recent estimates of late-type dwarf mass loss (Wood 2001; Wargelin & Drake 2002) using Lyα and X-ray studies have established upper limits on the M5.5V star Proxima Centauri of 6×10^{-14} $M_\odot$ yr^{-1} and $4 \times 10^{-15} M_\odot$ yr^{-1}. The mass outflow rate from the Sun itself has been supplanted by a more accurate measurement of $2 \times 10^{-14} M_\odot$ yr^{-1} (Wood et al. 2014). Again, if one uses the solar analogy to assess the amount of mass lost by the Sun in CMEs and flares, CMEs do not carry away much mass: data from the SOLWIND coronagraph suggested that even at solar maximum, no more than 16% of the solar wind mass loss rate was contributed by CMEs (Jackson & Howard 1993) and at solar minimum, with CME rates down by as much as a factor 5–10 compared to conditions at solar maximum, the CME contribution to mass loss rate would be even smaller (Mullan 2021). As revisions in our understanding of the solar wind took place, new insights into wind outflow rates in late-type dwarfs also emerged including the recognition of the so-called "hydrogen wall". Its importance merits further discussion as follows. As a star with a stellar wind is moving through the ISM, the H-rich wind outflow carves out a cavity around the star called its

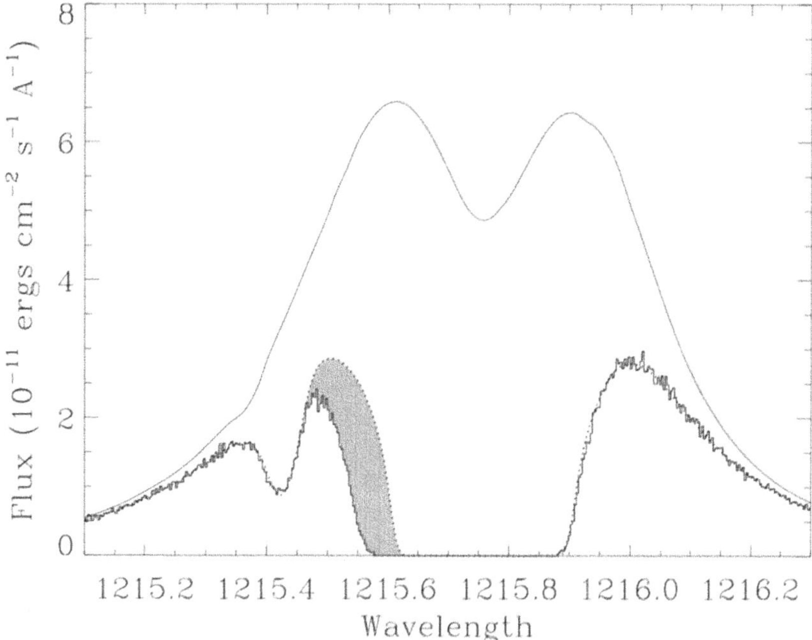

Figure 4.1. HST/GHRS Ly spectrum of Eri, showing broad H ɪ absorption at 1215.7 A? and narrow D ɪ absorption at 1215.4 A? The upper solid line is the assumed intrinsic stellar emission line, and the dotted line is the profile after ISM absorption alone, derived by forcing consistency between the ISM H ɪ and D ɪ absorption. The excess H ɪ absorption on the blue side of the line (shaded region) is astrospheric absorption known as the "wall". (Reproduced with permission from Wood et al. (2002). © 2002 The Astrophysical Journal. All rights reserved.)

"astrosphere" (Wood et al. 2002; Wood 2018). Charge-exchange reactions between a fully ionized stellar wind and the partially ionized warm gas in the ISM form a compressed region of hot neutral hydrogen atoms that are decelerated relative to the inflowing interstellar gas. If a sufficient column density of H is built up in the wall, then it becomes possible to detect the wall absorption in HST spectra of the Lyα line. From the wall line strength a mean mass loss rate can be derived for the star. This plowed up "hydrogen wall" produces a blueshifted absorption component in the stellar Lyα emission line that has been detected in HST spectra of at least 22 main sequence dwarf stars.

In Figure 4.1, the shaded region is the wall absorption detected in HST spectra of the Lyα line. Comparisons of the observed Lyα line profiles with theoretical models led to the first sensitive measurements of mass loss rates as low as $4 \times 10^{-15} M_\odot$ yr^{-1} in late-type main sequence dwarfs (Linsky & Wood 2004).

4.2 Characterizing the Population of Main Sequence–WD PCEBs

The SDSS spectroscopic catalog of WD–main sequence (WDMS) binaries: new identifications from DR 9-12 (Rebassa-Mansergas et al. 2014). Nebot Gomez-Morani et al. (2011) find that 21% to 24% of all SDSS WD + main sequence binaries, including through data release XII, have undergone CEE. Their report is consistent with published binary population models and HST high-resolution imaging of all such binaries unresolved from the ground. The largest and most homogeneous catalog of WD + MS binaries currently available is that from Rebassa-Mansergas et al. (2016), with a total number 3291 systems identified within the data release 12 of SDSS. The number of eclipsing systems among their sample is consistent with previous results indicating that ~10% of PCEBs—which make up roughly one fourth of WDMS binaries (Schreiber et al. 2010; Nebot Gomez-Morán et al. 2011; Rebassa-Mansergas et al. 2011)—are eclipsing systems (Parsons et al. 2013). From the sample of well-studied post-CE binaries (Schreiber & Gaensicke 2003), approximately half will evolve into CVs within a Hubble time.

4.3 Accretion of Outflows from Detached Main Sequence Dwarfs onto WDs in PCEBs

Accretion of matter by a hot WD in a close binary system usually occurs via an accretion disk, and one cannot directly observe the accreting photosphere. However, in detached red dwarf–WD spectroscopic binaries, accretion, if it happens at all, is thought to occur without the formation of an accretion disk. One envisages accretion rates in such binaries to be on the order of the solar wind loss rate convolved with the cross-section of the WD as it moves through the wind. Such accretion rates could be of the same order of magnitude as those experienced by single WDs encountering a dense interstellar cloud or passing through the intercloud medium (see Truran et al. 1977; Michaud & Fontaine 1984). Furthermore, any accretion of wind material by the WD component in a PCEB system can potentially shed light on the rate of wind mass loss by the late-type (typically dM) components of the system. As emphasized at the beginning of this chapter, the rate of outflow from an M dwarf star, either single or in a binary, remains poorly known despite the advances made from measurements of the "hydrogen wall" component of Lyα in the FUV. Indeed, the role of magnetic stellar winds from late-type components in compact binaries whether PCEBs or CVs, is the major mechanism, along with gravitational wave radiation, in draining orbital angular momentum, driving mass transfer in CVs and the timescale to Roche lobe overflow in PCEBs. Therefore, the study of accretion onto a WD from the mass loss of a late-type dwarf could ultimately shed new light on the winds and CMEs of late-type dwarfs in compact binaries as well as single late-type dwarfs.

4.4 The Ever-changing Saga of the Hot DAZ.9 WD in the PCEB Feige 24

The system Feige 24 is the most extensively studied PCEB over a broad range of wavelengths and is most illustrative of the challenges to understanding the source of metals in hot WDs. First of all, Feige 24's DAZ1 WD was the first and only known hot DA star to exhibit photospheric metals. Only one other object, the cool DA WD G74-7 (Lacombe et al. 1982) was known to have metals (Ca II) in the atmosphere. Feige 24 consists of an M dwarf and hot WD (dM4 + DAZ1), with a four day orbital period (Thorstensen et al. 1977; Sion & Starrfield 1984). Dupree & Raymond (1982) detected photospheric metal lines for Feige 24 utilizing high-resolution IUE spectra. Further observations (Dupree & Raymond 1983) confirmed the photospheric association for the N v, Si iv, and C iv resonance doublets from their higher velocities relative to the high-excitation (presumably circumstellar) lines, and the lower velocity ISM lines.

By the mid-1980s, Feige 24 was the only ($T > 30, 000$ K) DA WD known to exhibit photospheric metals (Dupree & Raymond 1982). The IUE spectrum revealing the highly-ionized metal lines is shown in Figure 4.2. The presence of these metal lines at their observed strengths indicates atmospheric abundances orders of magnitude

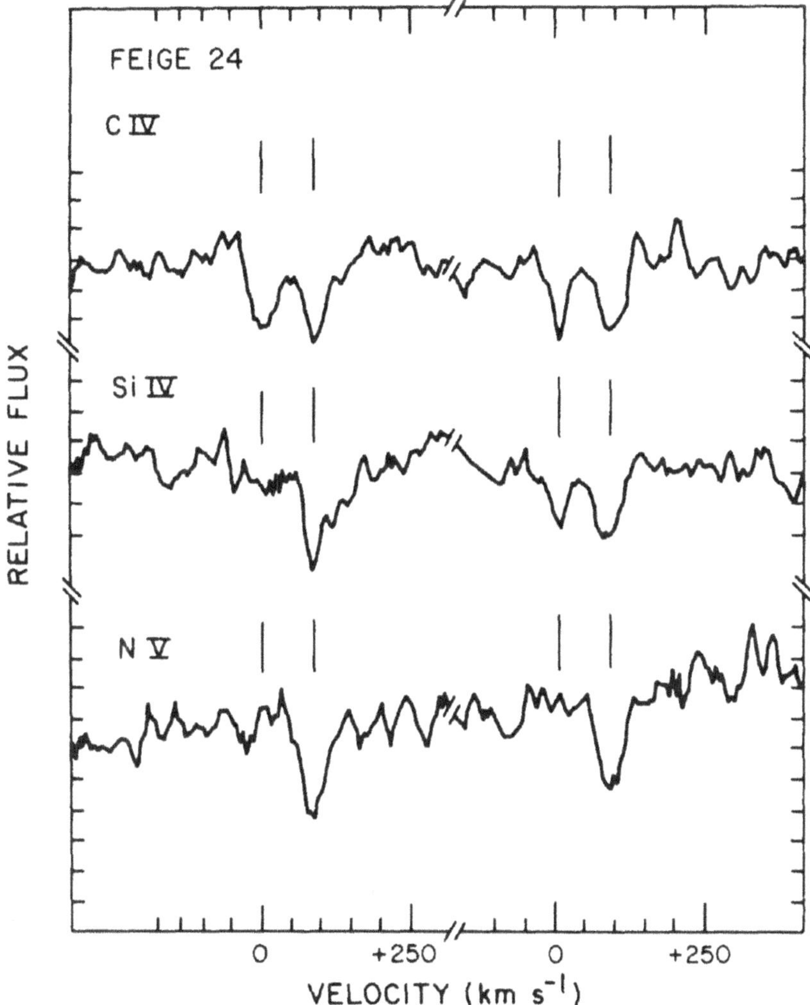

Figure 4.2. Highly-ionized absorption features in the IUE SWP spectrum of Feige 24. This spectrum has been smoothed with a 5 pixel average. Two absorption components (at velocities of 1–1 and 1–83 km s^{-1}) are labeled with vertical tick marks and clearly present in both members of the C Iv doublet (I548A and I550A). (Reproduced from Dupree & Raymond (1982). ©1982 The Astrophysical Journal. All rights reserved.)

larger than one would expect from a simple application of diffusion theory to the outer atmosphere of the WD (see Fontaine & Michaud 1979; Vauclair et al. 1979).

The first attempts to quantitatively investigate the source of the heavy elements in the atmosphere of the Feige 24 DAZ1 WD were two independent investigations reported by Sion & Starrfield (1984) and Wesemael et al. (1984). The two investigations differed in the way that diffusion was handled, especially the diffusion coefficients, how the accretion rates were estimated and how the abundances of carbon and nitrogen were determined.

Starting with the Sion & Starrfield (1984) analysis, they adopted the Chapman & Cowling (1970) diffusion coefficient for a fully ionized gas as well as accretion–diffusion equilibrium and employed the theoretical treatment of diffusion by Fontaine & Michaud (1979) and Vauclair et al. (1979), which are valid except for very small optical depths: $\tau \ll 1$. The possible prevention of rapid downward sinking of the C, N, and Si ions by meridional currents and turbulent motions were rejected for Feige 24 (see Tassoul & Tassoul 1983) since the WD is known to be a slow rotator (Sion & Wesemael 1983; Wesemael et al. 1984). This is not unexpected since no single WDs have Vsini $>$ 50 km s^{-1}. At the time of their investigation, radiative forces theory had not reached fruition. Sion & Starrfield (1984) were led to propose that accretion from the M dwarf companion at a separation of 8×10^6 km provides the enhanced photospheric metal abundances, a possibility first raised by Dupree & Raymond (1982).

Sion & Starrfield (1984) quantitatively predicted the abundances for C IV and N V corresponding to two different modes of wind accretion; the Bondi–Hoyle fluid rate and the Eddington geometric cross sectional rate. If accretion by the DAZ star is in steady-state equilibrium with diffusion, then

$$4\pi R_{\mathrm{wd}}^2 \rho V_{\mathrm{d}}\left(\frac{Z_i}{H}\right)_{M1} + \dot{M}\left(\frac{Z_i}{H}\right)_{M1} = \dot{M}\left(\frac{Z_i}{H}\right)_{M2} \qquad (4.2)$$

where $\dot{M}$ is the accretion rate, V_{d} is the diffusion velocity of a given ion, the subscripts 1 and 2 refer to the DAZ1 star and the dM star respectively, and (Z_i/H) denotes the metal to hydrogen ratio for metal element i. Because of the low mass of the red dwarf secondary star, the assumption of solar composition was assumed for its wind outflow with a mass loss rate of $10^{-13} M_\odot$ yr^{-1} by analogy with the solar wind. The diffusion velocities for a given ion can be obtained from expressions presented by Fontaine & Michaud (1979) and Vauclair et al. (1979). The diffusion velocity may be written

$$V_{\mathrm{d}} = \frac{Dm}{kT}(g_{\mathrm{rad}} + g_{\mathrm{GT}}) \qquad (4.3)$$

where g_{rad} is the radiative acceleration and g_{GT} is a combination of the inward acceleration of gravity and the thermal gradient. Since thermal diffusion has been shown to be negligible in WDs (see Fontaine & Michaud 1979), Sion & Starrfield (1984) ignored it, and g_{GT} depends upon the pressure gradient alone. The quantity D is the diffusion coefficient, m is the ion mass, and T is the temperature. For Feige 24, the models show that we can rule out envelope convection. Thus, the diffusion coefficient of Chapman & Cowling (1970) for a fully ionized gas is appropriate in PCEBs with hot hydrogen-rich DA WDs.

For diffusion of an ionized atom through an ionized medium, the diffusion coefficient (Chapman and Cowling 1970) is given by

$$D_{12} = \frac{3(2kT)^{5/2}}{16\left[(\pi m_1 m_2)/(m_1 + m_2)\right]^{1/2} n_1 Z_1^2 Z_2^2 e^4 A_1(2)},\qquad (4.4)$$

where subscripts 1 and 2 refer to the background element and trace element respectively, k is the Boltzmann's constant, T is the temperature, n_1 is the number density of the background gas, e is the electron charge, and $A_1(2)$ is a logarithmic term defined as

$$A_1\left(2\right) = ln\left(1 + X_D^2\right)\qquad (4.4a)$$

with

$$X_D^2 = \frac{16k^2 T^2 \lambda_D^2}{Z_1^2 Z_2^2 e^4}\qquad (4.4b)$$

and λ_D is the Debye length defined by

$$\lambda_D = \left(\frac{kT}{4\pi e^2}\sum_i n_i Z_i^2\right)^{1/2}\qquad (4.4c)$$

and the summation is over all charged particles i (see also Chapter 1, section 1.8).

All symbols are the same as defined in Fontaine & Michaud (1979). The parameters for the $0.6 M_\odot$ model for Feige 24, gave diffusion velocities of $V_d(\text{C IV}) = 6.8 \times 10^{-3}$ cm s^{-1} and V_d (N V) $= 6.7 \times 10^{-3}$ cm s^{-1}. Sion & Starrfield (1984) found the accretion rates for two different scenarios: (1) if the WD accretes by only its geometric cross-section (essentially the Eddington rate), for the radius of the DAZ1 star and the orbital separation, assuming the mass loss rate of $\sim 10^{-13} M_\odot$ yr^{-1} in analogy with the Sun, the DAZ1 star should be accreting at the rate $4 \times 10^{-20} M_\odot$ yr^{-1}. In view of the most recent solar wind mass loss rate of $3 \times 10^{-14} M_\odot$ yr^{-1} (Linsky & Wood 2014), the accretion rate would be roughly a factor of ten lower; (2) an accretion rate can be determined on the basis of what fraction of the stellar wind is gravitationally captured by the DAZ1 star (essentially the Bondi–Hoyle fluid rate)

$$\dot{M}_{acc} = \left(\frac{\pi r_a^2}{4\pi a^2}\right)\dot{M}_w\qquad (4.5)$$

where $\dot{M}_w$ is the wind mass loss rate from the M dwarf, a is the orbital radius and r_a is the accretion radius of the DAZ1 star;

$$r_a = \frac{2GM_{wd}}{V_{rel}^2}\qquad (4.6)$$

with $V_{rel} = V_w + V_{orb}$.

For Feige 24, the following parameters were adopted: orbital velocity $V_{orb} = 155$ km s^{-1}, the semimajor axis $= 8.98 \times 10^6$ km, and, wind velocity,

$V_w = 300$ km s^{-1}, in analogy with solar wind, to represent the wind velocity of the M dwarf companion. The sum of the masses of the M dwarf and WD mass was taken to be 0.6 Msun. Thus, the accretion radius of the DAZ1 star is 7.7×10^8 cm, and the fraction of the stellar wind captured is 1.8×10^{-3}. Thus, assuming the M dwarf has a mean mass loss rate comparable with the Sun, then the WD accretes at the rate $2 \times 10^{-16} M_\odot$ yr^{-1}. Assuming that this gas can reach the DAZ1 star, they used their numerical calculations of $\dot{M}$, their diffusion velocities for C IV and N V, and the solar (N/H) and (C/H) abundance ratios [log(C/H) $= -3.46$; log(N/H) $= -4.07$] to calculate the predicted abundances for carbon and nitrogen from Equation 4.1). If $M = 2 \times 10^{-16} M_\odot$ yr^{-1}, (C/H)$_{F24} = 9 \times 10^{-5}$ and (N/H)$_{F24} = 3 \times 10^{-5}$. If $\dot{M} = 4 \times 10^{-20} M_\odot$ yr^{-1}, then (C/H)$_{F24} = 2 \times 10^{-8}$ and (N/H)$_{F24} = 6 \times 10^{-9}$. At the higher accretion rate, the abundances considerably exceed a value of 10^{-3} solar. Their calculated abundances for carbon and nitrogen are not total abundances. However, in the absence of any other ionization states of nitrogen and carbon at the high effective temperature of Feige 24 ($\sim$65,000 K), N V and C IV should be the dominant ions.

In view of (1) the presence of photospheric metal lines in Feige 24, (2) the absence of absorption line of metals in other single hot DA WDs, (e.g., HZ 43) and (3) the unlikely possibility that the metal ions owe their presence in the atmosphere to radiative acceleration alone, and (4) the reasonable assumption of accretion–diffusion equilibrium (the rate of diffusion at the T_{eff} of the DAZ1 star is extremely short) they argued that accretion of material from the M dwarf companion is occurring at the present time.

The investigation by Wesemael et al. (1984) took a different approach to explore the source of metals in the atmosphere of the Feige 24 WD. They used newly determined diffusion coefficients by Paquette et al. (1986) and carried out H-rich model atmosphere fits to the IUE echelle spectrum of Feige 24 to determine the metal abundances. They calculated the accretion rates required to account for the observed abundances. The detected photospheric features and their line strengths (equivalent widths) are listed in Table 4.1, from Wesemael et al. (1984).

They also adopted the accretion–diffusion model of Vauclair et al. (1979) and Alcock & Illarionov (1980) which assumes a steady-state balance in the atmosphere between accretion and downward diffusion. They neglected any process that could prevent accretion onto the DAZ1 star. They assessed the rates of accretion that would

Table 4.1. Feige 24 Resonance Doublet Line Strengths and Metal Abundances

ION	λ_{Obs}	$W_{\lambda(A)}$	log N(Metals)/N(H)
C IV	1548.185	104	-6.4 ± 0.6
C IV	1550.774	108	
N V	1238.804	63.6	-5.3 ± 1.0
N V	1242.804	60.9	
Si IV	1393.755	97.9	-6.3 ± 0.9
Si IV	1402.770	75.6	

result from the geometric cross-section of the WD accretor intercepting wind outflow from the M dwarf. In that case, the accretion efficiency factor is of the order of

$$\left(\frac{R_{wd}}{2a}\right)^2 \sim 3 \times 10^{-7} \tag{4.7}$$

where R_{wd} and a are the WD radius and separation of the components, respectively. For gravitational capture of wind material from the M dwarf they computed an accretion efficiency factor of order

$$(M_{wd}/M_{rd})(R_{wd}R_{rd}/4a^2\gamma^2) \approx 2 \times 10^{-5} \tag{4.8}$$

where γ is the ratio of the wind velocity to the escape velocity from the dM star (which is very close to unity), the subscript dM refers to the red dwarf, and the subscript wd refers to the WD accretor. Finally, if fluid accretion is occurring onto the WD (in which case, the density of the wind must be high enough to enable the formation of a tail shock), the efficiency factor is

$$(M_{wd}/M_{rd})^2(R_{rd}^2/8a^2\gamma^4) \approx 8 \times 10^{-4}. \tag{4.9}$$

Like the work of Sion & Starrfield (1984), Wesemael et al. (1984) assumed a wind mass loss rate from the secondary of $\dot{M} = 10^{-13}M_\odot$ yr^{-1} in analogy with the Sun's outflow rate estimated at the time and derived an accretion rate onto the DAZ1 star which could be in the range $3 \times 10^{-20}M_\odot$ yr^{-1} to $8 \times 10^{-17}M_\odot$ yr^{-1}, depending upon whether the geometric (Eddington) rate or the fluid rate (Bondi–Hoyle) was adopted, respectively. Wesemael et al. (1984) noted that both rates were large enough to pollute the WD photosphere. Insofar as the origin of the heavy element ions in the atmosphere of the DAZ1 star, Wesemael et al. (1984) considered their results inconclusive as to whether accretion was occurring onto the WD and underscored the need for detailed calculation of radiative forces support in the atmospheres of hot WDs. This needed theoretical advance in radiative forces support would eventually arrive with the work of Chayer (1995).

For five more years, the origin of the metals in the WD atmosphere remained unresolved. However, in 1989, the Montreal group (Vennes et al. 1989) examined this problem with new model atmosphere modeling of high-resolution EXOSAT data (Paerels et al. 1986) obtained at extreme ultraviolet (EUV) wavelengths. In addition to showing that a stratified model atmosphere disagrees with the EXOSAT spectrum of Feige 24 due to the lack of the He II edge at 228A and the observed steep decline in the X-ray flux shortward of 250A, they explored radiative support as a means of levitating the observed metals in the FUV IUE spectra of Feige 24. They pointed out that IUE spectra covering the FUV (where the strong resonance doublets of N v, Si iv, and C iv predominate in hot stars), does not rule out the presence of other atmospheric metals in low abundance. Indeed, different metals in the EUV were detected in low abundances in the atmosphere of Feige 24, revealing that the model atmosphere fitting agrees satisfactorily with the observed EUV spectrum (Vennes et al. 1989). They acknowledged that although a proper theoretical framework for radiative forces was not yet available, their modeling provided solid evidence for the qualitative picture they described, namely that the EUV spectrum of Feige 24 can be explained by the "cumulative effects of a host of trace absorbers" in a hydrogen atmosphere. Although the source of the metals remained a mystery, they concluded that until improved, more realistic modeling of diffusion and radiative forces as well as weak wind outflow become available, the source of the metals must continue to be regarded as inconclusive.

The puzzle of the source of metals in the hot WD Feige 24 seemed to take an entirely new turn when the Hubble Space Telescope Imaging Spectrograph became operational. For the first time, it became possible to obtain high quality high-resolution observations of Feige 24 in the far-ultraviolet. Vennes et al. (2000) presented the first HST STIS spectra obtained at the orbital quadratures (phases 0.25 and 0.75) of the Feige 24 orbit. Their high quality spectra revealed many more absorption line features than could be detected in the dozens of IUE spectra. The presence of trace metals had already been established based upon the need for EUV/soft X-ray opacity revealed by the EXOSAT LETG data (Paerels et al. 1986), and modeled by Vennes et al. (1989) as described above. Dupuis et al. (1995) had extended the Vennes et al. (1989) analysis of the EUV Lyman Continuum data by including theoretical line profiles of C, N, O, Si, S, Fe, and Ni with variable abundances in their model atmosphere analysis. Thus, it became clear from these earlier studies that the presence of a host of trace metal species was present well beyond the features of C IV, N V, and Si IV first seen in the FUV spectra by Dupree & Raymond (1982). Vennes et al. (2000) used their HST STIS spectra plus 12 high dispersion IUE echelle spectra of Feige 24 to identify the following photospheric absorption features in addition to the N V (1238, 1242), C IV (1548, 1550), Si IV (1393, 1402) doublets. They detected C III (1174.75–1176.60), N IV (1718.500), O IV (1338.612, 1342.99, 1343.512), O V (1371.292), Si III (1206.510, 1206.533), S V (1501.799), Fe IV (1567.0–1569.0, 1608.0–1610.5), Fe V (1373.0–1381.0), Ni IV (1421.1–1421.5, 1534.5–1535.5), and Ni V (1243.5–1245.5, 1276.5–1277.5).

Armed with this rich data set of HST STIS spectra and their non-LTE models, Vennes et al. (2000) determined the $T_{\rm eff}$ of Feige 24 to be 57,000 K with $\log g = 7.5$. This result is displayed in Figure 4.3 where their non-LTE model fits to the individual ion species is displayed together with the Z/H abundance ratios shown below each feature.

The photospheric lines were affected slightly by a rotational velocity Vsini = 7 km s^{-1} which they accounted for in their abundance analysis. This abundance analysis of the complete list of detected photospheric absorption features revealed a preponderance of Fe, Ni, and O features over other ion species, and overall, they found C, N, O, and Ni to be underabundant relative to the prediction of diffusion theory, even with the first inclusion of selective radiative forces (Chayer 1995). The element Fe was found to be closer to the predicted abundances while Si was overabundant compared with the theoretical prediction. Thus, the takeaway from their study was that much remained to be done in reconciling the observed abundance patterns in hot DA and non-DA stars with the predictions of diffusion/radiative forces theory in its form at the time. While the authors did not rule out a role for accretion from the M dwarf companion or even exoplanetary debris (!) associated with this binary, it became apparent based upon Feige 24's similar abundance pattern to the DA1 star, G191-B2B, which has a very widely separated M dwarf companion, that accretion may be an unlikely source of the atmospheric metals. On the other hand, the discrepancies between the observed

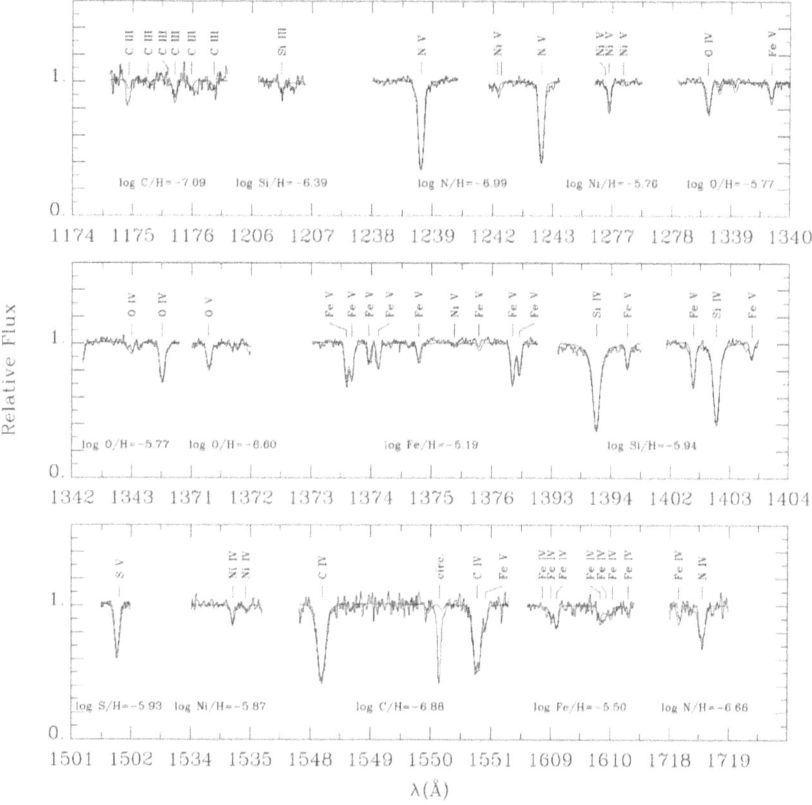

Figure 4.3. Summed HST STIS spectra and best non-LTE model fits to each detected (accreted or levitated) ion species in Feige 24. (Reproduced with permission from Vennes et al. (2000), their Figure 8. © 2000 The Astrophysical Journal. All rights reserved.)

abundances and theoretical predictions have not yet been resolved and the oft-invoked possible role of weak wind outflow in causing these disagreements is mentioned. However, at the time of this writing (2022), there has been no such theoretical modeling of weak wind outflow from a hot WD to determine its potential quantitative effects on the surface abundances.

Further studies of the DAZ1 star in Feige 24 were carried out by Kawka et al. (2008) using FUSE observations of Feige 24. Within the FUSE Range of EUV wavelengths, they reported the detection of 35 photospheric features including C III, O VI, P V, and Si IV in the spectra of Feige 24. The abundances of each element were determined from non-LTE model atmosphere line profile fitting to the FUSE spectra and are tabulated in Table 4.2 (from Kawka et al. 2008). It is gratifying that their abundance determinations for the photospheric metals in Table 4.2 are in close agreement with the HST STIS abundance measurements reported by Vennes et al. (2000). Using an effective temperature of $57,000 \pm 2000$ K, Kawka et al. (2008) obtained a mass of $0.58 \pm 0.05 M_\odot$ for the WD in Feige 24.

Table 4.2. Metal Abundances from Analysis of the
FUSE Spectra of Feige 24

C	-6.90 ± 0.06
N	-6.77 ± 0.10
Si	-6.20 ± 1.10
P	-6.20 ± -0.07
S	-7.11 ± -0.05
Fe	(-5.5) (Vennes et al. 2000)

At this juncture, the case for wind accretion from the dM companion as the source of metals in the photosphere of Feige 24's WD appears increasingly unlikely. Since Feige 24 is a wider binary than the other three PCEBs discussed above, and its closer resemblance to the abundance pattern and mix of ion species seen in G191-B2B (a very widely separated M dwarf + WD), accretion is virtually ruled out in favor of radiative forces levitation being the source of Feige 24's photospheric metals. This conclusion is further supported by rocket spectrometry of Feige 24 in which no photospheric helium was detected in the DA1 star in the EUV range of 220A to 250A where He II 237A and 243A should have been detected (Kowalski et al. 2011). However, on the basis of new evidence for accretion of M dwarf wind outflow in other PCEB systems, it may be premature to totally rule out a role for accretion in Feige 24. Indeed, a number of later investigations concern WDs in PCEBs which are so cool ($<20, 000$ K) that radiative forces levitation is not a factor in assessing the origin of absorption lines of metals in their photospheres. These cases are discussed in the following Section 4.9.

4.5 Beyond Feige 24: The Case for Accretion of M Dwarf Winds onto Non-magnetic WDs in PCEBs

The case for accretion onto the WDs in PCEBs was far from being decided when other PCEBs were observed in the FUV. These new possibilities that accretion could be an important source of metals in the atmospheres of hot WDs in PCEBs appeared with the opportunity to widen the exploration of atmospheric metals in hot WDs with the launch of the Extreme Ultraviolet Explorer (EUVE) spacecraft as part of the EUVE all sky survey. Vennes et al. (1997) observed the hot, hybrid composition DAOZ1 degenerate in the PCEB system RE101620-53 (= EUVE J10162053). They used model atmospheres computed assuming solar mixtures of helium and heavier elements distributed homogeneously over the surface. Vennes et al. (1997) found that the overall shape of the EUVE spectrum is well reproduced with hybrid composition models, supporting the idea that the WD in EUVE J1016 +2053 is accreting material in solar proportion. Thus, this system provides still another PCEB object that can serve as a laboratory for wind accretion and other physical processes in hot WDs. Interestingly, Vennes et al. (1997) also found a huge EUV flux variation with a periodicity of 57.3 minutes. They explain the EUV variation as due to the rotational modulation of surface abundance inhomogeneities by the DAOZ1 star. The EUVE spectrum on the WD manifests the effect of the opacity of a host of trace heavy elements on EUV energy distribution. These elements, in addition to the helium in this DAO1 star, include C, N, O, Si, S, and Fe in a hydrogen-dominated atmosphere. By assuming that the accreted heavy elements are distributed homogeneously over the stellar surface, Vennes et al. (1997) show that a steady-state equilibrium is never reached in the photospheric layers and the metal abundances are below the equilibrium values. Finally, they found that the helium abundance is low when averaged over the WD's surface and this lower He abundance constrains the accretion rate to low values between $10^{-19} M_\odot$ yr^{-1} and $10^{-18} M_\odot$ yr^{-1}. These values of the rate of accretion are substantially lower than the Bondi–Hoyle and comparable to (or lower than) the Eddington rate, underscoring that a simple accretion picture may not apply.

With the launch of HST, it became possible to obtain high-resolution FUV spectroscopy of hot WDs in PCEBs and look for evidence of accretion phenomena involving possible outflows in the dM companions. One of the first such HST investigations was carried out by Vennes et al. (1999). They used the HST Goddard High-Resolution Spectrograph (GHRS) to observe three PCEBs, EUVE J0720-317, EUVE J1016-053, and EUVE J2013+400. They determined the mass functions and mass ratios for the three systems and orbital elements but it was their detection of photospheric absorption lines in the three WD components that suggested accretion of dM wind material onto the WDs was likely occurring. This possibility arose from the helium and carbon abundances they measured in all three WDs. In one of the systems, EUVE J0720-317, they found further evidence of a variable helium abundance between 1996 September and 1997 January during which the He abundance decreased by an order of magnitude compared with measurement of the He abundance in 1994 January and February.

Dupuis et al. (1997) had measured an abundance of logHe/H = [4.5] based on a much earlier EUV spectrum obtained in 1995 December. They suggested that the variable He abundance may be indicating episodic accretion of the M dwarf wind with the increased helium abundance diminishing and diffusing downward following the accretion episode. The helium abundance in the other two PCEB WDs maintained a constant helium abundance of logH/He ~ -3. Vennes et al. (1999) concluded that the photospheric abundances of the three WDs revealed evidence of ongoing accretion from the red dwarf companion. They pointed to the large carbon and helium abundances, as well as long-term variability of the helium abundance in EUVE J0720-317 that accretion of dM wind outflow by the WDs in the three systems was likely occurring.

Further evidence of ongoing accretion onto the WDs in PCEBs was reported by Debes (2006) who investigated three PCEB systems containing M dwarfs and DAZ WDs to obtain accretion rates onto the WD from the M dwarf as the source of the photospheric calcium in DAZ stars. His analysis also included an important effort to infer the rates of the mass loss from the M dwarf companions which are very poorly known. This was done by computing the accretion rates onto the DAZ component needed to account for the strength of the Ca II absorption lines in the DAZ atmospheres.

Three of the six DAZs were PCEBs with previously measured orbital periods, primary, and companion masses. WD0419-487 (= RR Cae) is a 0.47 M WD with a $0.095 M_\odot$ dM companion having $P_{orb} = 7.3$ h; WD1026+002 is a $0.68 M_\odot$ WD with a $0.23 M_\odot$ dM companion having $P_{orb} = 14.33$ h; and WD1213+528 is a $0.63 M_\odot$ WD with a $0.32 M_\odot$ dM companion having $P_{orb} = 16.02$ h. Debes (2006) obtained the calcium abundances from the compilation of lines using a method similar to earlier studies, cited above, that assumed accretion–diffusion equilibrium. Debes (2006) adopted the diffusion timescales of Paquette et al. (1986) and the depths of the convection zones in the DAZ stars from the models of Althaus & Benvenuto (1998). For the wind outflow from the M dwarfs, he assumed spherically symmetric winds at the escape velocity from the M dwarfs and the accretion of this wind by the WDs was assumed to be by the Bondi–Hoyle mechanism. For the three PCEB systems, he obtained accretion rates in the range of $10^{-16} M_\odot$ yr^{-1} to $10^{-18} M_\odot$ yr^{-1} onto the DAZ WDs.

For deriving the wind mass loss rates for the M dwarfs, Debes (2006) derived the accretion rates onto the WDs adopting the Bondi–Hoyle fluid accretion rate and used the density of an assumed outflow at the dM surface, from which

$$\dot{M}_{rd} = \frac{q \dfrac{Ca}{H} R^2 v^4}{\tau_D \dfrac{Ca}{H}_{sun} G^2 M_{wd}} \tag{4.10}$$

where $\dot{M}_{rd}$ is the wind mass loss rate from the M dwarf companion, R_{rd} is the dM star's radius, v is the wind velocity, q is the depth of the DAZ star's convection zone, τ_D is the diffusion timescale of a Ca II ion, [Ca/H] is the Ca II abundance in the WD atmosphere and [C/H]$_{sun}$ is the solar Ca abundance, G is the gravitation constant and M_{wd} is the WD mass. Debes (2006) found dM mass loss rates two orders of magnitude smaller than the upper limit to the outflow rate of Proxima Centauri of $6 \times 10^{-14} M_\odot$ yr^{-1} to $4 \times 10^{-15} M_\odot$ yr^{-1} (Wargelin & Drake 2002). The unexpectedly low mass loss rates of the dM stars derived by Debes (2006) could be a result of the inapplicability of the Bondi–Hoyle fluid accretion scenario to these systems and the invalidity of using the solar analogy for the outflow. On the other hand, it cannot be ruled out that the wind outflow from late-type dwarfs is suppressed due to a change in the dynamo mechanism and a sharp break in the rotation-activity connection as noted by Mohanty & Basri (2003). Of course, in PCEB systems, it remains unclear how the magnetic fields and magnetized outflows may interact with the possibly magnetized WD companions. The most significant result of Debes (2006) is that it demonstrates that accretion onto WDs from the M dwarf outflows in the three PCEB systems can plausibly explain the observed Ca abundances in these three DAZ degenerates. Most importantly, the lower surface temperatures of the WDs in these three systems essentially rules out radiative forces levitation as the source of the photospheric calcium. This makes it even more likely that accretion of dwarf M outflow by the DAZ stars is the source of the metals. As shown by Zuckerman et al. (2003) accretion from the local ISM is unlikely to be a source of the metals although exoplanetary debris accretion cannot be entirely ruled out.

In the same program of FUSE observations by Kawka et al. (2008) summarized above for Feige 24, Kawka et al. (2008) obtained FUSE observations of the hot WDs in three other PCEBs, EUVE J0720-317, BPM 6502, and EUVE J2013+400. Within the FUSE Range of EUV wavelengths, they reported the detection of photospheric features in all three PCEBs. They detected photospheric absorption due to C III, and Si IV features in the photospheres of the hot WDs in EUVE J0720-317, and EUVE J2013+400. Their FUSE spectra of BPM6502 revealed absorption features due to C III, N II, Si III, Si IV, and Fe III. The authors used a grid of non-LTE model atmospheres using TLUSTY version 200 and SYNSPEC version 45 (Hubeny & Lanz 1995). From previous optical analyses by the same group (see Kawka et al. 2008), they obtained the following parameters for the hot WDs in the three systems: for BPM6502, they adopted $T_{\mathrm{eff}} = 19,960 \pm 400$ K, $\log g = 7.86 \pm 0.09$, for EUVE J2013+400, $T_{\mathrm{eff}} = 48,000$ K ± 900 K, $\log g = 7.69 \pm 0.09$, for EUVE J0720-317, $T_{\mathrm{eff}} = 52,400$ K ± 1800 K, $\log g = 7.68 \pm 0.01$. From their chemical abundance measurements, Kawka et al. (2008) derived the accretion rates needed to account for the helium and metals by assuming an equilibrium between accretion and diffusion (Fontaine & Michaud 1979; Sion & Starrfield 1984; Wesemael et al. 1984). In Figure 4.4, the FUSE abundance measurements as well as the T_{eff} and $\log g$ derived by Kawka et al. (2008) for the hot WDs in the three PCEBs are displayed.

For the cases of EUVE J0720-317 and J2013+400, with the presence of helium in an H-rich atmosphere and with heavy elements at ~1% solar, the accretion of wind outflow from the close late-type companion could reasonably be the source of the photospheric metals. For the hot WD in BPM 6502, the accretion rate onto the WD atmosphere based upon the carbon abundance $C/H = 0.01(C/H)_{\mathrm{sun}}$, was $1.1 \times 10^{-17} M_{\odot}$ yr^{-1}. With the accretion/diffusion equilibrium applied to EUVE J0720-317 and EUVE J2013+400, accretion rates onto both WDs was found to be $1.8 \times 10^{-19} M_{\odot}$ yr^{-1}, corresponding to their helium abundance and $\dot{M} = 3.4 \times 10^{-19} M_{\odot}$ yr^{-1} corresponding to their carbon abundance.

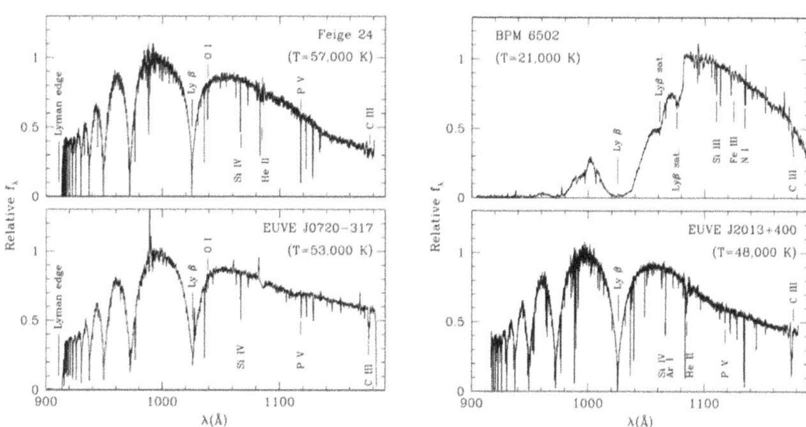

Figure 4.4. FUSE spectra of Feige 24, EUVE J0720317, BPM 6502, and EUVE J2013+400, revealing rich EUV spectra of accreted/levitated(?) heavy elements and ISM lines. (Reproduced with permission from Figure 1 in Kawka et al. (2008). © 2008 The Astrophysical Journal. All rights reserved.)

The aforementioned investigations starting with the work of Vennes et al. (1997, 1999), Debes (2006), and Kawka et al. (2008) presented strong evidence that accretion onto the WDs from late-type companion stars in PCEB is indeed occurring. Subsequently, a number of important new studies appeared, which also utilized the Bondi–Hoyle (fluid) accretion mechanism, the assumption of accretion–diffusion equilibrium, a constant accretion rate, the diffusion coefficients and diffusion velocities of the Montreal group, Vauclair et al. (1979), Alcock & Illiaronov (1980) and the most realistic model convection zone depths (see Althaus & Benveneto 1998), to determine the rate of accretion required to supply the observed elemental abundances in the atmospheres of the WDs in PCEBs.

New evidence continues to mount for wind accretion onto the WDs in PCEBs. Tappert et al. (2011) using higher resolution ground-based spectra than in an earlier paper (Tappert et al. 2007) on the PCEB system LTT560, and identified transitions of Mg, Al, Si, Ca, Sc, Mn, Fe, Co, and Ni, in the DAZ6.7 WD. Tappert et al. (2011) obtained a total of 32 Echelle spectra during a time interval of 7.3 hours covering a wavelength range of 3260–4528 Å, with resolving power ~70,000. This average spectrum has been produced by combining 38 individual spectra that had each been corrected for the radial velocity variations of the WD. Their study also revealed that a weaker component of H alpha emission was anti-phased with the stronger emission component that is associated with the (very) active M dwarf component, thus establishing the highly probable existence of a corona and hence temperature inversion well above the WD photosphere, presumably heated by accretion energy from the infalling material onto the WD surface. The model-fitted metal line profiles are displayed in Figure 4.5.

The accretion rate derived for LTT560's DAZ6.7 WD is $\sim 5 \times 10^{-15} M_\odot$ yr^{-1}. Pyrzas et al. (2012) analyzed the eclipsing PCEB, SDSS J121010.1+334722.9, containing a cool, low-mass, DAZ8.4 WD and a dM5 main sequence star. They detected photospheric absorption features due to Mg, Al, Si, Ca, Mn, and Fe. It is very significant to note that at the low surface temperature of the WD accretor, radiative forces support of the detected metals can be entirely ruled out unlike the hotter ($T_{\text{eff}} > 16,000$ K) accreting WDs in PCEBs where radiative levitation presents a competing interpretation of the presence of photospheric metals.

Moreover, Ribeiro et al. (2013) carried out an investigation of the PCEB system RR Cae which contains a DAZ6.9 WD. Like SDSS J1210, they detected a multitude of absorption lines due to the metals displayed in Table 4.3 from Ribeiro et al. (2012) along with their metal to hydrogen abundance ratios, the solar metals to hydrogen abundance ratios, the metal abundance ratio to the solar value and diffusion timescales in years, and the accretion rate in $M_\odot$ yr^{-1} corresponding to each detected photospheric metal.

From their analysis of RR Cae, they derived an average accretion rate of $7 \pm 2 \times 10^{-16} M_\odot$ yr^{-1}. This accretion rate is somewhat higher than the value derived by Debes (2006) but nonetheless within the uncertainties quoted in the two studies.

The conclusion one can draw from the above studies is that it appears very likely that accretion of wind outflow from the Roche Lobe-detached M dwarf companion stars onto the WDs in PCEBs is indeed occurring. The WDs in these PCEB systems

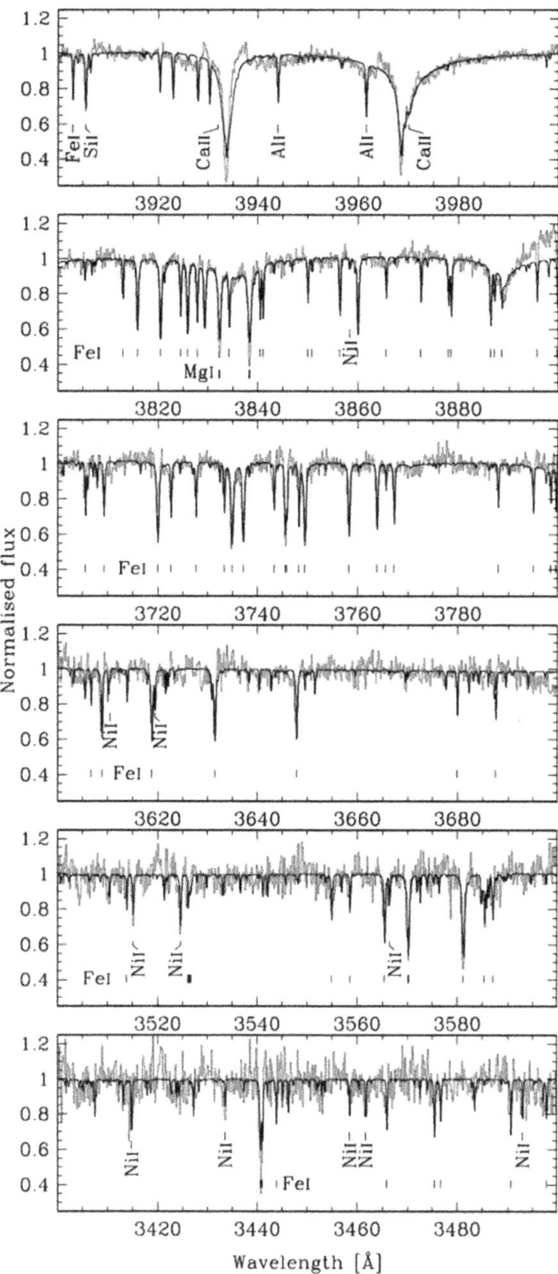

Figure 4.5. Numerous absorption lines of accreted metals in the average spectrum of the WD in LTT 560. The spectra were fitted with TLUSTY/SYNSPEC models to determine the metal abundances in the WD photosphere. The overplotted black line shows a best-fitting model atmosphere for TWD = 7500 K, $\log g = 7.75$. (Reproduced with permission from Tappert et al. (2011). © ESO.)

Table 4.3. Abundances of detected metals in the white dwarf spectra of RR Cae and the corresponding accretion rates, assuming accretion-diffusion equilibrium (Ribeiro et al. 2013).

X	$\log X/H$	$\log X_{sun}/H$	$\log X/X_{sun}$	τ_D	$\dot{M}$
Mg	7.1 ± 0.2	-4.42	-2.65	3.94	7.5×10^{-16}
Ti	9.8 ± 0.3	-6.98	-2.77	8.85	7.0×10^{-16}
Ni	8.5 ± 0.3	-5.75	-2.79	3.79	7.6×10^{-16}
Fe	7.2 ± 0.3	-4.50	-2.70	3.76	1.0×10^{-16}
ca	8.4 ± 0.3	-5.64	-2.77	3.93	5.9×10^{-16}
Al	8.2 ± 0.3	-5.53	-2.77	3.92	7.5×10^{-16}
Si	7.2 ± 0.3	-4.45	-2.76	3.95	5.7×10^{-16}
Cr	9.4 ± 0.3	-6.33	-3.11	3.83	3.3×10^{-16}

are predominantly DAZ stars with the majority observed thus far being cool enough that selective radiative forces levitation of metal ions is not a factor in competition with accretion of M dwarf wind outflow or CMEs. For the hotter DAZ stars, radiative forces levitation cannot be excluded and hence the interpretation of the source of the metals is rendered ambiguous and inconclusive at this point in time.

In a more recent EUV study bearing directly upon the role of radiative forces levitation in hot DA and DAZ WDs, Barstow et al. (2014) carried out an abundance analysis of the entire FUSE archive of 89 DA WDs covering the temperature range of 16,000 K to 77,000 K. Of the many DA stars in the FUSE sample, 56 stars showed no evidence of metals in the EUV range. The remaining 33 FUSE-observed DAZ stars did exhibit absorption features due to metals in the EUV. Of these DAZ stars, the measured abundances disagree with the theoretical predictions of radiative forces levitation, a result that has persisted since the first fully developed theoretical treatment of radiative forces levitation by Chayer (1995). The fact that both metal-free DA stars and DAZ stars overlap the same range of effective temperature casts doubt on radiative levitation unless for unexplained reasons, the sub-photospheric reservoir of metals differs between the metal-free hot DAs and the hot DAZ stars. Among the DAZ stars, debris disks comprise only 3% of DA degenerates hotter than 10,000 K while in the same temperature range, 20% of all DA stars are DAZ. 23% of the sample show evidence of heavy elements in the FUSE wavelength range. This fraction increases with T_{eff} but it is not clear by how much and what that implies.

As originally shown by Debes (2006), the cool DAZ stars in PCEBs are almost certainly accreting from M dwarf wind outflow but it is also possible that accretion of exoplanetary debris such as tidally disrupted asteroids, comets and other surviving planetesimal material could also be accreting onto these objects. This could arise from circumbinary exoplanetary material gravitationally disrupted by a planet(s) orbiting the M dwarf components. It is possible that such material vaporized by hotter WDs could be accreted in gaseous form and kept suspended against diffusion by radiative forces as hypothesized by Barstow et al. (2014).

4.6 Accretion onto Magnetic WDs in PCEBs

The PCEB systems that contain magnetic WDs are also of enormous importance for a number of reasons: they are presumed to be the progenitors of the magnetic CVs, the polar and intermediate polars; they present important tests of theories regarding the origin of the strong magnetic fields of the WD components; like their non-magnetic PCEB counterparts, they potentially can reveal the detailed physics of CEE and the factors that lead to mergers versus compact binary pairs. For example, is the magnetism of the WDs in CVs as well as the magnetism of high field single magnetic WDs associated with the CE process itself (Tout et al. 2008; Briggs et al. 2018)? At the lower end of the field strengths of WDs in magnetic PCEBs, the LARPs and PREPs (see Section 4.8 below) have very low accretion rates. Thus, the study of these objects can deliver important insights into the physics of magnetic accretion at the lowest regime of accretion rates and hence reveal the surface of the underlying magnetic degenerate itself, which is typically hidden by the accretion column light and polar accretion caps of the polars and intermediate polars accreting at much higher accretion rates. Finally, they are definitely accreting wind outflow from magnetically active M dwarf companions detached from their Roche lobes. By establishing their accretion rate and monitoring spectroscopic and photometric variations as well as through theoretical modeling, new insights into still poorly understood magnetic stellar winds and stellar activity of M dwarfs, may also be gained.

It would seem appropriate to begin detailed discussions of accreting magnetic PCEBs with perhaps their first recognized prototype, V471 Tauri (Young & Nelson 1972). The only problem with this designation is that V471 Tauri, to this day, remains unique! No other eclipsing, pulsing, accreting, magnetic WD hotter than 30,000 K is known to exist, thus far. The properties of this amazing PCEB are explored in the next section.

4.7 V471 Tauri, The King of Accreting Magnetic PCEBs

Among the PCEBs that were discovered with the SDSS survey, 10% contain magnetic WDs in close binary orbit with an M dwarf or K dwarf companion. These systems are detached like their non-magnetic counterparts but accretion of wind/flare/CME outflows by the magnetized WD could be occurring. Some fraction of these magnetic PCEBs will reach Roche lobe overflow within a Hubble time and hence be the progenitor systems of the magnetic CVs known as the polars or the intermediate polars. Among the magnetic PCEBs are objects known as low-accretion rate polars (LARPS) discovered by Schmidt et al. (2005). These are detached, compact systems in which the magnetic WD is accreting from the outflow of a main sequence-like dwarf. LARPS are discussed below in Section 4.8. First, there is an interesting PCEB that was discovered in 1971 by Nelson and Young that is regarded as the prototype of the PCEBs containing magnetic WDs and can provide insights into the physics of magnetic accretion in PCEBs.

The magnetic PCEB, V471 Tauri, is a short-period eclipsing binary in the Hyades star cluster (Nelson & Young 1970), whose stellar components are a K dwarf and a hot, magnetic DAZO2 WD. It is the prototype of the pre-cataclysmic binary systems

(Vauclair 1972; Paczynski 1976), which are detached close binaries containing a WD onto which mass transfer will be commence in less than a Hubble time (Schreiber & Gaensicke 2003). V471 Tau is also the prototype of the PCEBs. In the CE scenario (e.g., Iben & Livio 1993) the system originally had a much longer orbital period than its present 12.5 hours. When the more massive component reached the red giant (or AGB) stage, it underwent an episode of dynamically unstable mass transfer enveloping its main-sequence companion in a CE. The resulting frictional heating led the K2 dwarf to undergo orbital decay eventually leading to ejection of the CE due to frictional luminosity. Thus was born a compact binary containing the core of the red giant (now the magnetic DAZO2 WD) and the dK main sequence star. The membership of such a system in the Hyades cluster offers a treasure trove of fundamental information, because the distance, chemical composition, and current turnoff mass of the Hyades cluster is known accurately.

In previous work on V471 Tauri, O'Brien et al. (2001) determined a precise mass for the WD based upon their FUV spectroscopic analysis of 12 Hubble Space Telescope GHRS medium resolution spectra in the Lyα region of the WD's FUV spectra. The spectra were obtained at first and third quadrature in the binary orbit thus enabling the determination of a dynamical mass, a rarity for a nearby WD. Model fits to the Lyα absorption profile yielded a velocity semi-amplitude $K_{WD} = 164.0 \pm 3.5$ km s^{-1}. Using the known velocity semi-amplitude of the K dwarf, O'Brien et al. (2001) derived stellar masses of $M_{dK} = 0.93 \pm 0.04\ M_\odot$ for the K dwarf and $M_{WD} = 0.84 \pm 0.02\ M_\odot$ for the WD. Both component masses are considerably higher than estimates in most earlier investigations. Moreover, since the radius of the WD was previously determined for this eclipsing binary, the mass and radius of the WD is in excellent agreement with the theoretical mass–radius relation. O'Brien et al. (2001) found the orbital inclination (77.5° ± 0.7°), with a separation between the components of 3.30 ± 0.06 $R_\odot$, with the K dwarf underfilling its Roche lobe by 23%. With the newly determined orbital parameters from O'Brien et al. (2001), it was possible to constrain the CE efficiency parameter, (α_{CE}) to lie in the range of 0.1 < α_{CE} < 0.6. This parameter range implies a relatively high rate at which CE evolution creates PCEBs. O'Brien et al. (2001) also showed that the K dwarf is oversized for its mass and the WD is far too hot for its mass compared to other WDs in the Hyades cluster since the most massive WD in the cluster should be the oldest, and hence the coolest among the Hyad WDs. This led O'Brien et al. (2001) to posit that the WD is the product of a binary merger in a formerly triple system, which may also have had an important role in the origin of the WD's magnetism. Other important findings were (1) direct evidence of a CME from the K dwarf (Bond et al. 2001); (2) confirmation of the rotationally modulated magnetic accretion model for the origin of the 9.25 minute optical/far-UV (FUV)/X-ray oscillations (Sion et al. 1998); (3) detection of the first photospheric absorption line of a metal due to magnetic accretion (Sion et al. 1998); (4) the demonstration that the K dwarf is oversized for its mass and the WD is far too hot for its mass compared to other WDs in the Hyades cluster (O'Brien et al. 2001). Since the most massive WD in the cluster should be the oldest, and consequently the coolest, the high T_{eff} of the WD is puzzling. One possible explanation is that the WD is the product of a binary merger (Bond et al. 2001, O'Brien et al. 2001). Moreover, the WD component of V471 Tau exhibits variations at soft X-ray, EUV, and optical wavelengths at a period of 9.25 minutes. This variability is caused by rotational modulation of a magnetic WD, whose polar regions are darkened in the soft X-ray and EUV bands by accreted photospheric metals and helium (Clemens et al. 1992; Barstow et al. 1992), and brightened in the optical by UV flux redistribution.

Sion et al. (2012) carried out an archival study of the same HST data set of V471 Tauri with the GHRS for three key science objectives: (1) to determine K1 and the WD mass from observations at the quadrature orbital phases; (2) to look for the spectroscopic signature of medal lines accreted by the WD; and (3) to study the K2V star in which the WD was used as a beaming probe to study the K dwarf's mass loss, chromospheric and coronal structures at orbital phases near the ingress and egress of primary eclipse. Previous GHRS spectra covered only a 35 Å region centered on Lya, but they revealed a photospheric Si III (1206) absorption line modulated in strength on the X-ray rotational phase, such that at X-ray minimum (when the X-ray dark accretion pole faces the observer) the Si III absorption appears at maximum strength. This provided the first direct confirmation of the magnetic accretion model for the origin of the 9.25 minute X-ray/EUV/optical oscillations. The Si III detection (see Figure 1 in Sion et al. 1998) also marginally revealed Zeeman splitting into sigma + and sigma− components, as expected when viewing down the magnetic field lines at the time the pole is seen face-on. The observed Zeeman splitting indicates a polar field strength of 350 kG. It was the first time Zeeman splitting of a line due to metals has been seen in any magnetic WD, whether single or binary (Sion et al. 1998). They determined a silicon abundance of 0.1 solar (the first metal abundance determined for any magnetic WD) within the accreted Si spot, which covers about 40% of the visible hemisphere. Assuming that accretion and diffusion are in equilibrium, the low abundance implies an accretion rate four orders of magnitude lower than the Bondi–Hoyle rate that would occur in the absence of a magnetic field. The implied highly inefficient accretion strongly suggests that a magneto-centrifugal propeller may be operating. In the standard propeller formulation, e.g., Pringle & Rees (1972), a field of 350 kG is adequate for its operation in the case of V471 Tau.

Sion et al. (2012) found clear evidence of a secondary accretion pole shown in Figure 4.6 and rederived an effective temperature and spectroscopically derived surface gravity for the magnetic WD of $T_{\text{eff}} = 34, 100$ K $\pm$ 100 K, $\log g = 8.25 \pm 0.05$. In Figure 4.6, two phases of the rotational profile are displayed, computed from the STIS data with the rotational ephemeris HJED2450885.67019 +0.0064193953E due to Clemens et al. (1992). The first peak corresponds to the primary accretion pole (optical maximum, X-ray minimum) and the second peak to the clear FUV detection for the first time of the secondary accretion pole.

They detected a number of absorption features associated with metals accreted onto the photosphere of the magnetic WD. Among these features are C III (1175), Si (1206), C II (1335), Si IV (1393, 1402), and a feature at 1501 they identify as a blend of three Si III transitions. The C III (1175) and Si IV (1393, 1402) features are the strongest in the FUV spectrum and provide abundance determinations. All other absorption lines are either interstellar or associated with a region above the WD and/or with CME events illuminated as they pass in front of the WD. They derived radial velocities for all of the accreted metallic absorption features due to accretion and for

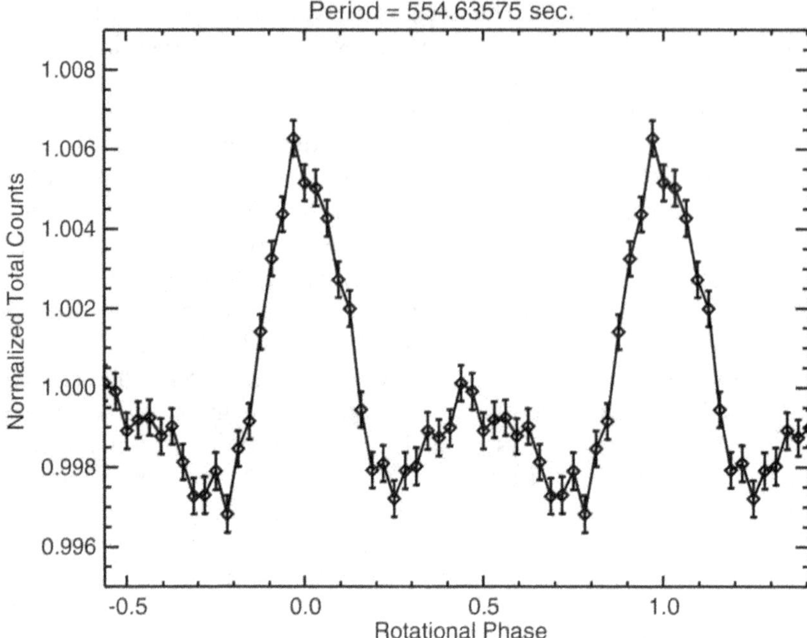

Figure 4.6. Two phases of the rotational profile showing the signature of the primary and secondary accretion poles. Ephemeris: HJED 2450885.67019+0.0064193953E (computed from the STIS data). (Reproduced with permission from Sion et al. (2012), their Figure 1. © 2012 The Astrophysical Journal. All rights reserved.)

the emission lines due to C IV (1548, 1550) and He II (1640). The two emission lines clearly indicate a shift as a function of orbital phase in the same direction as the WD but with velocities that are much smaller than the WD velocity. All of the absorption features are modulated on the 555 s rotation period of the WD with maximum line strength at rotational phase 0.0 when the primary magnetic accretion pole is exposed to the observer. The maximum absorption line strengths happen at minimum light in the soft X-ray and EUV light curves of the WD and with maximum light in the optical light curve, thus confirming the model in which the magnetic poles are darkened at short wavelengths by helium and metallic absorption and brightened in the optical by flux redistribution. In Figure 4.7, the rotational modulation of the resonance doublet Si IV (1393, 1402) is displayed. The maximum absorption strength of Si IV occurs at rotational phase 0.0, which is when the primary accretion cap is facing the observer (Figure 4.7; see also Figure 4.8 below for the rotational modulation of C III 1175.711 absorption line).

They determined the silicon and carbon abundances by fitting the detailed line profiles of Si IV and C III in order to determine the photospheric abundance of accreted silicon and carbon in the polar accretion region/spot. They adopted a two-component model in which a pure hydrogen photosphere covers 60% of the stellar surface visible at phase zero and an accreted cap covers 40% of the hemisphere. These surface areas are taken from the EUV light-curve analysis of Dupuis et al. (1992) and the analysis of Barstow et al. (1992). The abundances of Si and C were

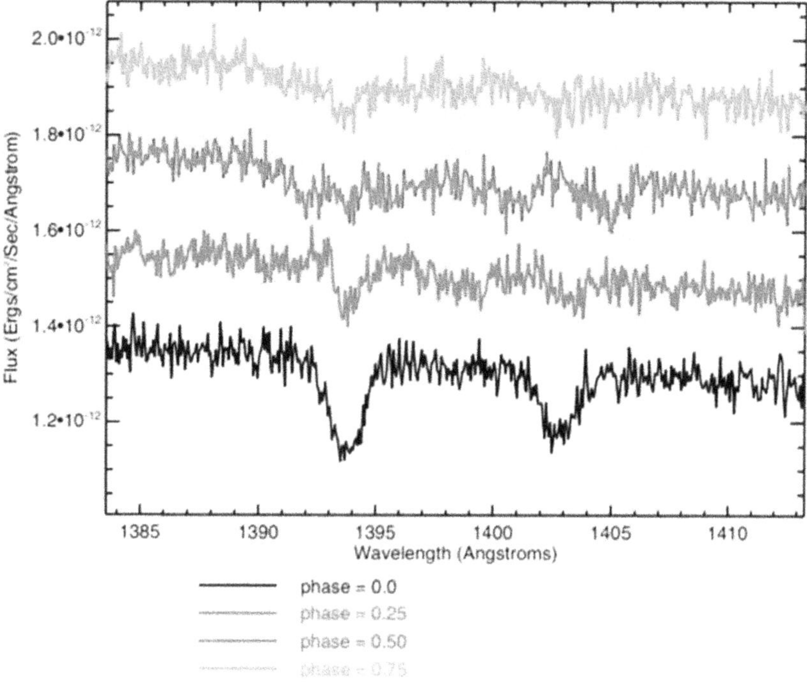

Figure 4.7. Variation of the S IV (1393, 1402) doublet as a function of X-ray rotational phase 0.0 when the primary accretion cap is facing the observer. In Figure 4.8, the C III (1175.711) absorption feature is displayed versus rotational phase revealing a strong absorption line at rotational phase 0.0. Note that there is no clearly detectable feature at phase 0.5 which corresponds to the secondary pole. (Reproduced with permission from Sion et al. (2012). © 2012 The Astrophysical Journal. All rights reserved.)

varied over the range $0.001 < (Si/H)/(Si/H) < 0.5$ while all of the other elements were fixed at very low abundances (<0.001).

In Figure 4.9, the model atmosphere line profile which best-fitted the Si IV (1393, 1492) is displayed.

Their derived Si abundances of $0.015 \times$ solar from the STIS spectra are a factor of seven lower than the Si abundance of $0.1 \times$ solar derived by Sion et al. (1998) for the Si III (1206) feature in the GHRS spectra. This lower Si abundance together with the flux levels of the STIS observations is likely indicating that V471 Tauri was less active during the STIS observations than the epoch when the GHRS spectra were taken. If the accretion rate and diffusion rate of Si and C are in equilibrium, the accretion rate required to provide the C and Si quantifies the efficiency of magnetic accretion in this system and is four orders of magnitude smaller than the Bondi–Hoyle fluid rate. This implies a magnetic-centrifugal propeller greatly lowers the accretion efficiency. Moreover, the accreted ions provide crucial information on element diffusion in the presence of a magnetic field.

It is puzzling however that the photospheric absorption features show no clear evidence of Zeeman splitting into sub-components. The Si IV and C III features are broad and flat-bottomed profiles with no detectable Zeeman splitting. However,

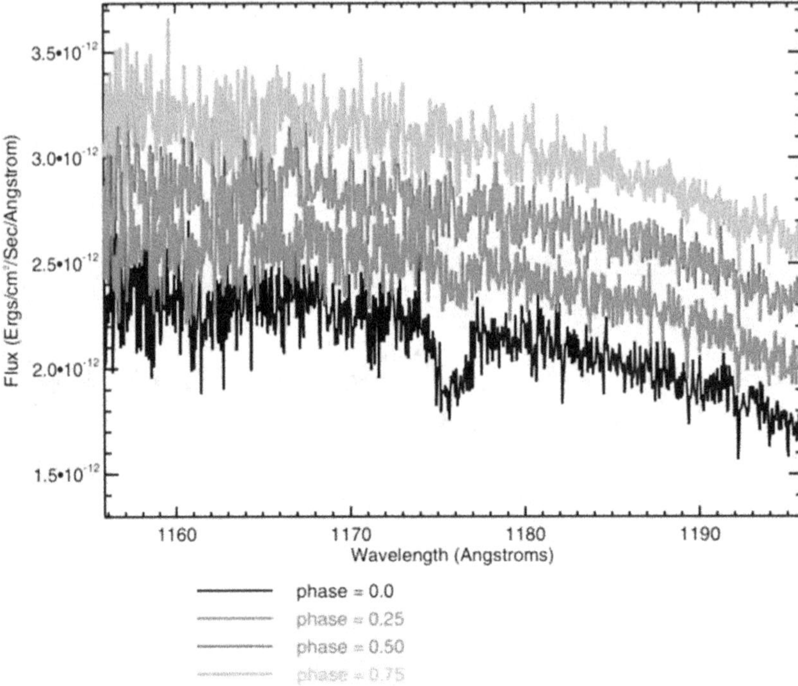

Figure 4.8. C III 1175.711 versus rotational phase with maximum strength at rotation phase 0.0. This feature is the strongest line absorption in the STIS spectrum. (Reproduced with permission from Sion et al. (2012). © 2012 The Astrophysical Journal. All rights reserved.)

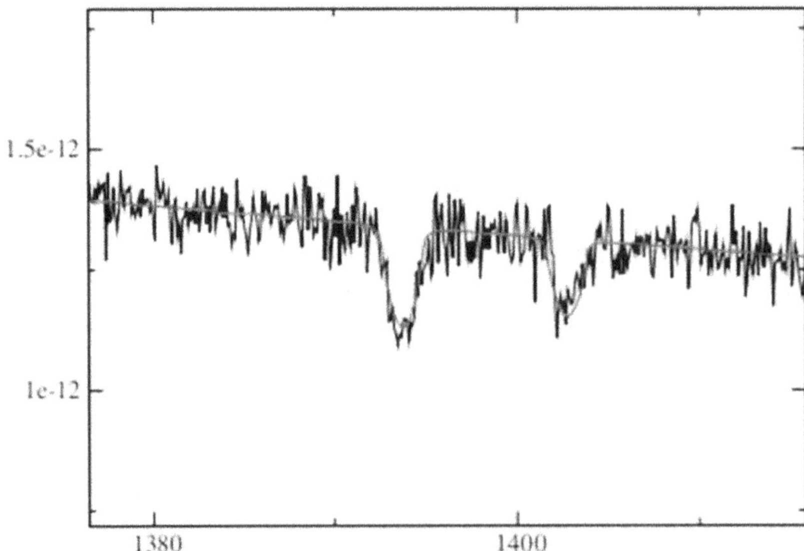

Figure 4.9. Best-fitting model fluxes to the Si IV line profiles in the STIS spectrum (see the text for details). (Reproduced with permission from Sion et al. (2012). © 2012 The Astrophysical Journal. All rights reserved.)

Sion et al. (2012) point out that it is likely that a field of a few 10^5 G, together with rotational broadening and magnetic field spread over the stellar surface, could easily explain the width, unresolved substructure, and the flat bottoms observed in these lines. It is also important to point out that the absorption lines show no evidence of a correlation between strength variations and orbital phase.

From the large abundances in the accretion spots, it is straightforward to derive the accretion rate needed to account for the silicon residing on the accretion spots. Sion et al. (2012) also assumed that the effect of a probable magnetic field is likely not strong enough to have a significant effect on the diffusion coefficient. Under these assumptions, they applied the gravitational diffusion developed by Paquette et al. (1986) and Dupuis et al. (1993) from their work on the accretion–diffusion equilibrium model for WDs. They derived the diffusion velocity of silicon ions in the atmosphere of V471 Tauri of about 0.1 cm s^{-1} and the e-folding time for gravitational diffusion is roughly 4.3 days. The diffusion timescale is therefore very short in the atmospheric zone and one expects that silicon would rapidly settle out of the atmosphere if accretion was not in a steady state with diffusion. Assuming that a steady-state is established, Equation (4) of Dupuis et al. (1993) which describes the steady-state condition for gravitational settling, can be applied with a constant accretion rate. This theory was developed in the context of cooler WDs with superficial convective zones. Here, they assumed that the atmosphere over the accretion spot has a constant silicon abundance and replaced M_{cz} by M_{atm} (total mass in the atmosphere) in the equation. The rate of accretion of silicon needed to provide this silicon abundance is about $7.2 \times 10^{-20} M_\odot$ yr^{-1} – 1 (or 4.5×10^6 g s^{-1}). Clearly, further detailed diffusion computations are badly needed to include the potential effect of the magnetic field on the accretion flow and on diffusion in the atmosphere.

The thermal history of the magnetic WD in PCEB V471 Tauri, with $T_{\text{eff}} = 35,000$ K, and $M_{\text{wd}} = 0.84 M_\odot$, has a cooling age of only 7.8×10^6 years compared with the mean cooling age of Hyades single WDs of 9×10^7 years. If its formation was essentially coeval with the single Hyad degenerates, then V471 Tauri is 13,000 K hotter than their average T_{eff} (19,680 K). This amount of heating by accretion is in the same regime as that experienced, on average, by WDs in CVs (Sion 1999). Comparing the positions of the Hyades degenerates in the HRD with a Bondi–Hoyle locus of constant accretion, as discussed by Castellani & Panagia (1971), there could be several possible physical explanations for the puzzling cooling history of the V471 Tauri degenerate. These include (1) the possibility of the V471 Tauri WD being the result of a stellar merger (Bond et al. 2001) as discussed earlier in this chapter; (2) an ancient nova WD still cooling; (3) steady nuclear burning providing its surface luminosity; and (4) compressional heating causing its observed T_{eff}. For steady hydrogen burning, an accretion rate of $\dot{M} = 10^{-11}\frac{L_s}{X_h} = 2 \times 10^{-12} M_\odot$ yr^{-1} is sufficient to account for the present T_{eff}. How much accreted mass/angular momentum is needed to spin the WD up to its present rotation (120 km s^{-1})? Substituting numerical values, one finds that, for an initial velocity of 0 km s^{-1}, the WD would have accreted $7.8 \times 10^{-3} M_\odot$ to be spun up to its present rotation rate. This is an upper limit since it assumes zero rotation initially. If the initial WD rotation was 50 km s^{-1}

(rapid for a field WD), then the only time the WD could be accreting at a rate similar to low state CVs is just after emergence from the CE when the K dwarf was bloated and could have filled its Roche lobe. The relaxation timescale for the bloated late-type dwarf following its emergence from the CE is of the order of 10^4 years (Hjellming & Taam 1991). It is during this phase that the WD could have accreted at a rate comparable to CV WDs.

4.8 Low-accretion Rate Polars (LARPS) and Pre-polars (PREPS)

Among the Hamburg–Schmidt (Hagen et al. 1995) and SDSS discovered CVs (Schmidt et al. 2005) there is a subset of compact binaries with magnetic WDs that Schwope et al. (2002) have coined low-accretion rate polars (LARPS). The low-accretion rates of these systems are orders of magnitude lower than the typical accretion rates of polars and intermediate polars. They were found in the same region of blue color as the quasi-stellar radio sources because of the large amplitude cyclotron humps characteristic of low-accretion rates and magnetic fields strength around 60 MG (Schwope et al. 2002). Cyclotron humps in the optical spectra of magnetic PCEBs are the primary means by which they are discovered and by which their field strengths and magnetic WD temperatures are determined from spectral fitting. Non-relativistic electrons gyrating in the WD field generate a power given by

$$\frac{-dE}{dt} = \sigma_t B^2 v^2 / c\mu_0 \tag{4.11}$$

where E is energy, t is time, σ_t is the Thomson cross-section, B is the magnetic field strength, v is the velocity perpendicular to the magnetic field, c is the speed of light and μ_0 is the permeability of free space. The broadening of cyclotron spectral features occurs due to imperfections (non-uniformities) in the magnetic field as the electrons propagate from one region of the field to another. The main cyclotron hump has the same frequency as the electron's orbit and the harmonic humps, due to the non-uniformities.

In Figure 4.10, cyclotron spectra are displayed for polars for a similar range of field strengths as the magnetic PCEBs (from Schwope 1996). While cyclotron emission can occur over the broad wavelength range from the FUV to the infrared, depending upon the field strength of the MWD, the magnetic PCEBs with the lowest field strengths will exhibit cyclotron emission at infrared wavelengths while the magnetic PCEB systems with the highest field strengths will display the cyclotron emission in the ultraviolet. Because magnetic PCEBs are identified predominantly in the optical, this explains the relatively narrow range of field strengths seen in the current census of PREPs/LARPs since in that narrow range of field strengths, strong cyclotron emission occurs in the optical.

The evolutionary association of LARPS to polars remains unresolved. Are LARPS the progenitors of polars or are they polars that have entered into deep low optical brightness states of very low accretion? The donor stars of LARPS are underfilling their Roche lobes, which implies that they could be still evolving toward Roche lobe overflow mass transfer.

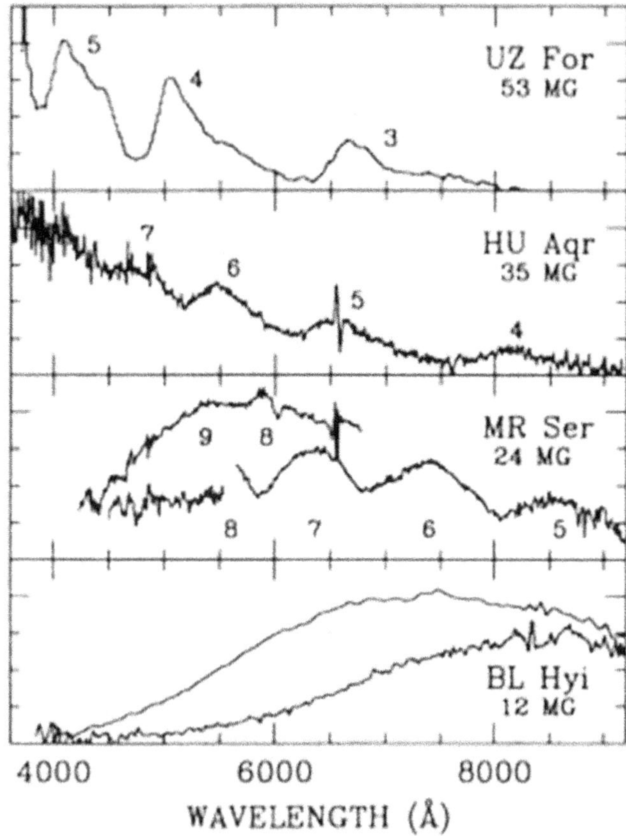

Figure 4.10. Cyclotron spectra of polars as a function of field strengths in the optical and infrared. The four objects that are displayed, BL Hyi, MR Ser, HU Aqr, and UZ For, have a range of field strengths similar to the range of field strengths of the magnetic PCEBs. The integer numbers of the cyclotron harmonics are shown. As the field strength increases, the cyclotron humps shift toward shorter wavelengths. (Reproduced with permission of Springer from Schwope (1996). Copyright © 1996, Kluwer Academic Publishers.)

The characteristics of these systems can be concisely summarized as listed below (Schmidt et al. 2005; Webbink & Wickramasinghe 2005; Schwope et al. 2009; Parsons et al. 2021):

(1) Very cool ($T \leqslant 10, 000$ K), intrinsically faint magnetic WDs.

(2) Roche lobe-detached M-dwarf donor companions.

(3) No evidence of states of high optical brightness.

(4) Accretion rates below even the threshold of gravitational wave emission in CVs with $\dot{M} \leqslant 3 \times 10^{-13} M_\odot$ yr^{-1} from cyclotron modeling.

(5) Strong, optical cyclotron emission features in the magnetic WD spectra.

(6) Weak or undetectable X-ray emission.

(7) Mode of accretion is by wind outflow from the M dwarf and capture by the WD very likely via a magnetic siphon (Webbink & Wickramasinghe 2005).

(8) Little evidence of photospheric Zeeman-split absorption lines.

(9) The accreting magnetic WD photosphere is virtually always directly observed.

It is useful to illustrate the observed properties of the magnetic PCEBs by selecting two systems, one near the high extreme of field strength, with an M dwarf donor SDSS J120615.73+510047.0 (Schwope et al. 2009), and the other system, NLTT5306 (Longstaff et al. 2019), at the low extreme of field strength with a brown dwarf donor. Both donors underfill their Roche lobes and neither system shows any evidence of an accretion disk or even a debris disk.

At one extreme, SDSS J120615.73+510047.0, consists of a late-type dwarf (dM3–dM4) donor star and a magnetic WD in binary with an orbital period between 200 and 270 minutes. The WD exhibits a single, strong cyclotron hump in the optical spectrum. The 9000 K magnetic degenerate with a field strength of 108 MG is accreting wind outflow from the secondary at the rate, $\dot{M} \sim 10^{-14} M_{\odot}$ yr^{-1} (Schwope et al. 2009). At the other extreme, Longstaff et al. (2019) present strong evidence that NLTT5306 is magnetically funneling a weak wind from its brown dwarf donor. The absence of any X-ray emission by SWIFT places an upper limit on the accretion rate of $2 \times 10^{-15} M_{\odot}$ yr^{-1} which is consistent with a magnetic field as low as 0.45 kG. If their model is confirmed, then this would be the first detection of wind outflow from a brown dwarf with implications for wind mass loss rates at the low end of the main sequence.

One of the most vexing problems in CV research is identifying the progenitors of the magnetic CVs, specifically the polars that are synchronized, have very strong magnetic field strengths, totally lack accretion disks and the intermediate polars that are asynchronous, have somewhat lower magnetic field strengths and truncated accretion disks due to magnetic pressure of the magnetic WD. This progenitor puzzle is particularly acute because recent complete survey studies reveal that as many as ~25% of the CVs contain magnetic WDs with MegaGauss field strengths (Parsons et al. 2021) yet the most current census of magnetic PCEBs at the time of this writing amounts to only 17 systems (Schwope 2021, Parsons et al. 2021).

It is important to note that the percentage of PCEBs with magnetic WDs is extremely small. It was first pointed out by Liebert et al. (2005, 2015) that the incidence of magnetism among detached WD plus main sequence star binaries is extremely low. Where are the young (i.e., hotter) magnetic WDs in detached binaries with non-degenerate companion stars? Indeed, out of thousands of PCEBs found in the SDSS, only a very small percentage are known to contain magnetic WDs (see Rebassa-Mangera et al. 2016). Since all PCEBs, whether containing non-magnetic degenerates or magnetic ones, went through the process of CEE, it is compelling to identify some factor(s) in the CE process that leads to a magnetic WD emerging. As stated above, the very origin of the strong magnetic fields of the WDs in magnetic PCEBs as well as the high field single magnetic degenerates has been theoretically attributed to the interaction of the WD and the co-rotating CE during the spiraling in process (Tout et al. 2008; see also Briggs 2018) leading to the origin of the high field single magnetic degenerates via merger. However, according to Parsons et al. (2021), the Tout et al. (2008) scenario has been thrown into question by Belloni & Schreiber (2020) who present several counterarguments to the Tout et al. (2008) mechanism, including recent population synthesis models that predict an unacceptably large and unobserved majority of CVs as strongly magnetic. Moreover, the

Tout et al. (2008) model would produce magnetic WDs with field strengths well below the observed field strengths in magnetic CVs (Belloni & Schreiber 2020). They also note the extraordinarily low number of very hot magnetic WDs in CVs although this paucity could be due to moderate suppression of accretion through interaction with the M dwarf magnetic outflow. This point will be discussed later in this chapter. In any event, it is clear that the progenitorship of the polars and the IPs remains a vexing problem, as discussed below.

Adding to the challenge of the troubling shortfall of potential progenitors for the magnetic CVs is the uncertainty of the true evolutionary status of LARPs. Are they indeed the genuine progenitors of polars or are they polars that have entered into prolonged deep low optical brightness states of very low accretion? The donor stars of LARPS are underfilling their Roche lobes, which implies that they could be still evolving toward Roche lobe overflow mass transfer. Anticipating the need to differentiate between the two possibilities, Schwope et al. (2009) renamed those systems that can be shown to have never achieved Roche lobe contact since their emergence from the CE, pre-polars (PREPs), while LARPs would designate those magnetic PCEBs that are currently in a deep low optical brightness state of accretion from a temporarily detached M dwarf donor star but the donor had reached Roche lobe overflow in the past. Both acronyms apply to a PCEB with Roche lobe-detached lower main sequence donor star in close orbit with a magnetic WD. Unfortunately, the two terms are sometimes confused or used interchangeably in the literature. In both LARPs and PREPs, accretion onto the magnetic WD is through wind outflow from the late-type M dwarf. The PREPs, due to their very low accretion rates and absence of ever having had Roche lobe overflow of the donor star, have suffered only very low levels of accretion heating. Hence, their surface temperatures reveal their cooling ages and therefore the elapsed time since they emerged from the CE.

Of these 17 systems, it is very likely that the vast majority are LARPs rather than PREPs but this remains to be determined definitively. All of the known systems are close to filling their Roche lobes by >80%. Of the relatively few magnetic PCEBs that have WD masses, they appear to be more massive than their non-magnetic PCEB counterparts. The accretion rates of the magnetic PCEBs are determined by setting the accretion luminosity equal to the cyclotron luminosity. Curiously, the accretion rates of the magnetic WDs in the magnetic PCEBs are higher than the accretion rates of the non-magnetic PCEBs but the surface temperatures of the non-magnetic WDs in the PCEBs are higher. This is likely due to the lower efficiency of magnetic accretion versus wind accretion onto a non-magnetic WD (Townsley & Gaensicke 2009). The reason for this lower efficiency probably stems from the wind outflow from the M dwarf companion being constrained by the magnetic field of the WD, resulting in less angular momentum being carried away (Webbink & Wickramasinghe 2002, 2005). The process that governs both the rate of accretion in magnetic PCEBs and the concomitant reduction in the rate of orbital angular momentum loss was first derived, on the basis of energy densities, by Webbink & Wickramasinghe (2005) and is termed a magnetic siphon. The importance of the magnetic siphon process merits a detailed quantitative examination in Section 4.9.

4.9 Webbink and Wickramasinghe (2005) Magnetic Siphon

For a detached binary with a magnetically active M dwarf and a magnetic degenerate with a field strength above about 50–60 MG (Li, Wickramasinghe and Wu 1995), the magnetic stellar wind of the M dwarf can be essentially "captured" by the much higher field strength primary companion, depending upon the orbital period and geometry. It is possible that the entire stellar wind of the M dwarf is brought into the primary degenerate's surface. As stated earlier, Webbink and Wickramasinghe (2005) explored the plausibility of this process quantitatively based upon energy densities. The fundamental importance of this mechanism (termed a magnetic siphon by R. Webbink) merits being highlighted here.

First, it is known that the M dwarfs in LARPS underfill their Roche lobes. Moreover, gravitational wave radiation sets a firm lower limit to the equilibrium mass transfer rate (Webbbink and Wickramasinghe 2005). For Roche lobe overflow, the constraint on the equilibrium accretion rate is given by

$$|\dot{M}_{\mathrm{eq}}| \geqslant \frac{M_{\mathrm{dM}}}{\tau_{\mathrm{GR}}}\left[\frac{2}{3} - \frac{M_{\mathrm{dM}}}{M_{\mathrm{wd}}}\right]^{-1}. \tag{4.12}$$

Since the low observed mass transfer rates in LARPs eliminate the possibility of steady accretion by Roche lobe overflow, then the case of non-equilibrium accretion states must predominate but on long timescales. Webbink and Wickramasinghe (2005) evaluated the growth/decay time scales for non-equilibrium mass transfer, as follows.

$$\frac{1}{\dot{M}}\frac{d\dot{M}}{dt} = \frac{1}{H}\left[\frac{dR}{dt} - \frac{dR_{\mathrm{L}}}{dt}\right] \tag{4.13}$$

where

$$\frac{dR}{dt} = \left(\frac{\zeta_{\mathrm{s}}R}{M}\right)\dot{M} + R\left(\frac{\partial \ln R}{\partial t}\right)_{\mathrm{M}} \tag{4.13a}$$

$$\frac{dR_{\mathrm{L}}}{dt} = \left(\frac{\zeta_{\mathrm{L}}R_{\mathrm{L}}}{M}\right)\dot{M} + 2R_{\mathrm{L}}\left(\frac{\partial \ln J}{dt}\right)_{\mathrm{M}} \tag{4.13b}$$

where J is the orbital angular momentum, $\zeta_{\mathrm{s}} = \left(\frac{\partial R}{\partial \ln M}\right)_{\mathrm{ad}}$ is the adiabatic mass–radius exponent of the donor star and $\zeta_{\mathrm{L}} = \left(\frac{\partial \ln R_{\mathrm{L}}}{\partial \ln M}\right)_{\mathrm{J}}$ is the corresponding Roche lobe mass–radius exponent in the absence of angular momentum losses. Setting $R \sim R_{\mathrm{L}}$ and collecting terms,

$$\frac{1}{\dot{M}}\frac{d\dot{M}}{dt} \sim \frac{R}{H}\left[(\zeta_{\mathrm{s}} - \zeta_{\mathrm{L}})\frac{\dot{M}}{M} + \left(\frac{\partial \ln R}{\partial t}\right)_{\mathrm{M}} - \frac{2d \ln J}{dt}\right]. \tag{4.14}$$

For the non-equilibrium case, with all terms in Equation (4.14) taken to be constant except for $\dot{M}$, then

$$t = \tau_{\dot{M}} \ln\left(\frac{\dot{M}}{\dot{M} - \dot{M}_{\mathrm{eq}}}\right) + \mathrm{const} \tag{4.15}$$

where $\tau_{\dot{M}}$ is the time needed for the Roche lobe radius to evolve through one pressure scale height in the absence of mass transfer and is given by

$$\tau_{\dot{M}} = \frac{H}{R}\left[\left(\frac{\partial \ln R}{\partial t}\right)_{\mathrm{M}} - \frac{2d \ln J}{dt}\right]^{-1}. \tag{4.16}$$

If $\tau_{\dot{M}} > 0$, then $\dot{M}$ relaxes to $\dot{M}_{\mathrm{eq}}$; if $\tau_{\dot{M}} < 0$, then $\dot{M}$ decays exponentially on the timescale $\tau_{\dot{M}}$.

For a late-type M dwarf donor star Webbink & Wickramasinghe (2005) point out that for a late-type M dwarf, the ratio of the pressure scale height H_{p} to the stellar radius, R_{dM}, is roughly 5e-5 implying that the characteristic relaxation time is 10^5 to 2×10^5 yr. On the other hand, they point out that the high space density of LARPs (a substantial fraction of the space density of polars) requires a mass transfer relaxation timescale $\tau_{\dot{M}} \gtrsim 5 \times 10^8$ yr. This ratio implies that the characteristic scale height of late M dwarf mass loss must be of order $\frac{H}{R_{\mathrm{L}}} \sim 0.5$, which is much larger than the $H_{\mathrm{p}}/R_{\mathrm{dM}}$ estimated above. From the size of the implied scale height, a corresponding kinetic temperature follows given by $T_{\mathrm{kin}} \sim 0.5 - 1 \times 10^7$ K.

By a happy coincidence, this temperature is of the same order as the coronal temperatures of the donor stars in PCEBs, the late-type active M dwarfs (Giampapa et al. 1996). Is the magnetic field of the WD sufficiently strong to capture the coronal outflow from its M dwarf companion? The field of, say, a $0.6M_\odot$ WD with surface field $B \sim 60$ MG, at a separation a $\sim 1R_{\mathrm{sun}}$ (corresponding roughly to a Roche lobe-filling M5 dwarf), is of order $B \sim B_{\mathrm{wd}} (R_{\mathrm{wd}}/a)^3 \sim 100$ G. The energy density of this WD field at the donor secondary is $U_{\mathrm{B}} = B^2/8\pi \sim 500$ erg cm^{-3} which is much greater than the energy density of the evaporative wind of an M dwarf, which is $U_{\mathrm{w}} \sim \rho c_{\mathrm{s}}^2 \sim \sum \dot{c}_{\mathrm{s}} \sim 0.2$ erg cm^{-3}.

It is thus physically plausible for the magnetic WD to capture the entire wind of the M dwarf donor even out to binary separations of $\sim 10R_{\mathrm{sun}}$ provided that the ambient field of the unspotted regions of the M dwarf remains small (Webbink & Wickramasinghe 2005).

4.10 What Is The Origin of the High Field Magnetic Degenerates?

Since all PCEBs, whether containing non-magnetic degenerates or magnetic ones, went through the process of CEE, it is compelling to identify the factor(s) in the CE process that leads to a magnetic WD emerging (Ferrario et al. 2020). Wickramasinghe et al. (2014) first showed that the strongest WD magnetic fields should result from the merger process of a binary occurring during the CE evolutionary stage. This process would also be expected to produce a magnetic WD having a significantly higher mass than the mass of isolated non-magnetic WDs. If a binary with a WD and a non-degenerate companion avoid merger and survive the CE phase and approach close to each other, a strong field in the WD would be generated. These systems could be the precursors of the PREPS and with time evolve into polars. This scenario receives some support from the observation that the PREPs have very short periods compared with the non-magnetic PCEBs as well as the fact that the PREPs have very high magnetic fields.

As stated above, the very origin of the strong magnetic fields of the WDs in magnetic PCEBs as well as the high field single magnetic degenerates has been attributed to the interaction of the WD and the co-rotating CE during the spiraling in process (Tout et al. 2008; see also Briggs et al. 2018) leading to the origin of the high field single magnetic degenerates via merger. As pointed out by Parsons et al. (2012), the Tout et al. (2008) scenario has been thrown into question by Belloni & Schreiber (2020) who present several counterarguments to the Tout et al. (2008) mechanism, including recent population synthesis models that predict an unacceptably large and unobserved majority of CVs as strongly magnetic. Moreover, the Tout et al. (2008) model would produce magnetic WDs with field strengths well below the observed field strengths in magnetic CVs (Belloni & Schreiber 2020). They also note the low number of very hot magnetic WDs in CVs although this paucity could be due to moderate suppression of accretion through interaction with the M dwarf magnetic outflow. In any event, it is clear that the progenitorship of the polars and the IPs remains a vexing problem but a promising new theory has emerged (Schreiber et al. 2021) that underscores the objections to the role of the CE and WD formation in the origin of the high magnetic fields.

The origin of magnetic fields in WDs remains a fundamental unresolved problem in stellar astrophysics. In particular, the very different fractions of strongly ($\geqslant 1$ MG) magnetic WDs in evolutionarily linked populations of close WD binary stars cannot be reproduced by any scenario suggested so far. Strongly magnetic WDs are absent among detached WD binary stars that are younger than approximately 1 Gyr. In contrast, in semi-detached CVs in which the WD accretes from a low-mass star companion, more than one third host a strongly magnetic WD.

Given the rarity of magnetic PCEBs and the large percentage of magnetic CVs among the general CV population, Schreiber et al. (2021) have explored binary evolutionary models of rotating WDs with carbon–oxygen cores, including in their models, the crystallization of the core and the interaction of the WD magnetic field and the donor late-type dwarf magnetic field. Their approach has yielded dynamo-generated strong magnetic fields (>1 MegaGauss) on the WD with the crystallizing C–O core, coupled with the WD rotation, driving the dynamo. Moreover, synchronization torques within the binary play a role in decreasing the loss of orbital angular momentum which would cause the magnetic CV donor star to detach from its Roche lobe for fairly short durations of time (the strongly magnetic PCEBs?). This new theory for the origin of the strong fields in magnetic WDs looks promising and will be thoroughly tested by independent investigations.

References

Alcock, I. 1980, ApJ, 235, 541

Althaus, L. G., & Benvenuto, O. G. 1998, MNRAS, 296, 206

Barstow, M. A., Schmitt, J. H. M. M., Clemens, J. C., et al. 1992, MNRAS, 255, 369

Barstow, M. A., Barstow, J. K., Casewell, S. L., Holberg, J. B., & Hubeny, I. 2014, MNRAS, 440, 1607

Barstow, M. A., Kleinman, S. J., Provencal, J. L., & Ferrario, L. 2020, IAU Proc. 357, White Dwarfs as Probes of Fundamental Physics: Tracers of Planetary, Stellar and Galactic Evolution (Cambridge: Cambridge Univ. Press)

Bauer, E. B., & Bildsten, L. 2019, ApJ, 872, 96

Belloni, D., & Schreiber, M. R. 2020, MNRAS, 492, 15

Bond, H. E., O'Brien, M.S., & Sion, E. M. 2001, AGM, 18S0913

Bond, H. E., Mullan, D. J., O'Brien, M. S., & Sion, E. M. 2001, ApJ, 560, 919

Briggs, G. P., Ferrario, L., Tout, C. A., & Wickramasinghe, D. T. 2018, MNRAS, 481, 360

Castellani, V., & Panagia, N. 1971, Ap&SS, 10, 122

Chamandy, L., Frank, A., Blackman, E.G., et al. 2018, MNRAS, 480, 1898

Chapman, D., & Cowling, T. B. 1970, The Mathematical Theory of Uniform Gases (Cambridge: Cambridge Univ. Press)

Chayer, P., et al. 1987, in Proc. Conf. Faint Blue Stars (Schenectady, NY: L. Davis Press), 653

Chayer, P. 1995, PhD thesis, Univ. de Montreal

Clemens, J. C., Nather, R. E., Winget, D. E., et al. 1992, ApJ, 391, 773

Debes, J. 2006, ApJ, 652, 636

Dupree, A. K., & Raymond, J. 1982, ApL, 263, L63

Dupuis, J., Vennes, S., Chayer, P., Cully, S., & Rodríguez-Bell, T. 1997, ASSL, 214, 375

Dupuis, J., Fontaine, G., Pelletier, C., & Wesemael, F. 1993, ApJS, 84, 73

Ferrario, L., Wickramasinghe, D., & Kawka, A. 2020, AdSpR, 66, 1025

Fontaine, G., & Michaud, G. 1979, ApJ, 231, 826

Giampapa, M. S., Rosner, R., Kashyap, V., et al. 1996, ApJ, 463, 707

Grichener, A., Sabach, E., & Soker, M. 2018, MNRAS, 478, 1818

Hagen, H-J., Groote, D., Engels, & Reimers, D. 1995, A&A, 111, 195

Heinenen, R. A., Saumon, D., Daligault, J., et al. 2020, ApJ, 896, 2

Hjellming, M. S., & Taam, R. E. 1991, ApJ, 370, 709

Hubeny, I., & Lanz, T. 1995, ApJ, 439, 875

Iben, I., & Livio, M. 1993, PASP, 105, 1373

Jackson, B. V., & Howard, R. A. 1993, SoPh, 148, 359

Kawka, A., Vennes, S., Dupuis, J., Chayer, P., & Lanz, T. 2008, ApJ, 675, 1518

Kowalski, M., Barstow, M. A., Wood, K. S., et al. 2011, ApJ, 730, 115

Lacombe, P., Liebert, J., Wesemael, F., & Fontaine, G. 1982, BAAS, 14, 915

Li, J K, Wickramasinghe, D T, & Wu, K W 1995, MNRAS, 276, 255

Liebert, J. 2005, AJ, 129, 2376

Liebert, J., Ferrario, L., Wickramasinghe, D. T., & Smith, P. S. 2015, ApJ, 804, 93

Linsky, J. L., & Wood, B. E. 2004, ASSL, 317, 1L

Linsky, J., & Wood, B. E. 2014, ASTRP Proc, 1, 43

Longstaff, E. S., Casewell, S. L., Wynn, G. A., et al. 2019, MNRAS, 484, 2134

Mohanty, S., & Basri, G. 2003, ApJ, 583, 451

Mullan, D. 2021, private communication

Nebot Gómez-Morán, A., et al. 2011, A&A, 536, A43

Nelson, B., & Young, A. 1970, PASP, 82, 699

O'Brien, M. S., Bond, H. E., & Sion, E. M. 2001, ApJ, 563, 971

Paczyński, B. 1976, in IAU Symp. 73, Structure and Evolution of Close Binary Systems, ed. P. Eggleton, S. Mitton, & J. Whelan (Dordrecht: Reidel), 75

Paerels, F. B. S., Bleeker, J. A. M., Brinkman, A. C., & Heise, J. 1986, ApJ, 309, L33

Parsons, S. G., Hill, C. A., Marsh, T. R., et al. 2012, MNRAS, 419, 817

Parsons, S. G., Gänsicke, B. T., Marsh, T. R., et al. 2013, MNRAS, 429, 256

Parsons, S. G., Gänsicke, B. T., Schreiber, M. R., et al. 2021, MNRAS, 502, 4305

Paquette, C., Pelletier, C., Fontaine, G., & Michaud, G. 1986, ApJS, 61, 197

Pringle, J. E., & Rees, M. J. 1972, A&A, 21, 1

Pyrzas, S., Gänsicke, B. T., Brady, S., et al. 2012, MNRAS, 419, 817

Rebassa-Mansergas, A., Nebot Gómez-Morán, A., Schreiber, M. R., Girven, J., & Gänsicke, B. T. 2011, MNRAS, 413, 1121

Rebassa-Mansergas, A., Ren, J. J., Parsons, S. G., et al. 2014, MNRAS, 458, 3808

Rebassa-Mansergas, A., Anguiano, B., García-Berro, E., et al. 2016, MNRAS, 463, 1137

Ribeiro, T., Baptista, R., Kafka, S., et al. 2013, A&A, 556, A34

Schreiber, M. R., Gänsicke, B. T., Rebassa-Mansergas, A., et al. 2010, A&A, 513, 7

Schmidt, G., Szkody, P., Vanlandingham, K. M., et al. 2005, ApJ, 630, 1037

Schwope, A. D. 1996, IAU Proc. 158, Cataclysmic Variable and Related Objects, ed. A. Evans, & J. H. Wood (Dordrecht: Kluwer) 189

Schwope, A. D., Brunner, H., Hambaryan, V., & Schwarz, R. 2002, in ASP Conf. Ser. 261, The Physics of Cataclysmic Variables and Related Objects, ed. B. T. Gaensicke, K. Beuermann, & K. Reinsch (San Francisco, CA: ASP), 102

Schwope, A. D., Nebot Gomez-Morán, A., Schreiber, M. R., & Gaensicke, B. T. 2009, A&A, 500, 867

Sion, E. M., & Wesemael, F. 1983, BAAS, 14, 915

Sion, E. M., & Starrfield, S. G. 1984, ApJ, 286, 760

Sion, E. M., Schaefer, K. G., Bond, H. E., Saffer, R. A., & Cheng, F. H. 1998, ApJ, 496, L29

Sion, E. M. 1999, PASP, 111, 532

Sion, E. M., Bond, H. E., Linder, D., et al. 2012, ApJ, 751, 66

Tappert, C., Gänsicke, B. T., Schmidtobreick, L., et al. 2007, A&A, 474, 205

Tappert, C., Gaensicke, B. T., Rebassa-Mansergas, A., Schmidtobreick, L., & Schreiber, M. R. 2011, A&A, 532, A129

Tassoul, M., & Tassoul, J. L. 1983, ApJ, 267, 334

Tout, C. A., Wickramasinghe, D. T., Liebert, J., Ferrario, L., & Pringle, J. E. 2008, MNRAS, 387, 897

Thorstensen, J.R., Charles, P.A., Margon, B., & Bowyer, S. 1977, PASP, 89, 623

Truran, J. W., Starrfield, S. G., Strittmatter, P. A., Wyatt, S. P., & Sparks, W. M. 1977, ApJ, 211, 539

Tutukov, A. V., & Yungelson, L. 1979, AcA, 29, 665

Vauclair, G. 1972, A&A, 17, 437

Vauclair, G., Vauclair, S., & Greenstein, J. L. 1979, A&A, 80, 79

Vennes, S., Thorstensen, J., & Polomski, E. 1999, ApJ, 523, 386

Vennes, S., Chayer, P., Fontaine, G., & Wesemael, F. 1989, ApJ, 336, L25

Vennes, D., Dupuis, J., Bowyer, S., & Pradhan, A. K. 1997, ApJ, 482, L73

Vennes, S., Polomski, E. F., Lanz, T., Thorstensen, J. R., Chayer, P., & Gull, T. 2000, ApJ, 544, 423

Verbunt, F., & Zwaan, C. 1981, A&A, 100, L7

Wargelin, B. J., & Drake, J. J. 2002, ApJ, 578, 503

Webbink, R., & Wickramasinghe, D. 2002, MNRAS, 335, 1

Webbink, R. F., & Wickramasinghe, D. T. 2005, in ASP Conf. Ser., 330, 137 The Astrophysics of Cataclysmic Variables and Related Objects, ed. J-M. Hameury, & J-P. Lasota (San Francisco, CA: ASP), 137

Wesemael, F., Henry, R. B. C., & Shipman, H. L. 1984, ApJ, 287, 868

Wood, B. E., Linsky, J. L., Müller, H.-R., & Zank, G. P. 2001, ApJ, 547, L49

Wood, B. E., Müller, H-R., Zank, G. P., & Linsky, J. L. 2002, ApJ, 574, 412

Wood, B. E. 2018, JPhCS, 1100, 012028

Wood, B. E., Müller, H-R., Zank, G. P., Linsky, J. L., & Redfield, S. 2005, ApJ, 628, L143

Wood, B. E., Meuller, H-R., Redfield, S., & Edelman, E. 2014, ApJ, 781, L33

Young, A., & Nelson, B. 1972, ApJ, 173, 653

Yungelson, Tutukov, & Livio, 1993, ApJ, 418, 794

Zuckerman, B., Koester, D., Reid, I. N., & Hünsch, M. 2003, ApJ, 596, 477

Accreting White Dwarfs

From exoplanetary probes to classical novae and Type 1a supernovae

Edward M Sion

Chapter 5

Accreting White Dwarfs in Symbiotic Variable Binaries

5.1 Introduction

Symbiotic variables are wide binary systems consisting of a red giant or red supergiant primary in mutual orbit with a hot accreting white dwarf (WD) companion. Their orbital periods typically range between hundreds of days to well over 1000 days. They are all close enough in separation that the WD accretes some fraction of the outflowing wind from the red giant primary. Most symbiotic systems contain a normal red giant with IR color indices corresponding to a red giant atmosphere having T_{eff} between 3000 and 4000 K. The vast majority of symbiotics (~80%) are classified S-type systems as they have little associated dust and the giant donors are spectral types G to M. The remainder (~20%), the D-type systems (D denotes dust) contain Mira variables, and their near-IR colors point to a combination of a reddened Mira variable and a warm dust shell. The observed spectra of symbiotic systems are composite, with the spectroscopic signature of both components covering the ultraviolet to the optical and revealing high excitation emission lines superposed on the optical and near-IR continuum of a red giant.

The optical–near-IR flux distribution component declines steeply into the UV reaching nearly zero continuum at ~2600 Å while the hot component contribution to the composite spectrum increases steeply into the far-ultraviolet (FUV). In Figure 5.1, a typical combination UV–optical–IR spectrum is displayed revealing the characteristic overall "smile"-shaped distribution of flux for the symbiotic variables CI Cygni and AG Dra.

5.1.1 Observational Evidence of Accretion Disks around the Accreting WD?

In many symbiotic systems, there is little evidence of an accretion disk surrounding the hot, accreting component while in a few systems, an accretion disk is evident during a quiescent period of lower accretion. In fact, there is no concrete evidence of a disk in any S-type symbiotic containing a WD. This also appears to be true for most of the D-type symbiotics. There is also no evidence for a FUV contribution by a nebular continuum

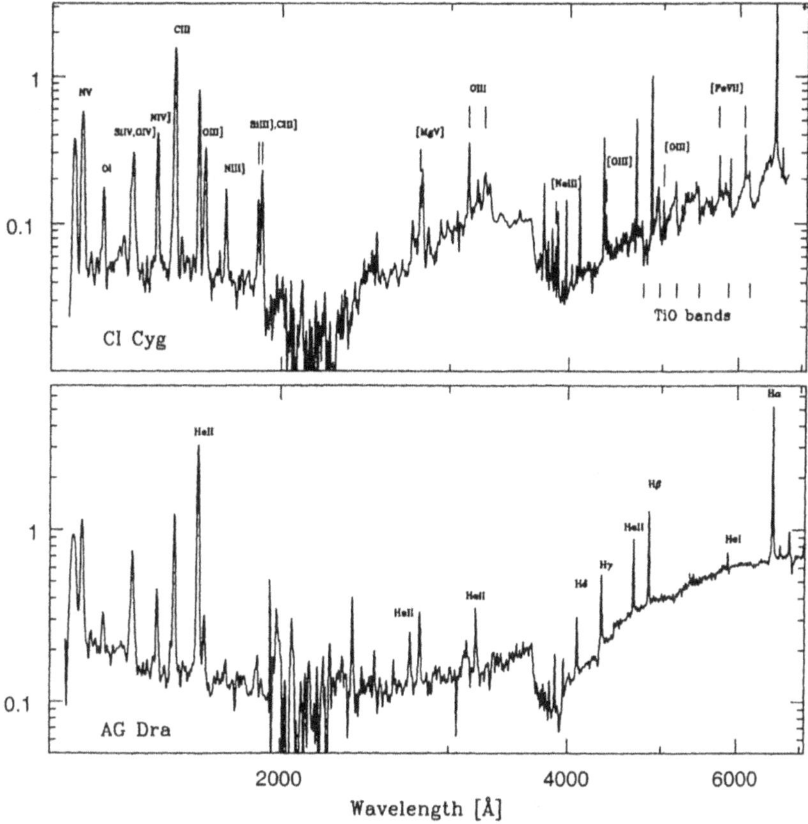

Figure 5.1. The optical UV to infrared spectral energy distribution of typical symbiotic variable, in this case, CI Cygni and AG Dra. Note the steep upturns at shorter wavelengths indicating the presence of an accreting hot source companion (a very hot WD) accreting from the wind outflow of a red giant. (Reproduced with permission from Mikolajewska (2003). © 2003. The Astronomical Society of the Pacific. All rights reserved.)

shortward of about 1600 Å. Therefore, the FUV continuum may be contributed virtually entirely by the radiation from the hot accreting WD up to at least 1600 Å. The presence of a hot accreting WD in symbiotic binaries offers several possibilities of studying the effect of wind accretion on the compact star. These include testing whether accretion disks form around the accreting compact component, the efficiency of wind accretion by the Bondi–Hoyle mechanism versus Roche lobe overflow by the giant donor star, and variations in the properties and structure of the hot WD in response to time-variable accretion. Among symbiotic variables two optical brightness phases are identified observationally, as follows.

5.1.2 The Hot Components during the Quiescent Phase

During quiescent phases the hot accreting WD component radiates energy at roughly a constant rate and unchanged spectral distribution. The intense radiation field of the accreting hot component photo-ionizes the giant's wind outflow, leading to a nebular continuum which contributes all the way down in wavelength to ~1600 Å, as well as several nebular emission lines at high excitation and ionization states. The inferred wind mass-loss rates from the giant donor during this quiescent stage is a few $\times 10^{-7} M_\odot$ yr^{-1}. This roughly corresponds to a blackbody pseudo-photosphere with a characteristic temperature of 10^5 K but with an effective radius an order of magnitude larger than expected for an accreting WD. Hence, it is unlikely that the actual accreting WD's surface can be detected. It is therefore virtually hopeless to detect photospheric absorption lines of ions accreting onto the actual photosphere, determine the WD's accreted chemical abundances or its rotational velocity.

The comprehensive study of over 20 symbiotic variables by Muerset et al. (1991) yielded the range of temperatures, luminosities, and masses of the hot components using their modified Zanstra methodology. The accreting WDs in symbiotic variables, with relatively few exceptions, are very hot (55, 000 K $\leqslant T_{\text{eff}} \leqslant$ 210, 000 K) (Muerset et al. 1991; see also Skopal 2005; Sion et al. 2017) with luminosities in the range $0.3 \leqslant L_\odot \leqslant 37, 400 L_\odot$ and radii of the hot components lying in the $0.002 \leqslant R_\odot \leqslant 0.8 R_\odot$. What seems now to be widely accepted is that the accreting WDs in symbiotic variables are largely low-mass WDs centered around 0.5 to $0.7 M_\odot$ (Muerset et al. 1991). In other words, symbiotics likely have lower WD masses than the average mass of accreting WDs in cataclysmic variables, which is $0.83 \pm 0.03 M_\odot$ (Zorotovic et al. 2011; McAllister et al. 2019). Many of these objects are extremely hot, high luminosity degenerate stars having low-WD gravities, extended envelopes, and thermonuclear reactions with possible steady-state hydrogen burning. In Figure 5.2, an HR diagram showing the locations of the hot components in S-Type and D-Type symbiotic variables is displayed in comparison with two theoretical evolutionary tracks for two different stellar masses. The evolutionary tracks in the HR diagram closely resemble the evolutionary paths of the central stars of planetary nebulae.

5.1.3 Observed Characteristics of the Active Phase

Typically during the active phase of a symbiotic variable, the spectral energy distribution of the accreting component manifests a significant change compared to the quiescent phase, along with a two to three magnitude brightening and spectroscopic evidence of high velocity wind outflow. The ionization state of the photoionized nebula changes as a result of this active phase wind outflow. The brightening referred to above remains of uncertain origin. However, it is thought to be caused by the release of gravitational energy due to an accretion event that triggered the active phase. A thermonuclear origin for these brightenings cannot be ruled out definitively at this time.

It is clear that the wind accreting hot component, whether fed through the Bondi–Hoyle mechanism or disk-fed by Roche lobe overflow of the giant star, demands fundamental insights into the process of wind accretion. Early estimates of wind accretion in symbiotics and the question of disk formation (e.g., Livio & Warner 1984) were followed by 2D and 3D hydrodynamic wind accretion simulations of increasing sophistication. These theoretical treatments of wind accretion onto the WD are discussed in the following section.

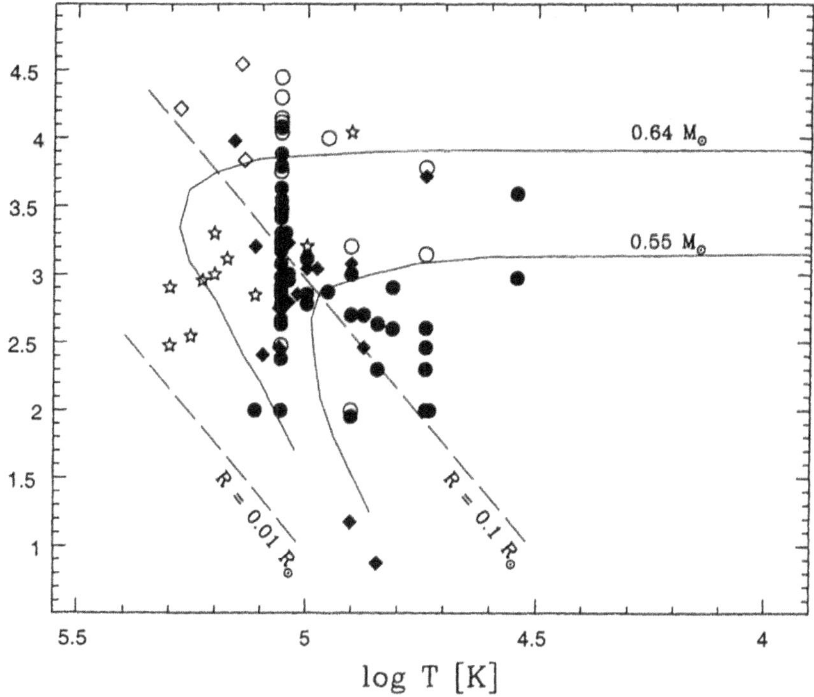

Figure 5.2. The hot components of symbiotic stars in the HR diagram. Data from Muerset et al. (1991) and Mikolajewska et al. (1997) are plotted as diamonds and circles, whereas filled and open symbols represent the Galactic S- and D-types, respectively. Stars correspond to the symbiotic stars in the Magellanic Clouds (Morgan 1992, 1996). The solid curves are the evolutionary tracks from Schonberner (1989), and the dashed lines correspond to lines of constant radius. (Reproduced with permission from Mikolajewska (2003). © 2003. The Astronomical Society of the Pacific. All rights reserved.)

5.2 Theoretical Studies of the Wind Accretion in Symbiotics

The initial approach to the questions of (1) the efficiency of wind accretion and, (2) whether disk formation occurs from wind accretion, used the Bondi–Hoyle accretion process (Bondi & Hoyle 1944) discussed in Chapters 2 and 4. This approach to the problem was implemented by Livio & Warner (1984). Recall that the rate of accretion is given by

$$\dot{M}_{bh} = \frac{4\pi G^2 M_{wd}^2 \rho_\infty}{(V_{rel}^2 + c_s^2)^{1.5}} \tag{5.1}$$

where M_{wd} is the WD mass, ρ_∞ is the density of the undisturbed medium, V_{rel} is the relative velocity of WD and gas, and c_s is the sound speed. This approach to the problem was implemented in a computationally convenient form by Livio & Warner (1984). The Bondi–Hoyle accretion radius of a WD accreter is

$$R_{acc} \sim 6.7 \times 10^{13} \left(\frac{M_{wd}}{M_\odot}\right)\left(\frac{V_{rel}}{20 \text{ km s}^{-1}}\right)^{-2} \text{cm} \tag{5.2}$$

where $V_{\text{rel}} = (V_{\text{wind}}^2 + V_{\text{orb}}^2)^{1/2}$ is the relative velocity of star and gas, with the implicit assumption that V_{rel} is highly supersonic. Moreover, the accretion radius of the WD defined above is an approximation subject to uncertainties discussed by Bondi & Hoyle (1944). If one assumes that the red giant wind has a constant velocity, then the accretion rate is expressed in a computationally convenient form, approximately as

$$\dot{M} \sim 5.6 \times 10^{-10} \left(\frac{M_{\text{wd}}}{M_\odot}\right)^2 \left(\frac{\dot{M}_{\text{wind}}}{5 \times 10^{-7} M_\odot \text{ yr}^{-1}}\right) \left(\frac{V_{\text{rel}}}{20 \text{ km s}^{-1}}\right)^{-3} \left(\frac{V_{\text{wind}}}{20 \text{ km s}^{-1}}\right)^{-1} \left(\frac{a}{10^{15}}\right)^{-2}$$

where a is the orbital semimajor axis and $\dot{M}_{\text{wind}}$, assuming that the wind is homogeneous, is the wind outflow rate from the red giant. The accretion luminosity generated by the Bondi–Hoyle fluid accretion process onto the WD is of order

$$L_{\text{acc}} \sim 4.7 \times 10^{33} \frac{M_{\text{wd}}^3}{M_\odot} \frac{\dot{M}_{\text{wind}}}{5 \times 10^{-7} M_\odot \text{ yr}^{-1}} \left(\frac{V_{\text{rel}}}{20 \text{ km s}^{-1}}\right)^{-3} \left(\frac{V_{\text{wind}}}{20 \text{ km s}^{-1}}\right)^{-1} \left(\frac{a}{10^{15}}\right)^{-2}$$
$$\left(\frac{R_{\text{wd}}}{10^9}\right)^{-1} \text{ erg s}^{-1}.$$

Crucial to whether a disk can form from Bondi–Hoyle accretion is the specific angular momentum gravitationally captured given by

$$j \sim 1.6 \times 10^{18} \eta \left(\frac{M_{\text{tot}}}{4M_\odot}\right)^{1/2} \left(\frac{a}{10^{15} \text{ cm}}\right)^{-3/2} \left(\frac{M_{\text{wd}}}{M_\odot}\right)^2 \left(\frac{V_{\text{rel}}}{20 \text{ km s}^{-1}}\right)^{-4} \text{ cm}^2 \text{ s}^{-1}$$

where η is a parameter dependent on the existing velocity and density gradients in the wind and is usually assumed to be close to unity. Using the above expression for the captured specific angular momentum, an estimate of the outer radius for the Keplerian disk to form around the WD is

$$R_{\text{disk}} \sim 2.0 \times 10^{10} \eta^2 \left(\frac{M_{\text{total}}}{4M_\odot}\right) \left(\frac{a}{10^{15} \text{ cm}}\right)^{-3} \left(\frac{M_{\text{wd}}}{M_\odot}\right)^3 \left(\frac{V_{\text{rel}}}{20 \text{ km s}^{-1}}\right)^{-8}.$$

The condition for a disk to form requires that $R_{\text{disk}} > R_{\text{wd}}$ and that the disk radius R_{disk} be smaller than the accretion radius, i.e., $R_{\text{disk}} < R_{\text{acc}}$. Livio & Warner (1984) underscored the importance of observationally establishing the existence or absence of accretion disks in symbiotic variables because this would be an important test of the validity of the Bondi–Hoyle accretion process applied to symbiotics. At the same time however, a critical need for more physically realistic theoretical treatments of the accretion process onto WDs in symbiotics existed. This needed progress soon followed in the form of fully multi-dimensional hydrodynamic simulations of accretion onto WDs in symbiotics that would answer whether or not an accretion disk can form from wind accretion in symbiotic systems, and yield the rate of mass accretion and the accretion efficiency (the fraction of the red giant wind that actually accretes onto the WD) as discussed below.

Although the initial impetus for multi-D hydrodynamic simulations of wind outflow from asymptotic giant branch (AGB) giants was driven by investigations of planetary nebula shaping and proto-planetary nebulae rather than accretion processes in symbiotic variables, the application of hydrodynamic codes to the wind accretion process in symbiotics soon followed. Specifically, the bipolar and asymmetric geometry of many planetaries suggested the presence of a binary companion, the idea being the presence of a detached binary companion to the AGB star's outflow naturally led to consideration of accretion of the outflow by the companion. Thus, hydrodynamic simulations of the accretion process and the question of whether or not accretion disks would form around the mass gainer component were directed to understanding symbiotic variables.

One of the first of such studies was carried out by Mastrodemos & Morris (1998). Although the accreting component in their smooth particle hydrodynamic simulations was a main sequence star, their results were directly relevant to symbiotic binaries since a mass losing giant component and a secondary component, a main sequence star, not a WD, was accreting a fraction of the giant's wind. Most importantly for symbiotic systems, their study revealed that stable, permanent accretion disks of different disk diameters form, as a function of wind velocity and component separation. These disks are geometrically thin but would be optically thick and resemble steady-state disks. Moreover, these disks would lead to significant accretion of mass and angular momentum by the accreting secondary component.

The wind models incorporated the basic ingredients of the wind including radiation pressure on dust grains, subsequent collisional momentum transfer from grains to gas, dust–gas viscous heating, and atomic and molecular cooling. Of course, the accreting component was represented by a hard surface with the low gravity of a main sequence star. Pulsations of the wind emitting late-type giant were not included in the simulation but the expected fluctuations in density and velocity in the wind due to pulsations would be expected to smooth out with increasing distance from the giant component. The boundary layer between the inner-most disk (where the motion is Keplerian) and star as well as the actual surface of the accreting component were not spatially resolved in these simulations given the much larger number of particles required and the concomitant exponentially large increase in the computing time. Instead, a free flow condition was imposed whereby particles that crossed a spherical surface centered on the main sequence star close to the main sequence star were removed from the computation, in effect having been accreted. However, while little was revealed about the details of actual accretion onto a realistic, accretion-heated secondary star with its radiation field and possibly wind outflow, this work established that stable accretion disks should be able to form from wind accretion in symbiotic systems.

In the wake of increasing interest in wind accretion studies using smoothed particle hydrodynamics (SPH), a new mode of mass transfer was presented by Psodialowski & Mohamed (2007). The potential theoretical breakthrough was stimulated by two key observations of the prototype Mira itself and its companion, presumed here to be a WD (see however Jura & Helfand 1984; Kastner & Soker 2004). Whether a WD or a main sequence dwarf, because of the large separation of the two components at 65 AU, it is regarded as a weak symbiotic. Surprisingly, despite the large separation, the two components are strongly interacting as evidenced by the first HST UV and Chandra X-ray observations (Karovska et al. 2004, 2005), which unexpectedly revealed mass outflow from Mira A to Mira B in this detached binary. Mira A is far from filling its Roche lobe because its radius is only between 2 and 3 AU depending on its pulsation phase while the Roche lobe radius of Mira A is ~25 AU. The key X-ray observation, with the Chandra X-Ray Observatory (Karovska et al. 2005), spatially resolved the X-ray emission from Mira AB and revealed that X-rays were being emitted from gas surrounding Mira A out to the Roche lobe itself. It is obvious that the slow (~20 km s^{-1}) wind material from Mira A was filling Mira A's Roche lobe!

This new process (Psodialowski & Mohamed 2007) was termed wind Roche lobe overflow (hereafter WRLOF) and will occur when the giant's wind acceleration region, parameterized by a wind acceleration parameter α, occurs at several stellar radii from the wind-emitting giant. If the wind is dust-driven as in the case of D-type symbiotics like Mira variables and the cool wind is very slow as observed in a number of systems, then the wind gas and dust fills the Roche lobe of the giant, unlike normal Roche lobe overflow when the giant's surface is in contact with the Roche lobe.

A few stellar radii away from Mira A, the wind temperature cools sufficiently that dust formation occurs. When this happens, it accelerates the outflow due to the effect of the radiation pressure on the newly formed dust grains. With the large separations of D-type symbiotic variables, the radius of the Mira's Roche lobe is typically much larger than the dust formation radius R_{dust}. Hence, the outflow from the Mira is relatively unaffected by the binary potential and the geometry of the outflow is expected to be roughly spherical. However, once R_{dust} is only a factor of a few of the Roche lobe radius R_{RL}, the wind outflow begins to manifest the effect of the binary potential and thus the onset of WRLOF.

The gas and dust flows to the WD at an enhanced rate, in effect, a gravitationally-focused flow of the gas and dust concentrated along the orbital plane of the binary. Psodialowski & Mohamed (2007) carried out SPH simulations using 10^5 particles and four cases of wind velocity; 20, 30, 40, and 50 km s^{-1}. Their starting model parameters were $1.4 M_\odot$ for the Mira and $0.6 M_\odot$ for the accreting component, with a binary separation of 11 AU corresponding to a binary period of ~9420 days. They included the radial pulsation period of the Mira component with a pulsation period of 332 days which facilitates the dust formation. The average wind mass-loss rate $\dot{M}_\mathrm{w}$ was ~$10^{-6} M_\odot$ yr^{-1}.

As was the case in the Mastrodemos & Morris (1998) SPH simulations, neither the surface of the WD nor the boundary layer between the disk and WD are resolved due to the need for a much larger number of particles and the prohibitively enormous increase in computational time. Instead, Psodialowski & Mohamed (2007) introduced a point sink term in the wind particles continuity equation that represents the gravitational potential well of the WD. As the wind particles approach the WD, they lose mass and are accreted (i.e., are removed from the simulation) when the mass of a given particle became 0.1% of their original mass.

For the higher velocity wind cases (30 km s^{-1}, 40 km s^{-1}, 50 km s^{-1}) the wind acceleration mechanism increased the velocity of the wind particles beyond the

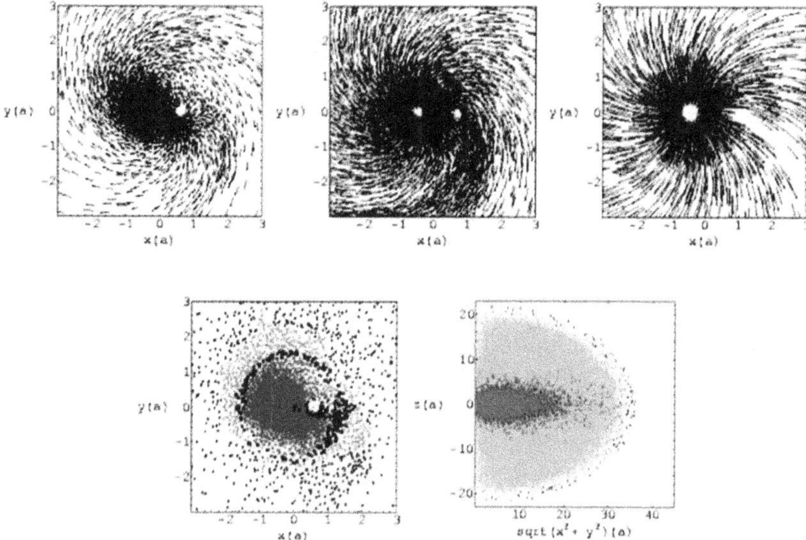

Figure 5.3. Top: The velocity field in the co-rotating center of mass frame of model A (left), B (middle), and D (right) in the orbital plane. The Mira is at (−.57, 0) and the WD is at (0.42, 0). Bottom (left): The wind particles in the orbital plane for model A. The triangles are particles bound to the primary, the stars are bound to the secondary, the boxes are bound to the system and the diamonds are unbound. Bottom (right): Particle positions in the meridional plane for model A (dark) and model D (light). (Reproduced with permission from Podsiadlowski & Mohamed (2007)).

escape velocity ($\sim$40 km s^{-1}) of the giant component. Thus, the fraction of the material lost by the primary that is accreted by the secondary is only $\sim$1.0%, which is close to the accretion efficiency of Bondi–Hoyle accretion ($\sim$0.7%). However, in model A, the slowest outflow velocity, the wind is concentrated in the Roche lobe of the primary and flows through the inner Lagrangian point (L1) toward the WD. In this case, the bulk of the wind material is accreted by the WD or remains bound to the system, and only a very small fraction of the wind is lost to the system.

In Figure 5.3, the velocity field of the SPH simulations of Psodialowski & Mohamed (2007) are displayed for wind velocities of 20, 30, and 40 km s^{-1}. The reader should focus on Model A shown top left with $V_{\rm wind} = 20$ km s^{-1} and the bottom left where the wind particles in Model A are shown in the orbital plane.

The key result from these simulations is that only for the slowest wind velocity, 20 km s^{-1}, is the wind contained within the giant's Roche lobe and flows through L_1 toward the WD. The accretion efficiency of WRLOF based upon the simulations of Psodliakowski & Mohamed (2007) is very high and the theoretically predicted accretion rate is a factor of 100 higher than the Bondi–Hoyle fluid rate, for the same set of input parameters. Indeed the accretion efficiency of Mira B by the Bondi–Hoyle mechanism is only $\sim$0.7%. The observational evidence supports a high degree of interaction between Mira A and Mira B. However, it is difficult to reconcile the high accretion rate onto the WD, Mira B, with the low temperature inferred for the WD by Sion (2003) and Sokoloski & Bildsten (2010). This troubling discrepancy will be explored further below. However, it is imperative to first examine the results of other independent SPH simulations on wind accretion in symbiotics carried out subsequent to the work of Mastrodemos & Morris (1998) and Psodliakowski & Mohamed (2007).

Following the SPH simulations of Psodialowski & Mohamed (2007), further hydrodynamic simulations of wind accretion targeted to symbiotic stars were carried out using an entirely different code. de Val-Borro et al. (2009) employed the FLASH code to explore the interaction of a slow AGB wind and an accreting WD companion with a two-dimensional simulation of wind accretion. The primary goals were to investigate disk formation and determine the resulting accretion rate onto the WD. For the accretion radius of the WD, they took it to be one-tenth of the radius of the WD's Hill sphere (see Chapter 3 for a discussion of the Hill sphere). The presence of mass outflows was studied as a function of mass-loss rate, wind temperature, and binary orbital parameters. Their code enabled a study of the stability of mass transfer in wind accreting symbiotic binary systems. The scheme they utilized provided higher resolution at the location of the WD including the formation of bow shocks and disk formation. These simulations yielded a quasi-steady flow of gas toward the WD and accretion disks with a range of disk diameters and component separations. Stable disks that formed were geometrically thin and had disk masses of $\sim 10^{-4} M_\odot$ when the wind acceleration occurred at several stellar radii. The accretion rates were found to be $\sim 10\%$ of the wind mass-loss rate from the AGB primary, enough to show that gravitationally-focused wind accretion, as found by Mastrodemos & Morris (1998) and by Psodliakowski & Mohamed (2007) further emphasized that wind accretion is a significantly important astrophysical process in symbiotic variables. Indeed, accretion disks were shown to be formed for wind outflow rates between 10^{-9} and $10^{-6} M_\odot$ yr^{-1} and component separations between 16 AU and 70 AU. Thus, the range of accretion rates and separations for which disk formation occurs would encompass the S-Type symbiotics as well as the D-types unless the relative absence of dust in the S-types inhibits the needed conditions for wind acceleration.

Mohamed & Psodliakowski (2012) continued their study of Mira-type (D-type) symbiotic binaries with further 3D SPH simulations of wind accretion, which again characterized the wind outflows to be highly aspherical and strongly focused toward the binary orbital plane due to the gravitational influence of the accreting WD. The authors ran two different simulations, the first one with a component separation of 20 AU, a wind velocity of 4 km s^{-1} and the dust formation radius larger than the Roche lobe of the giant, and a second simulation where the component separation was 60 AU, the same wind velocity but with the dust formation radius smaller than the Roche lobe of the giant. In the former case, the wind velocity of the outflow from the giant is well below the escape velocity from its Roche surface where $V_{\rm esc} = 14.4$ km s^{-1} at the Roche surface. Hence a high density region inside of the giant's Roche lobe forms which facilitates the WRLOF process. This matter can then flow through the L_1 "nozzle" and stream toward the WD forming an accretion disk.

In the second case, with $R_{dust} < R_{rl}$, the dust formation radius is ~10 AU compared with the Roche lobe radius ~25 AU. In this case, the dust formation is not affected by the presence of the WD gravity. Thus, the dust shell is almost spherical rather than the arcs of dust that resulted when $R_{dust} > R_{rl}$. The WD accretes ~10% of the wind outflow from the AGB star. When these dust shells are ejected they could lead to variations in the accretion rate onto the WD. In both simulations in their work, the accretion efficiency is greater than the Bondi–Hoyle accretion process by at least a factor of 10.

It is well known that symbiotic stars are regarded as one of the most promising candidates for being the single channel progenitors of the Type Ia supernovae (Patat et al. 2007, 2011). Specifically, attention has focused on the origin of narrow, variable absorption lines (e.g., Ca II H & K, Na D lines) exhibited by some Type Ia supernovae. These features suggest that the SNe Ia progenitors are surrounded by dense, circumstellar matter like supernova Type Ia 2006X (Patat et al. 2007, 2011). Similar spectral line variations have been detected in the symbiotic recurrent nova system RS Oph. This system undergoes thermonuclear outbursts every 20 years and, as with all known recurrent novae, contains a massive WD accreting at a high rate. Hydrodynamic evolutionary models by Yaron et al. (2005) and by Hillman & Kashi (2021) have shown that their sequences involving high mass WDs actually grow in mass after each classical nova outburst to reach the Chandrasekhar Limit and hence a Type Ia explosion. In a breakthrough study by Booth et al. (2016), SPH simulations of multiple mass transfer-nova cycles in RS Oph were carried out. Their simulations focused on studies of the large-scale structure of the circumstellar environment of recurrent novae, in an effort to further investigate the connection between recurrent novae and Type Ia supernovae. For the binary parameters, the red giant mass was 0.8 Msun and the WD mass 1.38 Msun, the orbital period P_{orb} = 453.6 days, the binary inclination 50°, and the semimajor axis = 1.48 AU. Their modeling included the quiescent mass-loss and simulated nova outbursts of RS Oph. The SPH code that they employed is GADGET-2 (Springel 2005).

Their approach included accretion onto the WD but at the resolution of the simulations, the accreting WD was represented as simply a sink for the particles. No mass-loss from the WD or the accretion disk was included. They found that the quiescent mass transfer produces a dense, equatorial outflow, focused toward the binary orbital plane (wind Roche lobe overflow), and an accretion disc forms around the WD. The collision of the simulated spherically symmetric nova outbursts with circumstellar matter like the aspherical nova shells of earlier novae and other circumstellar structures gives rise to a bipolar outflow, which closely resembles HST imaging of the 2006 outburst of RS Oph. The authors also generated synthetic absorption line profiles for comparison with observations of the 2006 outburst of RS Oph, which enables them to disentangle lines that are formed in the wind and lines that are given rise to by the novae outbursts, which are swept away by the supernova explosion.

In 2017, de Val-Borro et al. (2017) extended their results from their 2009 study with new hydrodynamic simulations. Instead of using the FLASH code, they chose a different source code, the PIERNIK magnetohydrodynamics code (Hanasz et al. 2010). This code incorporated a single-fluid stellar wind model in a Cartesian grid in a rotating frame of reference where both stars are at rest in the computational domain, to solve the equations of hydrodynamics in three dimensions. Their simulations used the equation of state for a polytropic fluid, taking into account the rotation of the symbiotic binary, given by

$$\omega = \left[\frac{G(M_1 + M_2)}{a^3} \right]^{0.5}.$$

The two stellar components were again treated as point masses and the secondary component, representing the WD, is treated as an absorbing sphere but the surface of the secondary, as in all other simulations to date, is not resolved.

One of the advantages of adopting this new code is that it can handle shock waves and flow discontinuities with high enough resolution that the use of artificial viscosity is avoided, which lends additional physical realism to their study. The authors compared their rates of accretion with the Bondi–Hoyle rate given by

$$\dot{M}_{\mathrm{bh}} = \frac{4\pi G^2 M_2^2 \rho_\infty}{(c_s^2 + v_\infty^2)^{1.5}}$$

where all of the symbols have their usual meaning (see Chapter 2) and accretion takes place on the downstream side of the shock (i.e., the tail shock behind the accretor). Defining the accretion efficiency f as

$$f = \frac{\dot{M}_{\mathrm{acc}}}{\dot{M}_{\mathrm{wind}}}.$$

Figure 5.4 displays the 3D model accretion efficiency compared with the Bondi–Hoyle rate $\dot{M}_{\mathrm{bh}}$.

The overall conclusion from the multi-D hydrodynamic studies is that for the slow, massive, radiatively driven winds of the AGB component, the flow between the components is preferentially concentrated along the binary orbital plane accompanied by the formation of a bow shock and accretion onto the WD via a tail shock. The accretion efficiency is significantly modified by the gravitational effect of the WD and amplifies the rate of accretion. This enhancement is as much as 5 to 20% or more of the wind mass-loss rate from the AGB star compared with the standard Bondi–Hoyle fluid prescription during the quiescence interval of a symbiotic. During an outburst period, the accretion rate is roughly 20% to 50% greater than the Bondi–Hoyle rate of accretion.

In general, the multi-D hydrodynamic simulations, even though the evolutionary simulations are restricted to a few Keplerian periods, clearly demonstrate the formation of accretion disks in the D-type symbiotics making it entirely plausible that accretion disks would also be expected to form in S-type symbiotics despite the fact that the primary components wind mass-loss rate is lower and the wind outflow not as massive.

Of course, the hydrodynamic evolution simulations are subject to a number of limitations. The prohibitively large computational demand restricts the model

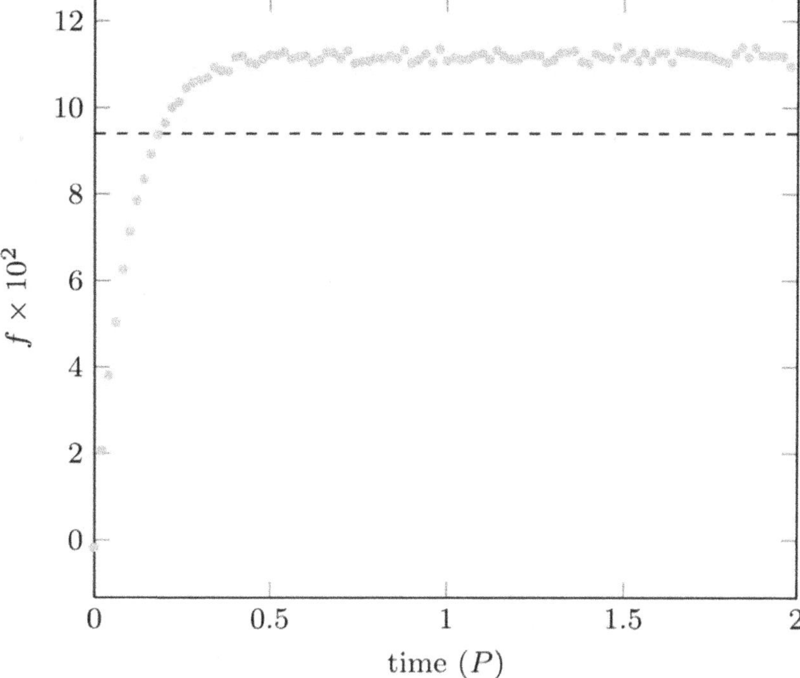

Figure 5.4. Time evolution of the accretion ratio, defined as the accretion rate divided by the mass-loss rate from the primary star. The green curve is the accretion efficiency of the 3D simulation. The dashed line shows the accretion rate, $\dot{M}_{bh}$, predicted by Bondi–Hoyle accretion. (Reproduced with permission from de Val Borro et al. (2017). Copyright © 2017, Oxford University Press.)

evolution time to be so short that it barely allows a steady-state flow to be established. Moreover, the wind acceleration mechanism is ad hoc without the inclusion of the requisite dust formation and radiative pressure. There is no treatment of the dust-driven acceleration ab initio.

From the perspective of accreting WDs and the microphysics of accretion onto the WD surface layers and in what ways the WD is affected, neither a boundary layer between disk and gas nor the WD surface layers are resolved in any of the hydrodynamic wind accretion simulations at the time of this writing. Therefore, while the simulations yield estimates of the efficiency of wind accretion and the rate of accretion relative to the Bondi–Hoyle process, they cannot provide information about angular momentum transfer vis-a-vis the rotational velocity of the WD, or chemical fractionation, chemical abundances or the mix of ion species in the accreting matter. FUV spectroscopy, specifically that extends down to the Lyman Limit, is the only hope at present for direct empirical determination of key parameters like the rate of accretion, the surface temperature of the accreting degenerate, the effects of heating and cooling of the WD, rotational velocities, and chemical abundances in the accreted atmosphere.

5.2.1 Far-Ultraviolet Spectroscopy of Accreting WDs in Symbiotic Variables

Since the nebular emission in symbiotics includes highly ionized ion species that require very high photoionization temperatures, it was clear that the derivation of hot component temperatures must rely primarily on ultraviolet radiation emitted shortward of Lyα (1216 Å). Fortuitously, the IUE spacecraft, launched in 1978, with its SWP camera for the UV wavelengths from 1170 Å to 2000 Å and its LWP camera covering 2000 Å to 3200 Å, was ideally suited to probe the properties of the hot components. The IUE observations obtained by a great many investigators comprise a vast archive of >1100 spectra obtained for 60 symbiotic variables, resulting in a large number of peer-reviewed, mainly observational, scientific articles and reviews. The stage was set for a deeper understanding of the nature of the hot components by the ground-breaking study of Webbink & Kenyon (1984) while in Europe, Harry Nussbaumer and his co-workers were also contributing new insights into the hot components using Zanstra methods. Prior to the launch of the IUE, relatively little was known about the fundamentally important temperatures, luminosities and masses of the accreting WDs in symbiotic variables. Landmark FUV studies by Webbink & Kenyon (1984) and Muerset et al. (1991) represented major advances in securing these fundamental parameters from the analysis of FUV spectroscopy. It was well known that if one assumed that the hot component radiated energy as a blackbody, radiation temperatures of $\sim 10^5$ K were obtained from the analysis of emission-line spectra.

Webbink & Kenyon (1984) used three flux components: (a) a cool giant; (b) the hot component; and (c) an ionized nebular component, which is a continuum source all the way into the UV arising from photoionization of the wind by the hot component. While the study of Webbink & Kenyon (1984) considered all the nebulae in symbiotics to be radiation bounded (i.e., insufficient UV photons from the hot component to ionize the nebula), Muerset et al. (1991) also allow for partially particle-bounded nebulae (i.e., there is enough nebular gas to absorb all of the UV photons and the nebula is fully ionized). Webbink & Kenyon (1984) utilize accretion disks as acceptable solutions in some cases while Muerset et al. (1991) ignored disks in their solutions and use the UV continuum radiation shortward of He II (1640Å) measured by its equivalent width while Kenyon & Webbink use the ionization edge of H I which they measured from the continuum radiation at 2600 Å. Both groups measure the temperatures of the hot components in symbiotics but a comparison of their hot component temperatures reveal differences by as much as 37,000 K in one case (AG Peg) to as little as 1000 K in another (V443 Her) with other systems in common to both groups show temperature differences between 5000 K and 14,000 K. The comprehensive study of over 20 symbiotic variables by Muerset et al. (1991) yielded the range of temperatures, luminosities, and masses of the hot components using their modified Zanstra methodology. As stated earlier, the accreting WDs in symbiotic variables, are generally very hot (55, 000 K $\leqslant T_{\text{eff}} \leqslant$ 210, 000 K), with radii of the hot components lying in the 0.002 $\leqslant R_\odot \leqslant 0.8 R_\odot$, and luminosities in the range 0.3 $\leqslant L_\odot \leqslant$ 37, 400 erg s^{-1}.

Given the plethora of FUV spectra in the IUE and HST archives along with ground-based optical and infrared spectrophotometry, there was relatively little multiwavelength accretion disk and white dwarf model fitting to the spectral energy distributions of symbiotic variables.

This all changed when Skopal (2005) introduced a comprehensive multi-wavelength approach that used theoretical fits to the observed spectral energy distribution from the infrared to the UV wavelength domains. Like Webbink & Kenyon (1984), Skopal (2005) characterized the infrared, optical, and UV continuum as consisting of the superposition of three different sources, a red giant, an accreting component (WD) and the continuum from a photoionized nebula. The total continuum flux was thus

$$F_\lambda = F_\lambda(\text{giant}) + F_\lambda(\text{WD}) + F_\lambda(\text{nebula}).$$

Skopal (2005) approximated the radiation from the giant by using synthetic red giant spectrum, covering the range of wavelengths normalized to the broad band [BVRI]JKLM photometric measurements, converted to flux points. The continuum contributed by the nebula included contributions from hydrogen and doubly ionized helium generated by free–free and free–bound transitions. Extensive details can be found in Skopal (2005). The analysis of the symbiotic system RW Hya by Skopal (2005) provides a good illustration of his multi-wavelength model fitting approach. RW Hya has never undergone a recorded outburst and has a light curve showing a wave-like variation versus orbital phase. The light curve is seen in the top panel of Figure 5.5. The observation that it is eclipsing indicates a high orbital inclination. There are numerous IUE spectra of the system in the MAST archive, which have been analyzed by many different authors. Skopal's (2005) analysis is perhaps the most detailed. He chose two sets of Short Wavelength Prime Camera (SWP) IUE spectra covering 1170 Å to 2000 Å and Long Wavelength Prime Camera (LWP) IUE spectra covering 2000 Å to 3000 Å. One SWP + LWP set was taken in 1987 at orbital phase 0.06, corresponding to optical light minimum and another set taken in 1979 at phase 0.32 corresponding to maximum optical light. The SWP continuum slope rises steeply toward shorter wavelengths as displayed in Figure 5.5 while the near-UV continuum is flat, demonstrating that a nebular continuum is contributing strongly to the LWP spectrum. Skopal's best fit to the optical maximum data yields a hot component temperature of 110,000 ± 30,000 K. His full multi-wavelength analysis is displayed in Figure 5.5.

5.2.2 FUSE Spectroscopy of the Hot Component in Symbiotic Variables

Up to now, virtually all of the temperatures and luminosities of symbiotic hot components have been derived using the modified Zanstra method to obtain radiation temperatures needed to account for photoionization of the observed emission lines (Muerset et al. 1991; Muerset & Nussbaumer 1994), or using black

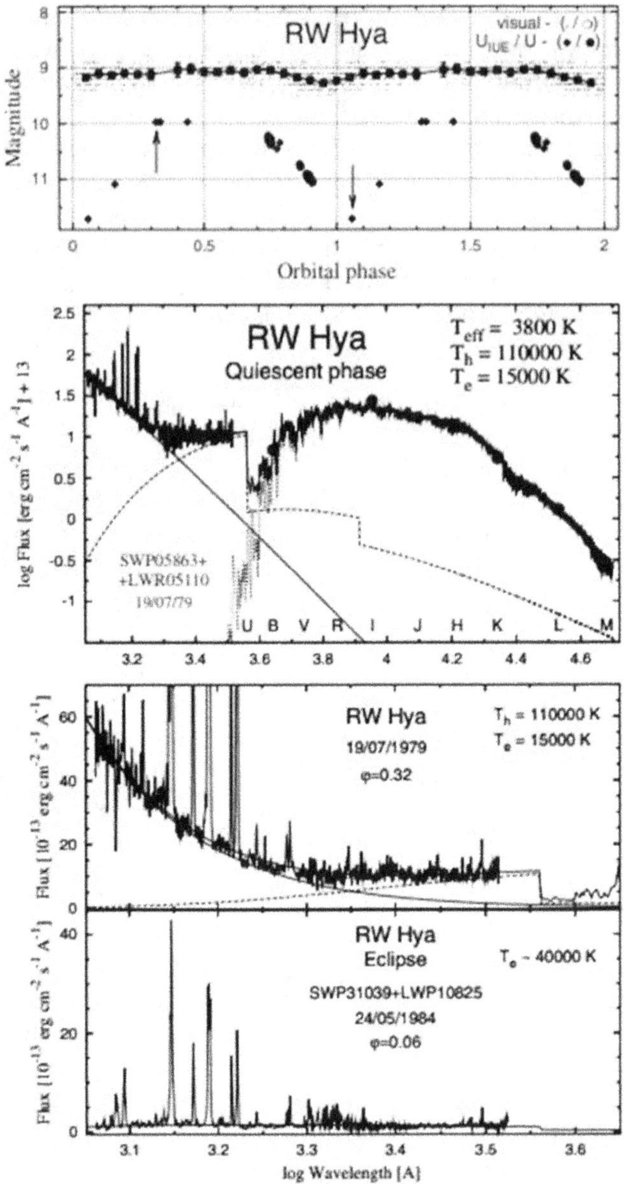

Figure 5.5. Top: Visual and U LCs of RW Hya (Skopal et al. 2002) with added U-magnitudes estimated from available IUE observations. The arrow marks position of the used spectrum. Middle and bottom: The SED near to the optical maximum shows typical characteristics of a quiescent phase with a high luminosity of the hot object, while during the eclipse only a very faint high-temperature nebular emission can be recognized. (Reproduced with permission from Skopal (2005). © ESO 2005.)

bodies or by simply assuming a photospheric temperature. This is understandable given the daunting complexity of symbiotic stars and the possible sources of FUV continuum radiation that may arise from that complexity. Nevertheless, many models have been proposed for the origin of the emission lines including the assumption that they all arise from photoionization. Moreover, the temperatures are derived for the most part from the He II (1640) recombination line and the H lines, but they yield different temperatures, with the temperature from the He II lines much higher than that from the H lines. The He II lines may arise from a wind or from wind collision shocks or coronae. Added to this uncertainty is the fact that for the very few systems analyzed up to now with actual non-LTE photospheric models (mostly post-novae), the model-derived temperatures for hot accreting WDs are a factor of two to three lower than the temperatures derived by Zanstra techniques.

An important archive of FUV spectra on symbiotic stars was obtained using the Far Ultraviolet Spectroscopic Explorer (FUSE) spacecraft. These spectra cover a wavelength range from 1180Å down to the Lyman Limit at 912Å. For observations of the hot accreting components of symbiotic stars, this offers a huge advantage because, unlike the IUE SWP spectral range, the FUSE spectra are unaffected by any UV contribution of the nebular continuum which arises from the photo-ionization of the red giant wind. In IUE and HST FOS, GHRS and STIS spectra, this nebular continuum tends to flatten their continuum slopes which, if not removed properly, leads to grossly underestimated effective temperatures and accretion rates. Moreover, in the short wavelength part of the FUV, the hot boundary layer (BL) between the star and disk may contribute to the FUV. Luna et al. (2013) have shown that most of the symbiotic systems they observed in the X-ray domain with the SWIFT spacecraft appear to have optically thick boundary layers that emit strongly in the FUV. Given this empirical result, then modeling of the FUSE spectra should, when possible, take into account the boundary layer between the hot star and disk in symbiotic stars. It is truly unfortunate that only a handful of symbiotics were observed with FUSE.

Sion et al. (2017) exploited the very limited FUSE archive of 5 symbiotic systems of which three of the FUSE spectra are discussed here: CQ Dra, RW Hya, and EG And. For CQ Dra and RW Hya, matching IUE SWP spectra and HST/GHRS spectra allowed an extension of the spectral coverage. There is no evidence that the S-type symbiotics are metal-poor, nor that their giant donor stars are s-process enhanced (Galan et al. 2016). This s-process enhancement is due to past mass transfer (when the present WD was on the AGB), and provides very strong indications that the WD mass should be at least $\sim 0.55 M_\odot$. CQ Dra has an outburst history and is clearly more interacting than RW Hya.

For those systems showing ellipsoidal light curves, especially in the red and near-IR range (where the red giant dominates the spectrum), it is very likely that the red giant is filling or nearly filling its Roche lobe. Therefore the mass transfer and accretion should be through L1, and hence an accretion disk could possibly be present. Their mass transfer/accretion rates should also be relatively high, $\sim 10^{-7} M_\odot$ yr^{-1} as revealed by numerical hydrodynamic simulations of mass transfer in such systems. In relation to this mass transfer, Shagatova et al. (2016) have found, using SWIFT X-ray observations, that the wind from the giants in S-type symbiotic stars are not spherically symmetric but is enhanced or focused in the orbital plane which raises the accretion efficiency onto the hot component, consistent with the WRLOF process of Psodialowski & Mohamed (2007).

These systems have been studied while in quiescence which optimizes the study of their hot components because their accretion rates and system luminosities tend to be lower relative to the WD photosphere light alone. The chief objective was to use non-LTE model accretion disks and model photospheres fitting to the FUV spectra to determine the accretion rates, accretion efficiency and the temperatures and luminosities of the hot components and ascertain whether their FUV radiation is from accretion light or WD photospheric light or both.

5.2.3 RW Hydrae

The S-type symbiotic system RW Hydrae (= HD 117970) has a primary component classified as a M2 III red giant. The inclination of RW Hya is high, just sufficient for the system to undergo eclipses while the orbital period is 370.2 days (Merrill 1950; Kenyon & Mikolajewska 1995). Published estimates of the WD mass range extend from 0.3 to $0.6 M_\odot$. The red giant mass is between 0.5 and $2 M_\odot$ (Schild et al. 1996; Kenyon & Mikolajewska 1995). The distance to RW Hya is based upon the corotation of the red giant (Schild et al. 1996) is 670 $\pm$ 100 pc. The total luminosity of the system at this distance is $\sim 700 L_\odot$. RW Hya exhibits ellipsoidal variability (Rutkowski et al. 2007) based upon the observed light curve. The distance set by the Roche lobe geometry, which requires a filling factor above 0.9, is then 1.7 kpc.

A noteworthy aspect of RW Hya for studies of the hot component is that there is little or no evidence for an accretion disk. Its multi-wavelength data was analyzed most recently by Skopal (2005) who dismissed an accretion disk but fit the FUV data with a hot blackbody SED at $T_{\mathrm{eff}} = 110,000$ K $\pm$ 30,000 K. Fortunately, a good FUSE spectrum of RW Hya is in the MAST archive which had never been analyzed and allowed a sampling of the FUV flux down to the Lyman Limit. The FUSE spectrum of RW Hya, with its flat continuum increasing toward shorter wavelengths, indicates that the hot component has a very high temperature, but the Lyman series is affected by ISM atomic H absorption lines. Single disk models require a very high mass accretion rate to fit the relatively flat continuum and yield a ridiculously short distance. It was therefore concluded that the disk is not a significant flux contributor to the FUV spectrum, and that the surface area of the hot component is much smaller than that of an accretion disk.

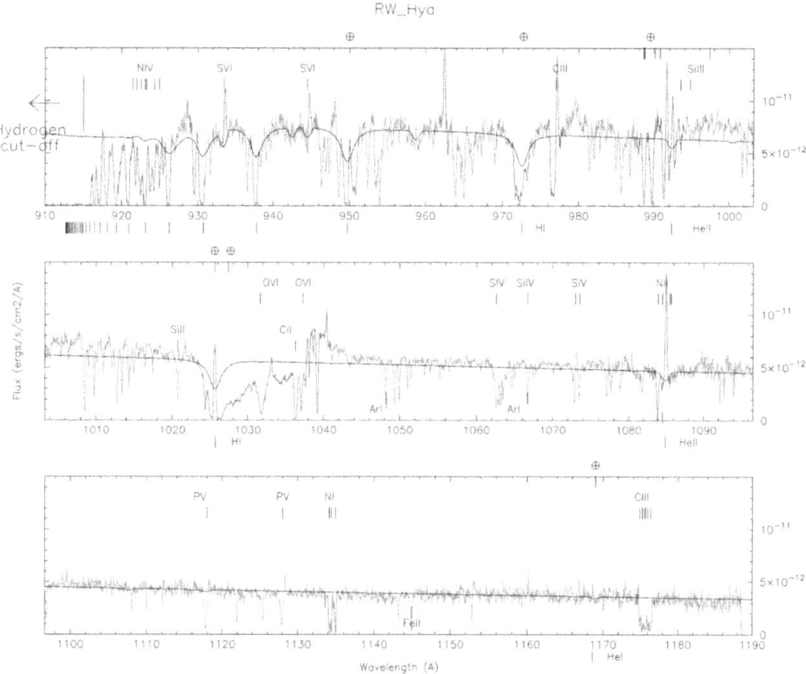

Figure 5.6. Single-temperature hot non-LTE WD model fit for a low-mass WD to the dereddened FUSE spectrum of RW Hya with a best fit corresponding to a WD with a surface temperature of 160,000 K and $\log g = 6.5$. Assuming a mass of $0.4M_\odot$ with a radius of $0.065R_\odot$ yields a scale-factor-derived distance of 811 pc, confirming that a low-mass WD with its larger radius and a very high temperature gives a distance to RW Hya that is within the error bars of the original Schild et al. distance. (Reproduced with permission from Sion et al. (2017). © 2017. The American Astronomical Society. All rights reserved.)

Sion et al. (2017) carried out single-temperature hot WD model fits for a low-mass WD to the FUSE spectrum of RW Hya and found a best fit corresponding to a WD with a surface temperature of 160,000 K with $\log(g) = 6.5$. The temperature and gravity are dictated by the shape of the wings of the Lyman series (where possible) and continuum. Assuming a mass of $0.4M_\odot$ and radius of $0.065R_\odot$ yields a scale-factor-derived distance of 811 pc confirming that a low-mass WD with its larger radius, and a very high temperature gives a distance to RW Hya which is within the error bars of the original Schild et al. distance.

This best-fitting non-LTE WD model atmosphere fit to the FUSE spectrum is shown in Figure 5.6. The luminosity of the accreting WD is 2 to $3000L_\odot$. Since the surface gravity of a WD is less accurately known from FUV spectra (Lyman electrons lie in the ground state and hence are least affected by pressure broadening), then if one assumes a $\log(g) = 6.8 - 7.0$, instead of 6.5, a luminosity $L_\mathrm{wd} = 700L_\odot$ is obtained. While a low mass (large radius) hot component seems consistent, it cannot be ruled out that the WD is undergoing hydrogen shell burning, and has a higher mass but an extended atmosphere whose radius is larger than that given by the WD mass–radius relation.

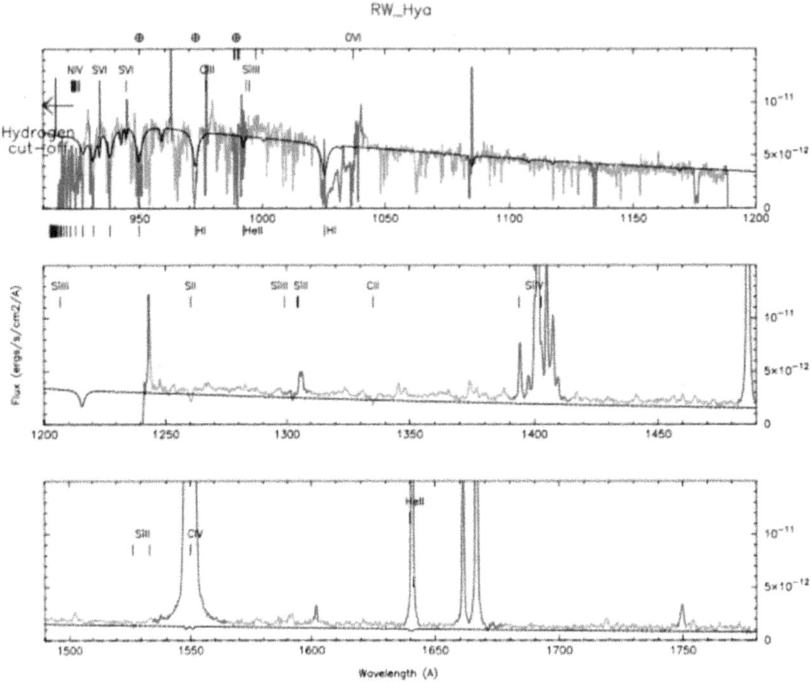

Figure 5.7. Combined dereddened FUSE+ HST GHRS spectrum of RW Hya modeled as in the previous figure. A very hot and very small mass WD provides the required flux. The fit reveals an accreting WD with $T_{wd} = 160,000$ K and $\log g = 6.5$. The two spectra were obtained at different epochs and with different instruments, and the HST spectrum has a flux slightly lower than in the FUSE spectrum. (Reproduced with permission from Sion et al. (2017). © 2017. The American Astronomical Society. All rights reserved.)

With the addition of an HST GHRS spectrum of RW Hya in the MAST archive that has a similar flux level to the FUSE spectrum, Sion et al. (2017) were able to extend the wavelength coverage from 1700A down to the Lyman Limit. After dereddening the spectra with $E(B - V) = 0.10$, they found basically the same solution for the combined spectrum (FUSE+GHRS) that they found for FUSE alone, that namely a very hot and very small mass WD can provide the flux needed. This single-temperature WD fit is displayed in Figure 5.7. When compared to the model, the HST GHRS spectrum has a flux slightly lower.

An accreting WD at this very high temperature almost certainly sustains thermonuclear burning of the accreted hydrogen and should be a supersoft X-ray source. The absorbing column of the ionized red giant wind in a symbiotic system is nominally more than sufficient to obscure X-ray emission from an accreting WD with shell burning. Thus, the absence of X-ray emission does not completely rule out the presence of thermonuclear shell burning in the hot component.

The recently published Gaia eDR3 parallax of RW Hya gives a distance of 1.5 kpc which exceeds the scale factor-derived best fitting distance of 0.811 kpc corresponding to Figures 5.6 and 5.7. By scaling the solution for the difference in temperature of the accreting WD corresponding to the two distances, the WD in

RW Hya would be ~ 50,000 K hotter at the Gaia eDR3 distance. Thus, at the Gaia distance the accreting WD in RW Hya has a surface temperature of ~210,000 K.

5.2.4 EG And

Observations of EG And suggest the wind from the giant is focused toward the orbital plane thus raising the accretion efficiency and rate of accretion onto the WD (Shagatova et al. 2016). The mass-loss rate from the giant is estimated to be a few times $10^{-6} M_\odot$ yr^{-1} with an inclination of $i = 80°$. The interstellar reddening is $E(B - V) = 0.05$ (Kenyon & Garcia 2016). The temperature (~75,000 K) and luminosity of the hot component in EG And were estimated using a modified Zanstra method while the orbital period, Porb = 482.6 days was re-determined by Fekel et al. (2000) who also determined a distance of 568 pc. van Leeuwen (2007) presented a distance of 513 ± 169 pc based upon a Hipparcos parallax.

The mass of the hot component $M_{wd} = 0.35 \pm 0.1 M_\odot$ (Kenyon & Garcia (2016). Sion et al. (2017) carried out a synthetic spectral analysis of the FUSE spectrum alone. They assumed a distance of 400–700 pc, a low WD mass between 0.35 and $0.6 M_\odot$ and orbital inclination of $i = 60°$, $75°$, and $81°$, and an interstellar reddening value of $E(B - V) = 0.05$. Their first fitting attempt involved a steady-state, optically thick accretion disk. However, the accretion disk fitting met with limited success. The disk fits are poor toward the short wavelength (Lyman Limit) end of the FUSE wavelength range. Moreover, the distance implied by the disk fits, the scale-factor-derived distance is far too close, even for $\dot{M} = 10^{-8} M_\odot$ yr^{-1}, $i = 60°$ and $75°$. At a high mass accretion rate, the disk becomes rather thick and with an inclination of $80°$, the boundary layer near the equatorial region of the star should be masked by the disk.

They concluded that a standard accretion disk cannot account for the FUSE spectrum of EG And. By fitting hot non-LTE WD photospheres to the FUSE data of EG And, the best fit was obtained for $\log(g) = 7.5$ and $T_{wd} \sim 80 - 95,000$ K. If $\log(g) < 7.5$, then the best-fitting temperature becomes lower. When they combine an optically thick disk with the WD model, there is an improvement over a disk-only fit but the WD + disk fit is clearly inferior to the WD-only model. The addition of a hot boundary layer to the WD (i.e., a two-temperature WD) introduces a degeneracy to the solution in that the contribution of small surface area BL with $T \sim 10^5$ K BL does not produce a noticeable change in the spectrum given that the WD itself has already a temperature close to 100,000 K (producing a flat spectrum). Also because of the high inclination, the contribution from a BL would be minimal. Thus, for EG And, it appears that the hot component can be identified with a hot, bare WD having a surface temperature of 80–95,000 K and $\log(g) = 7.5$. These derived parameters using synthetic spectral fitting are in good agreement with the results of Kenyon & Garcia (2016). Their best-fitting solution is displayed in Figure 5.8.

The recently published Gaia eDR3 parallax is in agreement with the scale factor-derived distance obtained from the best-fitting hot WD model displayed in Figure 5.8. The accreting WD in EG And has a surface temperature of 90,000 K and $\log(g) = 7.5$.

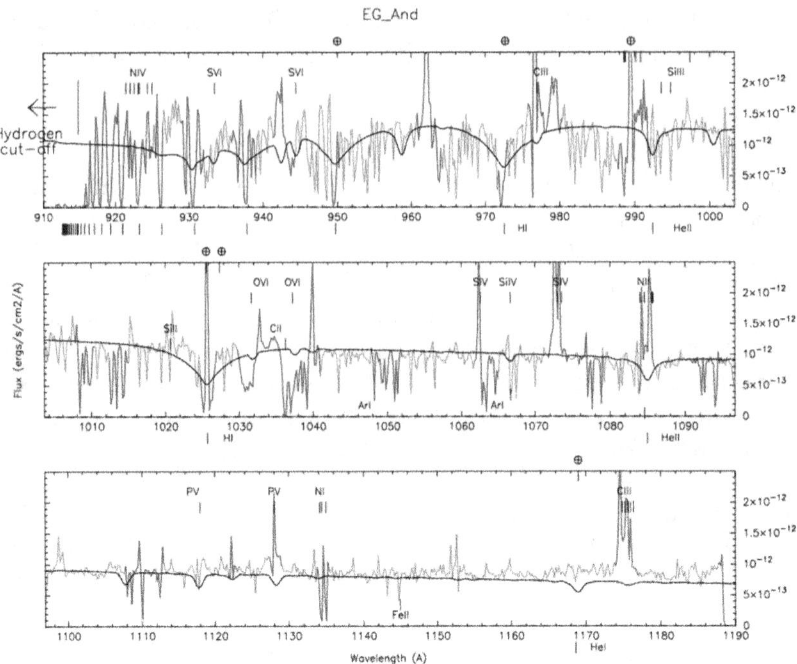

Figure 5.8. Best-fitting model solution for the dereddened FUSE spectrum of EG And. The hot component is revealed to be a hot, bare WD with a surface temperature of 90,000 K and $\log g = 7.5$. (Reproduced with permission from Sion et al. (2017). © 2017. The American Astronomical Society. All rights reserved.)

5.2.5 CQ Dra

CQ Dra was originally thought to be a cataclysmic variable in binary orbit with a red giant (Reimers et al. 1988), but was later re-classified as a symbiotic variable. However, as discussed below, the cataclysmic variable interpretation of the hot component may be correct after all. CQ Dra has a reliable HIPPARCOS parallax $\pi = 5.25 \pm 0.48$ mas (van Leeuwen 2007). The orbital period for CQ Dra was determined to be 1703 days by Eggleton et al. (1989) from radial velocity periodicity searches.

Sion et al. (2017) used an archival FUSE spectrum of CQ Dra in the MAST archive to carry out their investigation Lyman limit at 912A. The FUSE spectrum has a much lower continuum flux level than the IUE spectra and was obtained in a state of relative low luminosity. Curiously, the Lyman (series) profiles of H are not clearly seen, making it more difficult to identify a disk versus a WD. The spectrum does not show signs of interstellar absorption as seen in the FUSE spectra of the other three systems with FUSE spectra they considered except for atomic H (sharp Lyman series toward the shortest wavelengths). Some of the sharp emission lines are probably due to terrestrial airglow and sunlight reflected in the telescope, and some of the sharp, weak absorptions may also be either terrestrial (e.g., N I 1135) or interstellar (e.g., Fe II 1145). While identifying lines in the FUSE range via comparison with FUSE spectra of other systems with accreting WDs, they found that the FUSE spectrum of the hot WD in the dwarf nova RU Peg during

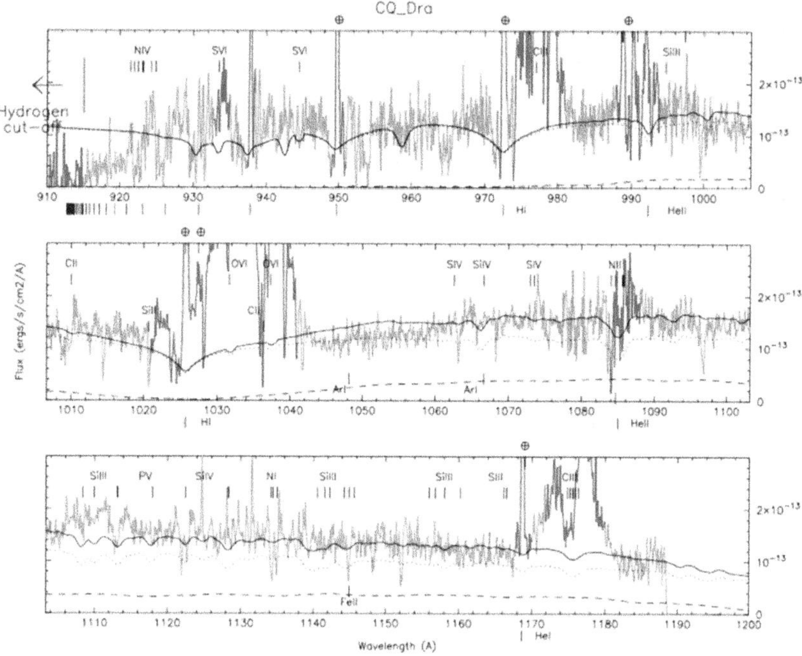

Figure 5.9. Best-fit theoretical disk+WD+BL spectrum (in black) to the dereddened FUSE spectrum of CQ Dra (in red). The WD has a mass of $0.8 M_\odot$ and a relatively low temperature of 20,000 K, with 120,000 K BL covering 4% of the WD surface. The disk has an accretion rate of $= 10^{-9.5} M_\odot$ yr 10^{-1} an inclination of 60°. The disk (dashed line) contributes 18% of the flux, and the WD+BL (dotted line) contribute the remaining 82%. The distance obtained from the fit is 164 pc. The known emission-line regions have been masked (in blue) and are not modeled. (Reproduced with permission from Sion et al. (2017). © 2017. The American Astronomical Society. All rights reserved.)

quiescence, is very similar to the FUSE spectrum of CQ Dra. There is also a similarity between the IUE spectra (e.g., SWP28355) of CQ Dra and the WD in the dwarf nova RU Peg. The IUE SWP spectra of CQ Dra in the archive are dominated by strong emission lines of C III (1175), weak N V (1240) emission, weak Si IV (1398) emission, and very strong C IV (1550) emission, with He II (1640) being notably absent. At first glance, these similarities would seem to add support to the possibility that the hot component in CQ Dra may indeed be a cataclysmic variable as Reimers et al. (1988) originally claimed, thus identifying CQ Dra as a triple system.

For the modeling of the FUSE spectrum, Sion et al. (2017) assumed a WD mass of $0.8 M_\odot$. Some parts of the FUSE spectrum show better agreement with a WD at $T_{\rm wd} = 30,000$ K while at the shortest wavelengths in the FUSE spectrum, there is better agreement with a WD surface temperature higher than 50,000 K. However, such a high temperature leads to an unacceptably large distance. When they fit only the FUSE spectrum with an accretion disk model, then for a distance of 100 pc (the Reimers et al. 1988 distance), they obtain $1 \times 10^{-9.5} M_\odot$ yr^{-1} for a disk inclination of 60°, as seen in figure 5.9. If $\dot{M} = 10^{-9} M_\odot$ yr^{-1}, the distance is of order 250 pc and the fit is poor. The only reasonable accretion disk fit to the FUSE spectrum of CQ Dra requires $\dot{M} = 1 \times 10^{-8} M_\odot$ yr^{-1} and corresponds to a distance of ~950 pc.

Hence, the very hot component of the system contributing flux to the shortest wavelengths of FUSE must have a very small size, smaller than the inner disk and smaller than the WD. This led to the consideration of a boundary layer (BL) contribution included in their modeling. The BL is included either as a direct optically thick component (at high mass accretion rate the boundary layer forms a hot equatorial spread layer on the WD surface (Inogamov & Sunyaev 1999; Piro & Bildsten 2004)), or as an optically thin hot component (at low-mass accretion rate; Narayan & Popham 1993) heating up the equatorial region of the WD through advection of energy (Abramowicz et al. 1995). Consequently, for either case, they assume that the WD equatorial region has an elevated temperature, and modeled the WD as a two-temperature component. The WD itself has a temperature that is moderate ($T < 30, 000$ K, so as to not contribute all of the flux) and the boundary layer has a temperature of the order of 100,000 K. If the visible part of the BL has a fractional area f, then the visible part of the WD has an area $1 - f$. To this two-temperature WD, they added an accretion disk and obtained the best fit to the continuum. The scale factor-derived distance yielded by the fit was reasonable. This three-component fit consisted of a WD with a temperature $T_{wd} \simeq 20, 000 \pm 3000$ K, a boundary layer of size $4 \pm 1\%$ and temperature of $120, 000 \pm 20, 000$ K, and an accretion disk having an accretion rate of $\dot{M} \simeq 10^{-10} M_\odot$ yr^{-1} and disk inclination $i \simeq 40° - 50°$. Most of the solutions were obtained for $\dot{M} = 10^{-10} M_\odot$ yr^{-1}, but some solutions were also obtained for $\dot{M} = 10^{-10.5} M_\odot$ yr^{-1} and $\dot{M} = 10^{-9.5} M_\odot$ yr^{-1}, therefore adding an error bar to the mass accretion rate $\log \dot{M} = -10.0 \pm 0.5 \, (\langle M_\odot$ yr$^{-1} \rangle)$. This best two component fit (WD + BL) to the FUSE spectrum of CQ Dra is displayed in Figure 5.9. Note that such a mass accretion rate is consistent with the low brightness state during which the FUSE spectrum was obtained. They also retrieved an IUE SWP spectrum (33521) of CQ Dra in the low state with a continuum flux matching the FUSE spectrum (in the spectral region where they overlap), and modeled the combined FUSE + IUE spectrum. They found that the IUE spectrum agrees well with their FUSE best-fit model, as shown in Figure 5.10, an indication that CQ Dra consistently comes back into the same low state.

The Gaia eDR3 parallax of CQ Dra yields a distance of 175 pc with an error-allowed range of 160 pc to 180 pc. The scale factor-derived distance corresponding to the best fitting multicomponent (WD + BL + disk) model shown in Figure 5.9 corresponds to 164 pc, which is within the error range of the Gaia eDR3 distance to CQ Dra.

5.2.6 The Weak Symbiotic System Mira AB

Mira AB, is a wide binary system consisting of the prototypical pulsating (pulsation period of 332 days), AGB (red supergiant) star, Mira A, with a companion star Mira B. Due to its very wide separation (wider than typical of symbiotic variables), and the level of interaction between A and B, Mira AB is regarded as a weak symbiotic binary-star system Mira AB (o Ceti). The orbital period is 497.9 years (Prieur et al. 2002) with an orbital inclination of 112° and the separation between the two components is ~65 AU. Mira B accretes from the wind outflow of Mira A. While Mira B has been widely

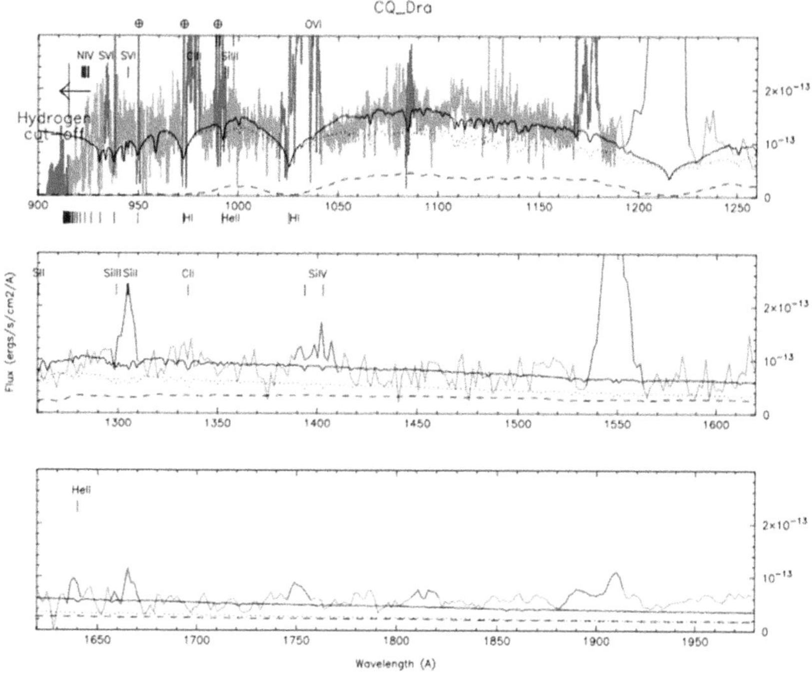

Figure 5.10. Same model as shown in Figure 5.9 but now it fits the combined FUSE+IUE SWP (33521) spectrum of CQ Dra. The fit to the continuum is relatively good all the way down to ~1800 Å. (Reproduced with permission from Sion et al. (2017). © 2017. The American Astronomical Society. All rights reserved.)

identified as being a WD (e.g., Warner 1972, Reimers & Cassatella 1985), the nature of Mira A's companion is uncertain. Indeed, Mira B's classification has intrigued and troubled investigators for the past 50 years. In two independent investigations, Jura & Helfand (1984) and Kastner & Soker (2004) presented observational evidence based upon X-ray data supporting the identification of Mira B as a main sequence star rather than a WD. Jura & Helfand (1984) use Einstein X-ray observations to argue that the observed X-ray luminosities are lower than expected for an accreting WD ($<10^{-3}L_\odot$) and that the total bolometric luminosity of Mira B is more akin to what is expected from accretion onto a low-mass main sequence dwarf. Kastner & Soker (2004) utilized XMM-Newton observations of Mira B to propose that the X-ray spectrum and luminosity of the Mira AB system suggest that the accreting component is either a late-type, pre-main-sequence star or late-type, $0.5M_\odot$, magnetically active main sequence star (spun-up by accretion) than being an accreting WD. They argue that the X-rays in Mira AB could arise from magnetospheric accretion of Mira A's wind by a main sequence star or from coronal activity associated with a main sequence star.

In the FUV spectral region, which is free of contamination from Mira A, the nature of the line spectrum and continuum energy distribution provide important clues on the nature of Mira B. Bochanski & Sion (2001) analyzed numerous IUE archival spectra of Mira B, in the wavelength range shorter than 2600 Å (i.e., no flux contribution from Mira A), with high gravity photosphere and accretion disk

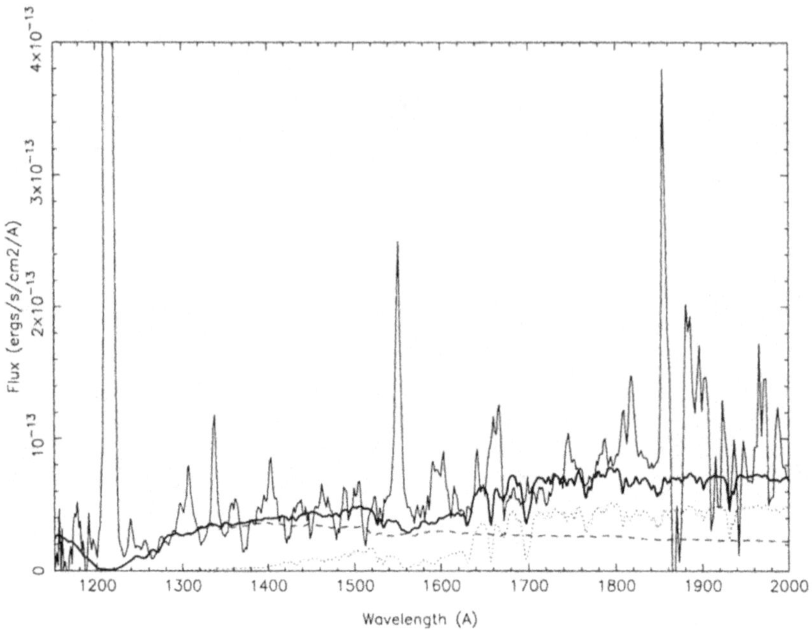

Figure 5.11. A best-fitting WD plus accretion disk fit to an archival IUE SWP spectrum of the WD component, Mira B in the weak symbiotic variable system Mira AB scaled to Mira AB's distance of 107 pc. The dashed line denotes the flux component due to an optically thick, steady-state accretion disk with inclination $i = 81°$ and accretion rate $\dot{M} = 3 \times 10^{-11} M_\odot$ yr^{-1} for a WD mass of $0.6 M_\odot$. The dotted line represents the flux component due to a cool WD with $T_{\mathrm{eff}} = 9000$ K, $\log g = 8$ (corresponding to $0.6 M_\odot$). The disk flux accounts for the significant continuum flux toward shorter UV wavelengths while the cool WD provides the needed UV flux toward longer UV wavelengths. The crossover between the two flux components occurs at approximately 1650 Å. (Reproduced with permission from Sion (2003). © 2003. The Astronomical Society of the Pacific. All rights reserved.)

models with vertical structure. An optically thick accretion disk model viewed at high inclination with an accretion rate of $10^{-10.5}$ Mo yr^{-1} reveals large disagreement with the observed continuum. Higher accretion rates, as proposed by Reimers & Cassatella (1985), are even worse. Bochanski & Sion (2001) find that the Lyα region and FUV continuum out to 2000 Å is best fit by a WD with a surface temperature of 9000 K, as displayed in Figure 5.11. Using the Hipparcos distance, re-processed by Knapp et al. (2003), to Mira AB (107 pc), their estimate of the observed FUV luminosity agrees with the bolometric luminosity of a 9000 K WD. If this temperature is maintained by accretion, then the required accretion rate is only $5.45 \times 10^{-12} M_\odot$ yr^{-1}. If accretion is not a factor in retarding the effect of thermal cooling on the WD temperature, then the cooling age of the WD is 854 Myr, with a mass of $0.6 M_\odot$. If their model fitting is correct then one can draw important implications for wind accretion efficiency and disk formation at large binary separation.

Sokoloski & Bildsten (2010) obtained ground-based observations of Mira B at the Lick Observatory to monitor the optical brightness variation and their associated timescale. They discovered that the amplitude of the aperiodic optical variations changed on timescales of minutes to tens of minutes. They find that the amplitude of the optical variations is ~two-tenths of a magnitude. The timescale of variations being so short point to an accreting WD, not an accreting main sequence star. The accretion rate they inferred ($\sim 10^{-10} M_\odot$ yr^{-1}) is similar to the accretion rate of WDs in dwarf novae during quiescence. It is high enough to be consistent with the low X-ray emission if Mira B accretes from a disk via a boundary layer girdling the WD since an optically thick accretion disk also radiates strongly in the FUV and EUV. The low WD temperature they present is consistent with the expected compressional heating from the weight of the accreted gas on the WD. Their interpretation of the nature of Mira B is supported by the disk and WD model fitting of Bochanski & Sion (2001; see also Sion 2003) whose best fit to the IUE FUV spectrum of Mira B is a combination fit of a cool WD and a low accretion rate of $3 \times 10^{-11} M_\odot$ yr^{-1} as displayed in Figure 5.11.

5.3 The Outbursts of Symbiotic Variables

The physics underlying the luminosity variations and outbursts of symbiotic variables is either accretion-driven gravitational energy release as gas accretes onto the WD secondary component or that the energy releases are due to thermonuclear fusion processes on the WD, also driven by accretion. What seems now to be widely accepted is that the accreting WDs in symbiotic variables are largely low-mass WDs centered around 0.5 to $0.7 M_\odot$ (Muerset et al. 1991). In other words, symbiotics likely have lower WD masses than the average mass of accreting WDs in cataclysmic variables, which is $0.83 \pm 0.03 M_\odot$ (Zorotovic et al. 2011; McAllister et al. 2019). This comparison excludes the masses of the WDs in the symbiotic novae which almost certainly have higher masses than ordinary symbiotics and suffer classical nova thermonuclear explosions (cf. Yaron et al. 1995). These objects will be discussed in Chapter 7 on classical novae.

We also know that the surface temperatures of symbiotic variable WDs are much higher ($>10^5$ K) than the WDs in cataclysmic variables which have surface temperatures in the range 10^4 K $< T_{\rm eff} < 50,000$ K (Sion 1985, 1999). It may follow that their core temperatures are higher than the core temperatures of CV WDs, perhaps because accretion switched on when the WD was younger and retarded its thermal cooling, possibly even re-heating the core after it had started cooling down during the onset of wind-fed or disk-fed accretion. The initial temperature of the core at the onset of accretion is subject to which of the numerous formation channels that are outlined by Iben & Tutukov (1996). If the WD core is hot (e.g., $>5 \times 10^7$ K), the WD mass is low, the surface temperature $>10^5$ K and the accretion rate is high ($>10^{-8} M_\odot$ yr^{-1}) and variable, then the evolution of the accreting WD and its outbursts will be affected by its initial core thermal structure (Iben & Tutukov 1996). The thermonuclear explosions will be less energetic than a classical nova explosion. While quasi-static and hydrodynamic evolution of accreting WDs with low WD masses have been followed by Hillman & Kashi (2021), Shara et al. (1993), Cassisi et al. (1998), and Piersanti et al. (1999, 2000, 2002), the WD cores are initially cold ($T_{\rm core} < 20 \times 10^6$ K), the accretion rates are, in all but the Hillman & Kashi

(2021) case, constant, and modulation of the accretion rate by outflowing wind from the hot WD and variations in the wind outflow rate of the red giant are not included.

The Hillman & Kashi (2021) study specifically examined Bondi–Hoyle accretion onto a massive WD in a symbiotic variable over the lifetime of a symbiotic in order to compare the energetics of a symbiotic nova explosion with a classical nova explosion in a CV over the same evolutionary time interval as the symbiotic. Surprisingly, they found that the symbiotic nova explosion was more powerful than the energy of the classical nova explosion in the CV case. In any event, one expects that the thermonuclear shell flashes on accreting WDs in symbiotic systems lead to far less mass ejection and nuclear energy release than occur in classical nova explosions. There is even a case of a disk instability driven accretion event that appeared to trigger a thermonuclear outburst in Z Andromeda (Sokoloski et al. 2006). Realistic simulations (e.g., the MESA code) of the structure and evolution of an accreting, low-mass, hot WD with initially hotter carbon–oxygen cores and helium cores over the estimated lifetime of a symbiotic variable with time variable accretion are needed to yield insights into the luminosity variations and outbursts in normal symbiotic variables. The physics of this topic will be discussed in detail as part of Chapter 7 on classical novae.

References

Abramowicz, M. A., Chen, X., Kato, S., Lasota, J-P., & Regev, O. 1995, ApJL, 438, L37

Bochanski, J., & Sion, E. M. 2001, BAAS, 198, 1208

Bondi, H., & Hoyle, F. 1944, MNRAS, 104, 273

Booth, R., Mohamed, S., & Podsiadlowski, P. 2016, MNRAS, 457, 822

Cassatella, A., Holm, A., Reimers, D., Ake, T., & Stickland, D. J. 1985, MNRAS, 217, 589

Cassisi, A., Iben, I., & Tornambe, A. 1998, ApJ, 496, 376

de Val-Borro, M., Karovska, M., & Sasselov, D. D. 2009, ApJ, 700, 1148

de Val-Borro, M., Karovska, M., Sasselov, D., & Stone, J. M. 2017, MNRAS, 468, 3408

Eggleton, P., Bailyn, C., & Tout, C. A. 1989, ApJ, 345, 489

Fekel, F. C., Joyce, R. R., Hinkle, K. H., & Skrutskie, M. F. 2000, AJ, 119, 1375

Galan, C., Mikolajewska, J., Hinkle, K. H., & Joyce, R. R. 2016, MNRAS, 447, 592

Hanasz, M., Kowalik, K., Woltanski, D., & Pawlaszek, R. 2010, in EAS Publications Series, Vol. 42, Extrasolar Planets in Multi-Body Systems: Theory and Observations, ed. K. Gozdziewski, A. Niedzielski, & J. Schneider, 275

Hillman, Y., & Kashi, A. 2021, MNRAS, 501, 201

Iben, I., & Tutukov, A. 1996, ApJS, 105, 145

Inogamov, N. A., & Sunyaev, R. A. 1999, AstL, 25, 269

Jura, M., & Helfand, D. 1984, ApJ, 287, 785

Karovska, M., Wood, B., & Marengo, M. 2004, RMxAA, 20, 92

Karovska, M., Schlegel, E., Hack, W., Raymond, J. C., & Wood, B. E. 2005, ApJ, 623, L137

Kastner, J., & Soker, N. 2004, ApJ, 616, 1188

Kenyon, S., & Mikolajewska, J. 1995, AJ, 110, 391

Knapp, G. R., Pourbaix, D., Platais, I., & Jorissen, A. 2003, A&A, 403, 993

Livio, M., & Warner, B. 1984, Obs, 104, 152

Luna, G. J. M., Sokoloski, J. L., Mukai, K., & Nelson, T. 2013, A&A, 559, 6

Mastrodemos, N., & Morris, M. 1998, ApJ, 497, 303

McAllister, M., Littlefair, S. P., Parsons, S. G., et al. 2019, MNRAS, 486, 5535

Merrill, P. 1950, ApJ, 111, 484

Mikolajewska, J., Acker, A., & Stenholm, B. 1997, A&A, 327, 191

Mikolajewska, J. 2003, ASP Conf. Proc. 303, Symbiotic Stars Probing Stellar Evolution, ed. R. L. M. Corradi, R. Mikolajewska, & T. J. Mahoney (San Francisco, CA: ASP) 9

Mohamed, S., & Psodliakowski, P. 2012, BaltA, 21, 88

Morgan, D. H. 1992, MNRAS, 258, 639

Morgan, D. H. 1996, MNRAS, 279, 301

Muerset, U., Nussbaumer, H., Schmid, H. M., & Vogel, M. 1991, A&A, 248, 458

Muerset, U., & Nussbaumaer, H. 1994, A&A, 282, 586

Narayan, R., & Popham, R. 1993, Natur, 362, 820

Patat, F., Chandra, P., Chevalier, R., et al. 2007, Sci, 317, 924

Patat, F., Chugai, N. N., Podsiadlowski, Ph., et al. 2011, A&A, 530, A63

Piersanti, L., Cassisi, A., Iben, I. Jr+, & Tornambe, A. 1999, ApJ, 521, L59

Piersanti, L., Cassisi, A., Iben, I., & Tornambe, A. 2000, ApJ, 535, 932

Piersanti, L., Cassisi, S., Iben, I. Jr+, & Tornambe, A. 2002, in AIP Conf. Proc. 637 (Melville, NY: AIP), 99

Piro, A. L., & Bildsten, L. 2004, ApJ, 610, 977

Prieur, J. L., Aristidi, E., Lopez, B., et al. 2002, ApJS, 139, 249

Podsiadlowski, P., & Mohamed, S. 2007, BaltA, 16, 26

Reimers, D., & Cassatella, A. 1985, ApJ, 297, 275

Reimers, D., Griffin, R., & Brown, A. 1988, A&A, 193, 180

Rutkowski, A., Mikolajewska, M., & Whitelock, P. A. 2007, BaltA, 16, 49

Schild, H., Muerset, U., & Schmutz, W. 1996, A&A, 306, 477

Schoenberner, D. 1989, IAU Symp. 131, Planetary Nebulae (Dordrecht: Kluwer) 463

Shagatova, N., Skopal, A., & Carikov, Z. 2016, A&A, 588, 83

Shara, M., Prialnik, D., & Kovetz, A. 1993, ApJ, 406, 220

Sion, E. 1985, ApJ, 297, 538

Sion, E. 1999, PASP, 111, 532

Sion, E. 2003, in ASP Conf. Proc. 303, Symbiotic Stars Probing Stellar Evolution (San Francisco, CA: ASP), 193

Sion, E., Godon, P., Mikolajewska, J., Sabra, B., & Kolobow, C. 2017, AJ, 153, 160

Skopal, A., Vanko, M., Pribulla, T., et al. 2002, CoSka, 32, 62

Skopal, A. 2005, A&A, 440, 995

Sokoloski, J., Kenyon, S., Espey, B., et al. 2006, ApJ, 636, 1002

Sokoloski, J., & Bildsten, L. 2010, ApJ, 723, 1188

Springel, V. 2005, MNRAS, 364, 1105

Van, Leeuwen, F. 2007, A&A, 474, 653

Warner, B. 1972, MNRAS, 159, 95

Webbink, R., & Kenyon, S. 1984, ApJ, 279, 252

Yaron, O., Prialnik, D., Shara, M. M., & Kovetz, A. 1995, ApJ, 623, 398

Yaron, O, Prialnik, D, Shara, M M, & Kovetz, A 2005, ApJ, 623, 398

Zorotovic, M., Schreiber, M., & Gänsicke, B. T. 2011, A&A, 536, A42

Accreting White Dwarfs
From exoplanetary probes to classical novae and Type 1a supernovae
Edward M Sion

Chapter 6

Accreting White Dwarfs in the Helium-rich AM Canum Venaticorum Systems

6.1 Introduction

The cataclysmic variables (CVs) with the shortest known orbital periods are the AM Canum Venaticorum systems with orbital periods in the range of 5.4 minutes $\leqslant P_{\mathrm{orb}} \leqslant$ 66 minutes. They are among the most exotic interacting binaries known. These ultra-compact binaries consist of semi-degenerate helium donor stars or degenerate helium-core donors filling their Roche lobe and transferring helium-rich gas to a helium-rich white dwarf (WD) accreter. The transferred helium gas is typically of such a high degree of purity that the helium to hydrogen number ratio, $N_{\mathrm{He}}/N_{\mathrm{H}} > 10^4$ makes them among the purest helium objects known in nature. The primary accreting degenerates are not unlike DB WDs. Their extremely short orbital periods place them among the strongest sources of low-frequency gravitational waves. The formation of these binaries remains poorly understood but not surprisingly their ultra-compactness requires, for the majority of such systems if not the entire class, two episodes of common envelope evolution. Current theories of the formation of these bizarre, extremely compact binaries are discussed in Section 6.3 below. Excellent reviews of these systems have been published by Solheim (2010) and Ramsay et al. (2018).

The AM CVn binaries were first discovered on the basis of their optical spectra which were devoid of hydrogen and closely resembled normal DB WDs. Since the first ultra-short binary to be identified was the H-rich dwarf nova WZ Sge, the most exotic aspect of AM CVn was that it turned out to be a CV with no hydrogen detected whatsoever. While DB WDs were known, it seems that nobody envisioned a CV with no hydrogen but it was known that an H-rich CV, WZ Sge also had an ultra-short period. The path to their discovery, the realization of their importance and even the time frame closely parallel each other. For that reason, it is worthwhile to follow, in tandem, the observational and theoretical advances that led to their critical scientific importance.

doi:10.1088/2514-3433/ac930cch6   6-1

Early on, WZ Sagittae with outbursts in 1913 and 1946 was classified as a recurrent nova primarily because of its large-amplitude outbursts. It was first observed photometrically by M. F. Walker in 1954 who found that the star exhibited maxima about 40 minutes apart and noted that its colors were very similar to those of Nova DQ Her 1935. Jesse Greenstein carried out spectroscopic observations which revealed that at minimum, WZ Sagittae exhibited very broad hydrogen absorption lines of a WD. In 1961, WZ Sge was observed by Krzeminski (1962) and by Robert Kraft who observed WZ Sge with the prime-focus spectrograph of the 200 inch (5.1 m) reflector. Krzeminski & Kraft (1964) identified it as a spectroscopic binary with a period of about only 81 minutes. This established WZ Sge as an ultra-short period H-rich CV with an orbital period of 81 minutes. Six years later, Paczynski (1967), in a landmark paper, recognized WZ Sge as a system evolving by virtue of angular momentum loss due to gravitational wave emission and hence would pose a crucial test of the theory of general relativity. Vila (1971) computed evolutionary models of binaries with different WD masses and H-rich degenerate donors applicable to ultra-short period H-rich CVs. As stated earlier, the history of the identification of AM CVn in many ways, reflects a similar discovery story.

The discovery spectrum of Humason–Zwicky 29 (aka HZ 29; Greenstein & Mathews 1957) obtained with the 5.1 m telescope on Mt. Palomar, is displayed in Figure 6.1 and is dominated by very broad shallow absorption lines of He I. A few years later, small amplitude photometric variability was detected by Smak (1967)

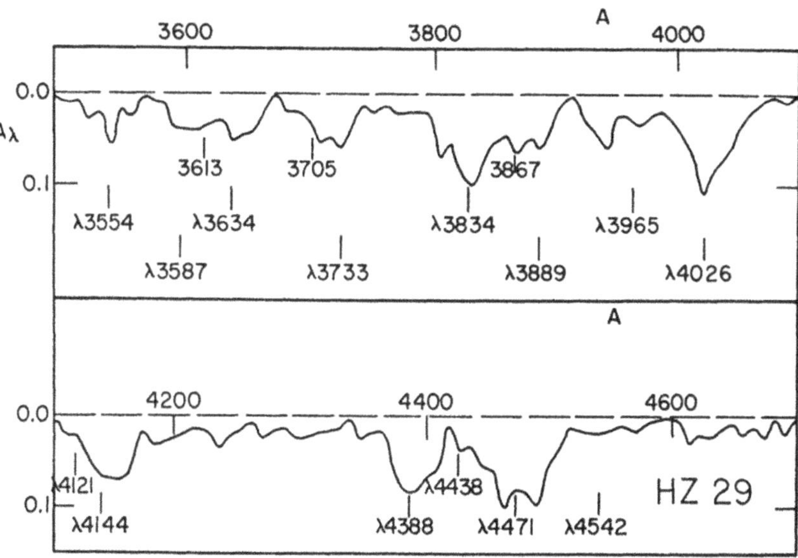

Figure 6.1. The discovery spectrum of HZ 29 (= AM CVn) the prototype of the entire class of helium-rich CVs. The profiles of the absorption lines in HZ 29, based upon on five spectrographic plates that were obtained with 200″ Hale Reflector on Mt. Palomar by Jesse Greenstein and Mildred Mathews. The vertical dashes and wavelengths correspond to lines of He I, which dominate the spectrum; the spectrum is entirely devoid of hydrogen lines. The spectrum most closely resembles a DB WD. (Reproduced with permission from Greenstein & Mathews (1957); their Figure 4. © 1957. The American Astronomical Society. All rights reserved.)

who found variability on a very low level ($\sim$0:04 mag) along with a double-humped light curve with a periodicity of 1051 s $\pm$ 0.015 s (Smak 1967, Ostriker & Hesser 1968). Five years later, Faulkner et al. (1972) proposed a model for the He-rich CV HZ29 in which a $\sim$0.041$M_\odot$ degenerate helium donor, filling its Roche lobe, was transferring helium-rich gas to a helium-rich WD primary.

It was quickly realized that these systems are an extreme counterpart of the compact binary WZ Sge and normal H-rich CVs but their orbital periods are far shorter than the shortest known orbital period of the H-rich systems which at present is 75 minutes. In addition to transferring nearly helium pure gas from the donor star, and the accreted material being nearly pure helium, their angular momentum loss due to gravitational wave emission rapidly lengthens their orbital periods and causes their mass-transfer rates to decline toward the longest known orbital periods characterizing this subclass, at present $\sim$67 minutes. The optical brightness variations displayed by AM CVn systems are similar to the H-rich CVs. These optical brightness states include high states, low states, dwarf nova-like outbursts and rare (due to their limited CV lifetimes) helium thermonuclear runaways triggered by helium accretion. For example, the helium nova V445 Puppis (Nyamai et al. 2021 and references therein) may have been spawned by a precursor AM CV system. In any case it is clear that the AM CVn systems can be categorized into similarly defined groups as the H-rich CVs based upon orbital period, accretion rate, and mass transfer driven by angular momentum loss.

6.2 Characterizing the Observed Behavior of AM CV Systems

The current census of known AM CVn systems is 56 (Ramsay et al. 2018). Of this sample, 18 systems have no recorded outbursts while 37 systems have had recorded outbursts. Following their discovery, extensive optical spectroscopic and photometric studies revealed rapid light variations (flickering) with amplitudes of $\sim$0.01 to 0.04 magnitudes. AM CVn systems like V803 Cen revealed emission line-dominated spectra in low states and absorption line-dominated spectra in high brightness states. The former applies to faint optically thin disks at low accretion rate and the latter to a bright optically thick disk at higher accretion rates. An example of the high states and low states represented in optical spectra is seen in Figure 6.2. This behavior occurs in AM CVn systems with orbital periods in the middle range and resembles a similar behavior seen in H-rich CVs like dwarf novae. Of course, if GWR, angular momentum loss and orbital period were the only factors in the observed behavior of AM CVn systems, then they would differ little from that of H-rich CVs. However, the differences seen in AM CVn systems relative to H-rich CVs clearly point to other factors that must be considered such as the primary and secondary component masses, the donor temperatures and donor compositions.

They divide into three groups with some overlap: (1) There are eight AM CVn systems at the shortest orbital periods (5.3 min$<P_{\rm orb}<$ 23 min) and highest accretion rates with no recorded low states. These objects are the helium CV counterparts (e.g., AM CVn itself) to the H-rich CVs known as the UX UMa nova-like variables. They are characterized by high mass-transfer rates and hot, luminous optically thick

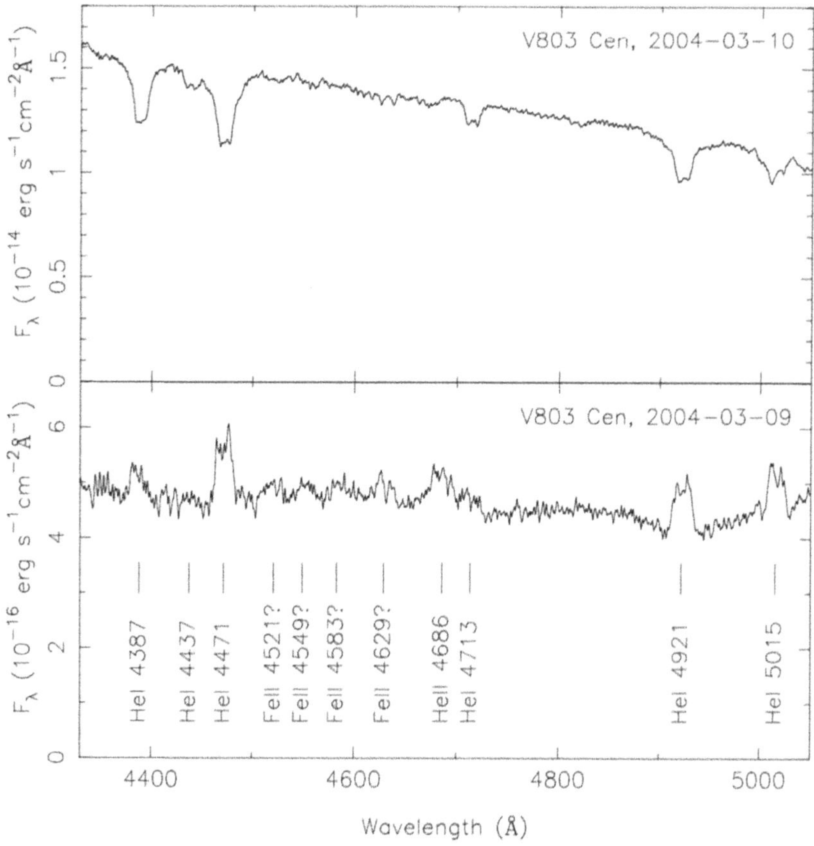

Figure 6.2. New Technology Telescope spectra of V803 Cen taken one day apart. The bottom panel shows the low state ($V = 17.2$) with emission lines and the top panel shows the high state ($V = 14$) with absorption lines. (Reproduced with permission from Roelofs et al. (2007); their Figure 2. Copyright © 2007, Oxford University Press.)

disks, sometimes with wind outflow. (2) Those AM CVn systems in the intermediate orbital period range 24 min$\leqslant P_{orb} \leqslant$ 50 min have lower mass-transfer rates and suffer instabilities in their disks. They exhibit dwarf nova-like outbursts, including "normal" outbursts and superoutbursts, with three to four normal outbursts occurring in between superoutbursts. The normal outbursts are seen mostly at the short period end of the above intermediate orbital period range and have durations between 1 and 1.5 days. The superoutbursts have a larger amplitude, 3 to 5 magnitudes above quiescence and recur in the range of ~40 days to several years (Levitan et al. 2011, 2015). One such system's light curve, PTF1J071 912.13+485 834.0, is displayed in Figure 6.3 where both normal outbursts and superoutbursts are seen. These systems are the helium-rich analogs of the H-rich dwarf novae. While the recurrence times of superoutbursts in these systems are similar to the WZ Sge H-rich CVs, the latter do not exhibit normal outbursts between superoutbursts. (3) At the longest orbital periods (lowest accretion rates), there are eight AM CVn systems with $P_{orb} \geqslant$ 46 minutes with the very lowest accretion rates and lowest optical brightness

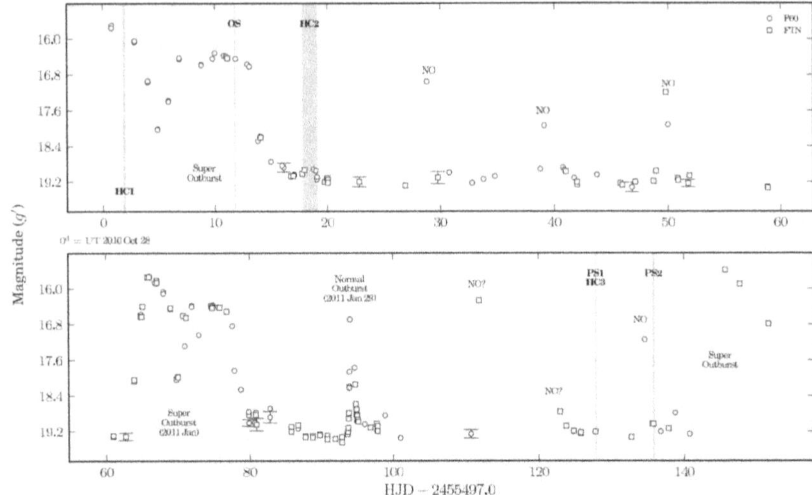

Figure 6.3. Light curve of PTF1J071 912.13+485 834.0. (Reproduced from Levitan et al. (2011). © 2011. The American Astronomical Society. All rights reserved.)

which have never been observed in outburst (e.g., GP Com) and appear to be in permanent low states. The two best examples of the spectra of the long orbital period systems are GP Com itself and V396 Hya as shown in Figure 6.3. The only analogs to these among the H-rich CVs are magnetic CV objects like those polars that appear to be in permanent low states (e.g., VV Puppis). However, there are ten systems which have exhibited outbursts even with $P_{orb} > 46$ min so it is clear that there is considerable overlap among the above three groups defined by P_{orb}.

For the systems that exhibit outburst behavior regularly, Levitan et al. (2015) showed that for 15 AM Cvn systems with several observed outbursts, the detailed analysis of their light curves utilized quantitative outburst criteria to better characterize the identification of outbursts. From this analysis, they derived the following empirical relationships between outburst properties and orbital period. For the recurrence times of outbursts, they obtained

$$P_{rec} = \left(1.53 \times 10^{-9}\right)P_{orb}^{7.5} + 24.7.$$

For the amplitudes of the outbursts, they derived

$$\Delta_{mag} = 0.13P_{orb} - 0.16$$

and for the duration of outburst, they obtained

$$t_{dur} = \left(2.53 \times 10^{-6}\right)P_{orb}^{4.54} + 10.6.$$

Statistical tests (p-values) for all three outburst properties revealed that the recurrence times and outburst durations are strongly correlated with P_{orb} while the outburst amplitude is only slightly less correlated. Since these relationships were based upon the analysis of only 15 systems, these outburst properties versus orbital period will very likely be further refined.

6.3 Theoretical Interpretation of the Observed Behavior of AM CVn Outbursts

The rough similarity of the observed behavior of outbursting AM CVn systems to the behavior of H-rich dwarf novae raised the question of whether the thermal-viscous disk instability mechanism operative in H-rich dwarf novae also apply to AM CVn systems. The answer to this important question was provided by Smak (1983) who constructed helium accretion disk models with vertical structure which undergo thermal instability within the temperature range corresponding to the ionization of helium. Using the alpha-disk approximation, with $\alpha = 0.1$, he showed that the models revealed the familiar S-shaped relationship between surface density Σ and effective temperature T_{eff} resulted as displayed in Figure 6.4 for a WD mass $= 1.0 M_{\odot}$. The points A and B correspond to the ionization of helium at temperatures $T_A = 4.1$ and $T_B = 3.95$, which are higher than the H-ionization case, as expected. The only two AM CVn systems known at the time were AM CVn and GP Com, neither of which were observed to undergo outbursts although AM CVn with its high accretion rate and luminous optically thick helium disk could be seen as in a state of "permanent outburst". Nevertheless, the known sample of AM CVn systems began to grow and soon a number of systems showing outburst behavior were observed and analyzed.

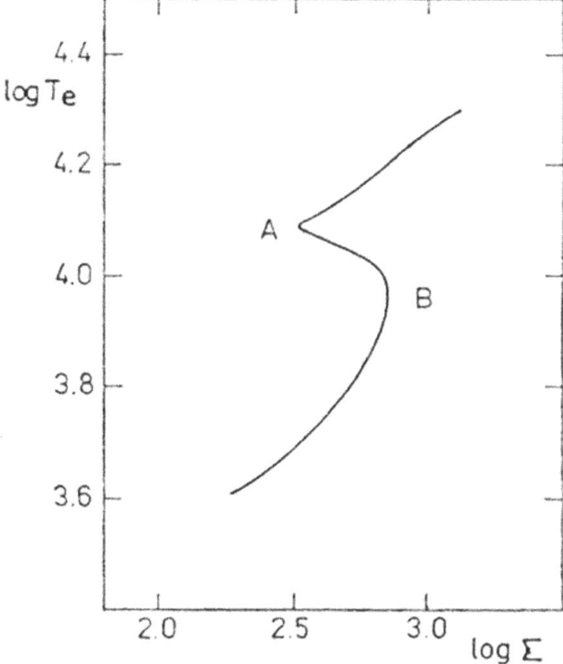

Figure 6.4. The Σ versus T_{eff} relation for a helium accretion disk, a WD mass $M_{\mathrm{wd}} = 1.0 M_{\odot}$, the α viscosity parameter $= 0.1$ and a disk radius $R = 1 \times 10^{10}$ cm. The upper curve is the stable branch and the lower part is the unstable branch. (Reproduced with permission from Smak (1983); his Figure 1. © Copernicus Foundation for Polish Astronomy.)

The increase in the sample size of AM CVns and the greater range of observed system behavior required more sophisticated and extensive grids of helium accretion disk models. In particular, Tsugawa & Osaki (1997) further investigated the application of the disk instability model to AM CVn systems by calculating a grid of helium disk models for a range of WD masses and alpha-viscosity parameters. They confirmed that helium accretion disks are thermally unstable with the S-shaped form found by Smak (1983). They also argued that the observed large-amplitude photometric variations (SOBs) in AM CVn stars can be explained by the thermal-tidal instability model of helium accretion disks.

However, new models of helium accretion disks and outburst mechanisms presented by Kotko et al. (2012) challenge the validity of the thermal-tidal instability model put forth by Tsugawa & Osaki (1997). They explored the disk instability model and included the effect of mass-transfer rate variations, additional sources of the disc heating, and the primary's magnetic field on the light curves of outbursting AM CVn systems. They reproduced various features of the outburst cycles in the light curves of AM CVn stars. They also presented arguments supporting the enhanced mass-transfer rate, as being due to variable irradiation of the donor star but of course the source of the irradiation remains unknown. Since the largest outbursts in AM CVn systems are the superoutbursts, they stressed the importance of the enhanced mass-transfer rate as the essential factor in accounting for the AM CVn star outbursts over and above the thermal-tidal instability mechanism of Tsugawa & Osaki (1997). From the perspective on the accreting helium-rich WD, these superoutbursts are key events that will heat the WD and eventually trigger explosive behavior.

In the same year that Levitan et al. (2015) derived the empirical correlations between the orbital period P_{orb} and outburst recurrence times, outburst duration and outburst amplitude, Cannizzo & Nelemans (2015) undertook a theoretical investigation using the disk instability model to constrain the outburst properties discussed in Levitan et al. (2015). Most importantly, Cannizzo & Nelemans (2015) showed that using the observed range of outburst properties for AM CVn systems as a function of orbital period, they constrained the mass transfer versus P_{orb} relation. The relation Cannizzo & Nelemans (2015) derived is

$$M_{\odot} \sim 5 \times 10^{-9} M_{\odot}\ \mathrm{yr}^{-1}\left(\frac{P_{orb}}{1000\ \mathrm{s}}\right)^{-5.2}$$

where the exponent on the right-hand side has an exact value which depends on the structure of the donor star and masses of the two components. They reported that this relation is consistent with the empirical recurrence time versus P_{orb} relation of Levitan et al. (2015).

6.4 The Formation of AM CVn Binaries

The ultra-compactness of AM CVn binaries and their extreme helium richness relative to the orbital separations of the H-rich CVs suggest evolutionary formation scenarios that differ substantially from the formation of the H-rich CVs. Indeed, three different evolutionary channels leading to the emergence of AM CVn binaries have been proposed. These three channels are potentially distinguishable from each other based upon the abundance of nitrogen to carbon and nitrogen to oxygen ratios in the chemical composition of the Roche lobe-filling donor star. The composition of the donor star potentially reveals its chemical composition from the analysis of the chemical abundances of C, N, and O of the freshly accreted material on the WD companion. These abundance ratios depend sensitively on the different levels of CNO burning and helium burning in the donor star (Nelemans et al. 2010). The three channels and their evolutionary stages leading to the formation of AM CVn systems are schematically illustrated in Figure 6.5. Channel I involves not one common envelope stage like the formation of H-rich CVs but rather two common envelope episodes needed to shrink the orbits to their observed compactness. This channel results in a double WD binary in which one of the WDs fills its Roche lobe and transfers helium to its Roche lobe detached WD companion (Tutukov & Yungelson 1979). The abundance ratios characterizing this channel would be N/C > 100, N/O > 10.

A second possible channel (II) also involves two common envelope stages but the resulting AM CVn system has a Roche lobe-filling semi-degenerate helium star (Iben & Tutukov 1991). In this case, the helium star donor is expected to manifest N/C < 10, N/O < 15. A third possible channel (III) would be identifiable if hydrogen lines are detected since this would indicate the AM CVn binary is the progeny of a CV with an evolved donor star (Podsiadlowski et al. 2003). All three of the above formation channels are color-coded and displayed in Figure 6.6.

6.5 The Masses, Effective Temperatures, Chemical Abundances and Rotational Velocities of Accreting WDs in AM CVn Binaries

Fundamental parameters characterizing the accreting WDs in AM CVn binaries (WD masses, effective temperature, chemical abundances, and rotational velocities) are becoming increasingly available due to major developments on several fronts. First, the availability of accurate distances thanks to Gaia DR1, DR2, eDR3, and DR3, enable ever more precise model fitting to optical and far-ultraviolet spectra, yielding robust model-derived parameters. Second, the number of known AM CVn binaries has grown to 56 systems of which a substantially increased number have been discovered at the intermediate orbital periods that undergo dwarf nova-like outbursts (high states followed by low states) as well as numerous systems at the longest orbital periods and lowest accretion rates (like GP Com). It is these systems that offer the best chance to study the underlying accreting WD, potentially exposed, during the quiescence between AM CVn outbursts and superoutbursts or in the low brightness states of systems with the longest orbital periods like GP Com that are not known to suffer outbursts. Third, the number of known eclipsing AM CVn systems had increased dramatically thanks to ZTF. These systems offer the most accurate masses for the accreting AM CVn WDs. Since the accretion disks and accreting WD photospheres have their Planckian peaks in the far-ultraviolet (FUV), and other

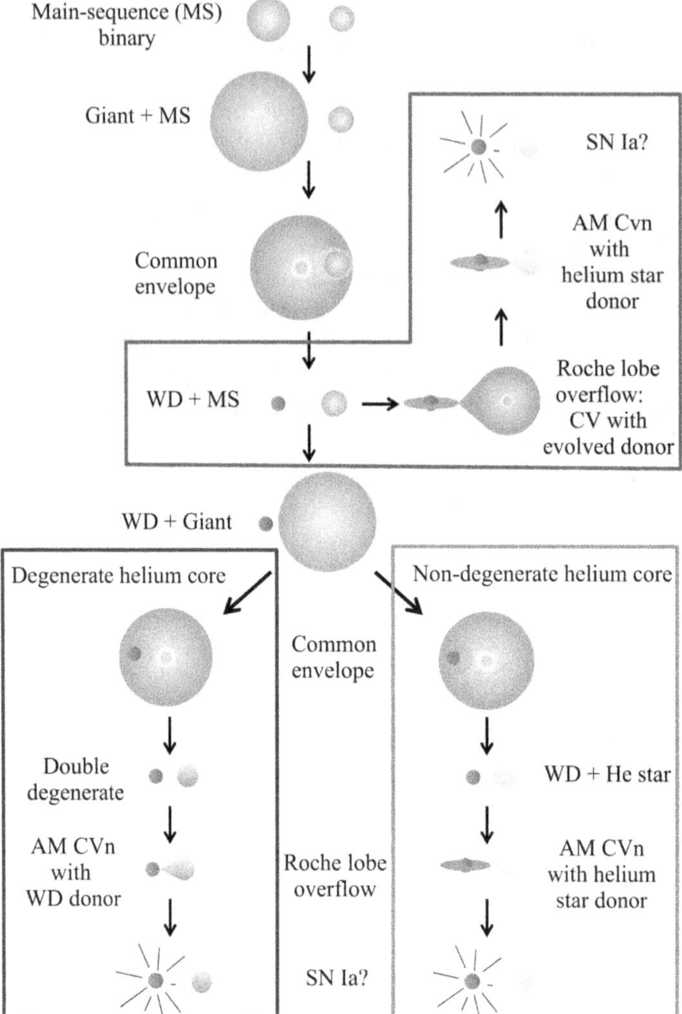

Figure 6.5. The proposed ultraviolet observations will provide firm constraints on the formation channels of AM CVns. Understanding the relative importance of these channels is critically important for the verification of the LISA space mission, as it will allow to establish the low-frequency foreground noise from Galactic binaries. AM CVns are thought to descend from wide binaries that underwent two common envelope phases. If, at the moment of the onset of the second of these phases, the core of the companion is degenerate, then a double degenerate is formed. In the opposite case, the binary emerges from the common envelope as a WD plus a He star. Driven by angular momentum loss due to gravitational wave radiation, the donor fills its Roche lobe and the system becomes a AM CVn with a WD (brown branch) or a He star (red branch) donor. Alternatively, if a CV donor has undergone significant nuclear evolution by the time the binary comes into contact, then the system can evolve to short periods, becoming a AM CVn (blue branch). The final fate of these systems is not known but they are among the most promising SN Ia progenitor candidates. In AM CVns, the direct detection of the donor stars is impossible due to their intrinsic faintness. However, their composition can be inferred from observations of the accreting WD, as its photospheric composition reflects that of the material accreted from its companion. (Reproduced with permission from Pala et al. (2022). Copyright © 2021, Oxford University Press.)

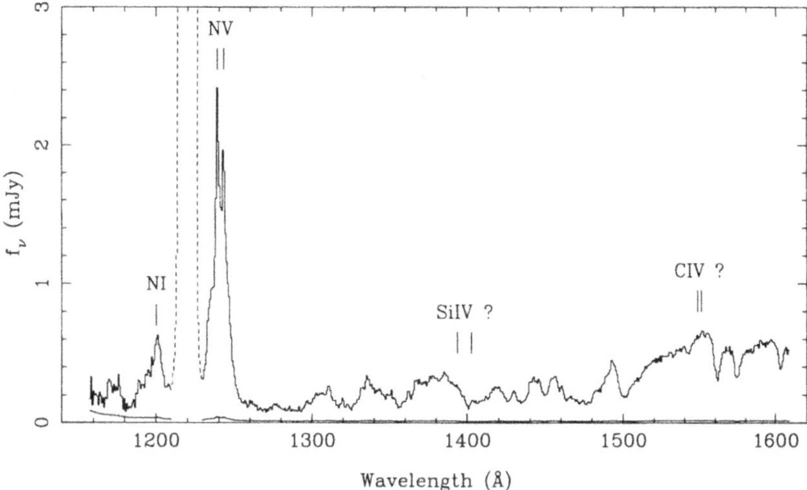

Figure 6.6. The average spectrum of GP Com taken with HST on 1993 February 23. The geocoronal Lyα line, plotted as a dashed line, reaches a level of about 12 mJy. The rest wavelengths of Si IV and C RV are marked for reference and not to suggest detection of these lines. The line at the bottom of the figure represents the 1σ uncertainties on the data. (Reproduced with permission from Marsh et al. (1995), their Figure 1. Copyright © 1995, Oxford University Press)

Table 6.1. AM Canum Venaticorum Binaries Observed with IUE.

Name	P_{orb}	M_1	M_2	i	$E(B-V)$	d (pc)
ES Cet	621	<0.8			0.025	350
GP Com	2794	0.5–0.68	0.009–0.012		0.026	75
HP Lib	11.03	0.49–0.8	0.48–0.088	26–34	0.149	197
CR Boo	1471	0.67–1.10	0.044–0.088	30	0.023	337
V803 Cen	1612	0.78–1.17	0.059–0.109	12–15	0.120	347

sources of light (e.g., the hot spot, the cool donor, gas streaming) do not contribute appreciably at FUV wavelengths, orbiting FUV, EUV, and X-ray telescopes are the instruments of choice to gain the deepest insights into the accreting WD. Therefore, a summary is provided.

6.5.1 International Ultraviolet Explorer (IUE)

The rich IUE archive contained the first-ever FUV spectra of AM CVn binaries but unfortunately, the small telescope and the faintness of most of the targets, led to low quality, noisy data with only a few exceptions. Nonetheless, it became apparent that only a few, if any systems, would be awarded coveted observing time with HST. This led Sion et al. (2011) to obtain whatever FUV information was attainable with the IUE telescope. The systems selected are tabulated in Table 6.1 along with the parameters known or estimated at the time primarily from the review by Solheim (2010). Except for ES Ceti and GP Com, the systems in table 6.2 underwent high states and low states with the latter state being most likely to reveal the underlying accreting WD.

Their analysis was carried out using the code BINSYN (Linnell & Hubeny 1996), which takes into account the donor companion star, the shock front which forms at the disk edge, and the FUV and NUV energy distribution. The distance of each system was its parallax-derived value and adopted appropriate values of orbital inclination and WD mass from Solheim (2010). We find that the accretion-heated "DO/DB" WDs are contributing significantly to the FUV flux in five of the systems (ES Ceti, CR Boo, V803 Cen, HP Lib, GP Com). In three of the systems, GP Com, ES Ceti, and CR Boo, the WD dominates the FUV/NUV flux. They presented model-derived accretion rates which agree with the low end of the range of accretion rates derived by earlier authors from blackbody fits over the entire spectral energy distribution.

One noteworthy result was that the accreting WD in ES Ceti is surrounded by a small disk and therefore is not undergoing direct impact accretion (Sion et al. 2011). The empirically determined outer accretion disk radius is much smaller than the tidal truncation radius (Warner 1995). Figure 6.7 shows the orbital plane view of the system with the accretion disk marked by the diagonal line region. The mass transfer stream is shown, terminating on the accretion disk rim. Continuation of the stream shows that, in the absence of an accretion disk, the stream would miss the WD if one neglects stream spreading (Lubow & Shu 1975, 1976).

Recently, on the basis of Magellan and VLT spectroscopic and spectropolari-metric observing campaigns focused on ES Ceti, Bakowska et al. (2021) presented evidence of double-peaked emission, characteristic of an accretion disc, with an additional component associated with the outermost disc, rather than a direct impact, which is broadly consistent with S-wave emission from the gas stream or disc impact region. This confirms beyond any doubt that 620 s is the orbital period of ES Cet. They find no significant circular polarization (below 0.1%). The trailed spectra show that ES Cet's outer disc is eclipsed by the mass donor, revealing at the same time that the photometric minimum coincides with a previously unrecognized eclipse. This confirms the original finding by Sion et al. (2011) that based upon the ballistic trajectory of the gas stream, ES Ceti is NOT a direct impact accretor but rather accretes helium via a small disk.

6.5.2 Hubble Space Telescope

For H-rich cataclysmic Variables, the launch of HST with its higher resolution spectrographs and large 2.3 m aperture was a real milestone. It enables the first spectroscopic determinations of the chemical abundances, rotational velocities, gravitational redshifts and higher precision surface temperatures of the underlying accreting WDs in dwarf novae during quiescence and nova-like variables in their low states. However, the number of AM CVn systems observed with the HST spectrographs has been very small due in large part to a focus on understanding the structure and evolution of far more numerous and greatly varied subclasses of the H-rich objects. At the time of this writing, there have been only four AM CVn binaries

Table 6.2. Deeply Eclipsing AM CVn Binaries.

System	P_{orb}	q	i	T_{wd}	Mass ratio	M_d	R_{wd}	R_d
ZTFJ1637 + 49	61.5	0.026 ± 0.010	82.7 ± 0.09	11.2 ± 0.3	0.90 ± 0.05	0.023 ± 0.008	0.009 ± 0.001	0.068 ± 0.007
ZTFJ0003 + 14	55.5	0.021 4 ± 0.010	86.5 ± 2.0	11.6 ± 0.8	0.79 ± 0.11	0.017 ± 0.011	0.010 6 ± 0.001	0.057 ± 0.012
ZTFJ0220 + 21	53.5	0.017 4 ± 0.005	85.3 ± 2.1	14.2 ± 1.0	0.83 ± 0.07	0.014 ± 0.006	0.010 ± 0.001	0.054 ± 0.007
Gaia14aae	49.7	0.029 0 ± 0.000 6	86.27 ± 0.10	17.0 ± 1.0	0.872 ± 0.007	0.025 3 ± 0.000 7	0.009 24 ± 0.000 09	0.060 3 ± 0.000 3
ZTFJ2252?05	37.4	0.034 ± 0.006	87.0 ± 1.0	15.2 ± 0.9	0.76 ± 0.05	0.026 ± 0.008	0.010 ± 0.001	0.049 ± 0.004
ZTFJ0407?00	35.4	0.024 ± 0.004	86.5 ± 0.7	17.4 ± 1.2	0.79 ± 0.06	0.019 ± 0.003	0.010 ± 0.001	0.044 ± 0.002
YZ LMi	28.3	0.041 ± 0.002	82.6 ± 0.3	17.0 ± 1.0	0.85 ± 0.04	0.035 ± 0.003	-0.047 ± 0.001	

The first column is the system name, 2nd column: P_{orb} in minutes, 3rd column: the mass ratio, 4th column: the inclination angle in degrees, 5th column: the estimated temperature of the primary WD, 6th column: the WD primary mass in solar masses, 7th column: the mass of the secondary donor star in solar masses, 8th column: the radius of the primary WD in solar radii, 9th column: the radius of the donor secondary star in solar radii.

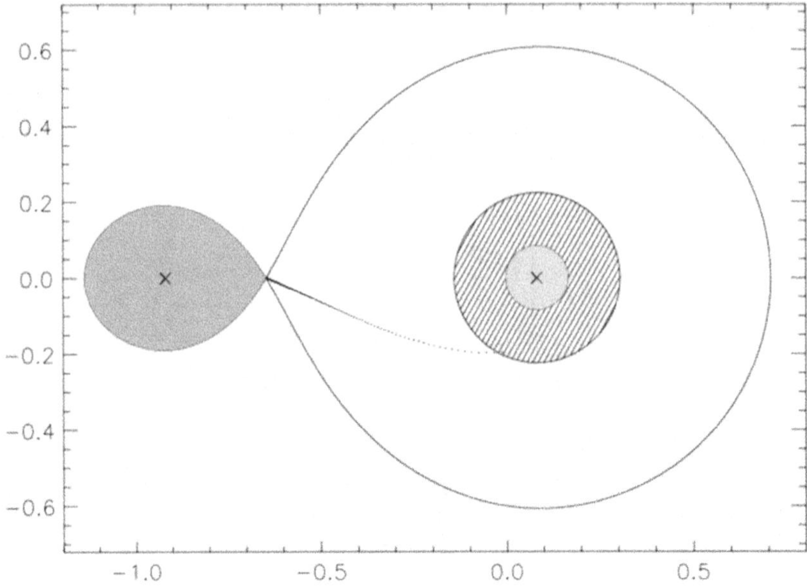

Figure 6.7. Orbital plane view of the AM CVn system ES Ceti. The mass transfer stream intersects the rim of the accretion disk. Continuation of the mass transfer stream trajectory misses the WD by a small amount. (Reproduced with permission from Sion et al. (2011); their Figure 3. © 2011. The American Astronomical Society. All rights reserved.)

observed with HST AM CVn, CE 315, GP Com, and CP Eri. Of the four, CE 315, and GP Com appear to be in long standing low optical brightness states except for flares in the case of GP Com. Neither system has any recorded outbursts while CP Eri exhibits high states and low states. In all three of these systems, the underlying accreting WD may outshine an optically thin accretion disk. On the other hand, AM CVn remains in a high optical brightness state analogous with the UX UMa H-rich nova-like variables in which these systems appear to be in permanent outburst. The findings from the analysis of the Hubble FUV spectra are summarized below.

6.5.2.1 CE 315
CE 315 (= V396 Hya) was discovered spectroscopically by Ruiz et al. (2001), who noted a blue continuum with He I and He II emission features but no hint of any hydrogen. They noted that the optical spectrum was very similar to GP Com and that the emission lines were triple-peaked with the central peak forming near the WD. They found that the He emission lines had very broad widths of 2000 km s^{-1}. They derived a period of 3906 s making it the longest period AM CVn system. They estimated that the primary mass was $0.77 M_\odot$ and the donor secondary mass $\sim 0.017 M_\odot$ with a mass ratio of 0.022. Like GP Com, CE 315 reveals the N/C anomaly with very strong N V emission seen in its HST spectrum and weak or absent C IV in emission which is indicative of CNO processing (Gänsicke et al. 2003). The optical spectra of the spectroscopic twins, GP Com and V396 Hya, are comparatively displayed in Figure 6.8.

6.5.2.2 AM CVn

By the time of HST Cycle 8 when CP Eri was observed with HST STIS, the only other AM CVn that had been observed spectroscopically with HST was AM CVn itself which remains in a high brightness state accreting at a high rate with a luminous, optically thick accretion disk dominating the system light. Wade et al. (2007) obtained STIS spectra of AM CVn in time-tag mode covering the FUV and NUV bandpasses over a time baseline of two HST orbits. The identified absorption lines were the same as previously detected with IUE but were at higher spectral resolution and phase-resolved with the opportunity to search for short timescale variations. Blue-shifted absorption lines due to the well-known zero volt resonance doublets typically seen associated with wind outflow, N v (1238, 1242), Si iv (1394, 1402), C iv (1548, 1550) were detected as well as the UV recombination lines due to He ii (1640) and N iv (1718). The authors also detected weak redshifted emission from N v and He ii. Over the two HST orbits that were covered, the authors detected a 20% decline in light from the system as well as flickering resembling dwarf nova oscillations on a timescale less than a minute. They utilized helium disk models from two different disk methods (blackbody disk and helium model atmospheres) but the two types of disk models gave inconsistent best-fit solutions. Therefore, their analysis could not deliver a robust determination of AM CVn's accretion rate which was their most important objective.

6.5.2.3 GP Com

The first two exposed accreting WDs to be spectroscopically identified were GP Com (Lambert & Slovak 1981) and CP Eri (Sion et al. 2006). Starting with GP Com (aka Giclas 61-29), Lambert & Slovak (1981) obtained a single IUE spacecraft observation on 1980 July 31 through the large aperture with the short-wavelength prime camera (SWP09641) with an exposure time of 3 hours. The spectrum revealed emission lines of N v and He ii but notably absent was C iv at (1548, 1551Å) which is ubiquitously seen in FUV spectra of CVs. The authors concluded that the absence of C iii 1909 ruled out nebulosity as the origin of the emission lines and instead suggested that they form in an accretion disk.

A much higher quality FUV spectrum that was time-resolved was obtained with HST by Marsh et al. (1995) and is displayed in Figure 6.8. Surprisingly, the N v emission feature manifested three remarkable flare events occurring during a total of 13 hours of FUV observation. Whether the flare event was due to actual mass ejection via a wind or was associated with irradiation of the accretion disk remains unclear. In addition, Marsh et al. (1995) found from optical data that the heavy elements are under-abundant relative to nitrogen. Their measured silicon to nitrogen ratio from the HST STIS spectrum confirmed the underabundance metals with respect to nitrogen seen in the optical with FUV data. After removing the flare spectrum from the data and phase-folding on the 46 minute orbital period, they saw the S-wave behavior of the UV lines that is seen in GP Com's optical spectra.

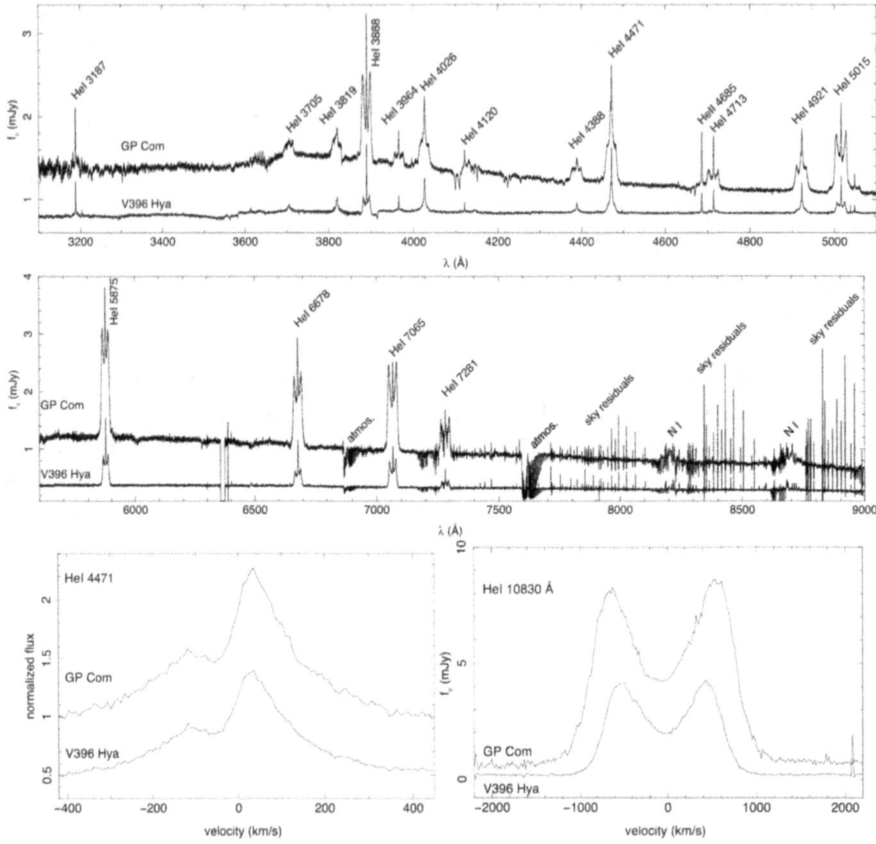

Figure 6.8. Gaussian smoothed average spectrum of GP Com and V396 Hya obtained with VLT/X-Shooter. Helium emission lines are indicated. The lower left panel shows the central-spike feature with its blue-shifted forbidden component in He I 4471 Å observed with UVES. The lower right panel shows the strongest helium lines in the NIR arm observed with X-Shooter. The narrow spikes at wavelength >7500 Å are residuals from the nights sky line removal. (Reproduced with permission from Figure 1 in Kupfer et al. (2016). Copyright © 2016, Oxford University Press.)

6.5.2.4 CP Eri

CP Eri has an orbital period of 28.73 minutes, an optical brightness range of 16.5 in outburst and 19.7 in quiescence with broad shallow optical He I absorption in outburst and, in quiescence, double-peaked He I and He II emission features as well as an Si II emission line in the optical (Abbott et al. 1992). Abbott et al. (1992) did not detect hydrogen in either high- or low-state optical spectra.

Sion et al. (2006) obtained two HST Space Telescope Imaging Spectrograph (STIS) spectra of CP Eri on 2000 September 11. Unlike the optical spectrum of CP Eri seen in quiescence, the HST FUV spectrum contains many absorption features, including a strong feature at Ly Alpha, strong C III (1175), C II (1335), Si II (1260, 1265), C I (1270), O I, Si III + O I (1300), C I (1356, 1490, 1657), Si II (1526, 1533), plus moderately strong emission features at Si IV (1393, 1402), a possible N V (1238, 1242) feature, and possible He II (1640) emission.

The rather deep absorption line near 1216 could not be due to He II at the T_{eff} of the WD that is indicated by the continuum slope and by the low-ionization metal line profiles. Hence, unless the hydrogen absorption line is totally of interstellar origin, which is unlikely given its breadth, there is a possibility that it is due at least in part to photospheric H I Lya. Therefore, the authors chose hybrid composition "DBA" atmospheres in which the dominant element is helium, with hydrogen far less abundant than helium.

Helium accretion disk models by Nasser et al. (2001) were compared with the HST STIS spectrum of CP Eri over the effective wavelength range of the STIS spectrum, 1150–1716. The helium accretion disks are steady state, non-LTE models, which are more appropriate for the high state of AM CVn systems. However, as a first approximation to the actual accretion disk during a low state, Sion et al. (2006) applied the optically thick models to CP Eri's spectrum since realistic optically thin helium disk models are not yet available. Their disk model grid covered inclination angles of 30° and 45°, He/H by number of 10^3 to 10^5, fixed WD mass of $0.6M_\odot$ ($\log g = 8$), metal abundances $Z = 0.001$ with CNO abundances of 3900, $1.5 \times$ solar, respectively, and outermost annulus (disk radius) of 8 to 15 WD radii. All of the five best-fitting helium accretion disk fits yielded distances of 1.34, 1.04, 1.21, 1.27, 1.15 kpc, which are close to the Gaia distance of CP Eri. The best-fitting accretion disk model has He/H = 1000, (including Fe), a disk inclination angle of 45° and a reduced $\chi^2 = 2.416$. Curiously, the disk fits are improved considerably below 1350 Å if Fe is over-abundant because of the large number of low-excitation Fe lines whose collective absorption eats away at the continuum and broadens line profiles.

Although the disk inclination is relatively low at 45°, the possibility of an Fe curtain above the disk plane cannot be ruled out. Indeed, Godon & Sion (2021) have shown that an Fe curtain can be present in some H-rich dwarf novae, even at lower inclinations. The best-fitting helium accretion disk model to the HST STIS spectrum of CP Eri is displayed in Figure 6.9.

Sion et al. (2006) also explored the possibility that the STIS spectrum of CP Eri is in its low brightness state, and like the FUV spectra of the shortest period H-rich dwarf novae during quiescence, is dominated by the accreting WD with only a minor contribution from an accretion disk. Assuming that the profile is entirely H I, the authors kept the gravity fixed at $\log g = 8$ with various He/H ratios from 10^2 to 10^5, metal abundances $Z = 0.5, 0.1, 0.05,$ and 0.005, and temperatures in the range $15,000 < T_{eff} < 20,000$ K. They found that the optimal He/H ratio needed to replicate the profile is 10^3. This ratio is smaller than the He/H = 10^4 stringent He/H (by number) ratio characterizing the DB WDs where in order for Balmer lines not to be detected in their optical spectra, H/He by mass fraction $\geqslant 10^{-5}$ (which they are not). The best-fitting DBAZ photospheric spectrum of CP Eri is displayed in Figure 6.10.

Their analysis suggested that CP Eri may not be a typical AM CVn system in that they find a significant abundance of H and a higher metallicity compared with other

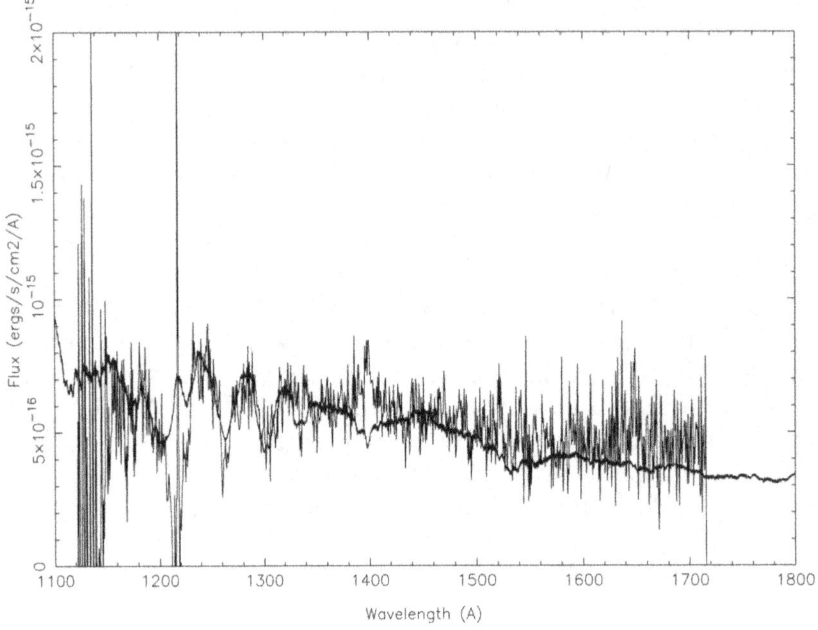

Figure 6.9. Flux distribution, flux versus wavelength, for the best-fitting helium accretion disk model compared with the HST STIS spectrum of CP Eri. The model disk parameters are metallicity, $Z = 0.001$, inclination $i = 45°$, central WD $\log g = 8$, and an accretion rate $= 3 \times 10^{-10} M_\odot$ yr^{-1}, scaled to a distance of 1.04 kpc. (Reproduced with permission from Sion et al. (2006), their figure 1. © 2006. The American Astronomical Society. All rights reserved.)

AM CVn systems such as the well-studied object GP Com, which is metal-poor, has shown little evidence of any H, and is always seen in a low state. It is obviously important to explore whether the abundance of H from our UV analysis would lead to detected H features in the optical spectrum. While a reexamination of the optical quiescent spectrum of Groot et al. (2001) suggests a possible hint of very weak H emission features in the optical low-state spectrum (see Figure 1 in Groot et al. 2001), much higher signal-to-noise ratio optical spectra are clearly needed. In any case, the metallicity they derived is consistent with the Groot et al. conclusion that CP Eri has higher metallicity than GP Com and CE 315, implying that it is likely not a Population II object.

Since CP Eri's accretor is the only WD in an AM CVn so far with a directly determined photospheric H abundance, we cannot compare with any other AM CVn cases. Therefore, analyses of other exposed WDs in these objects are clearly needed. At the time of this writing, the author is a member of an approved HST Large GO proposal in HST Cycle 29 that will include the first FUV spectra for 15 AM CVn binaries. These new HST COS spectra will deliver new insights into the physical processes affecting the helium-accreting WDs in AM CVn systems.

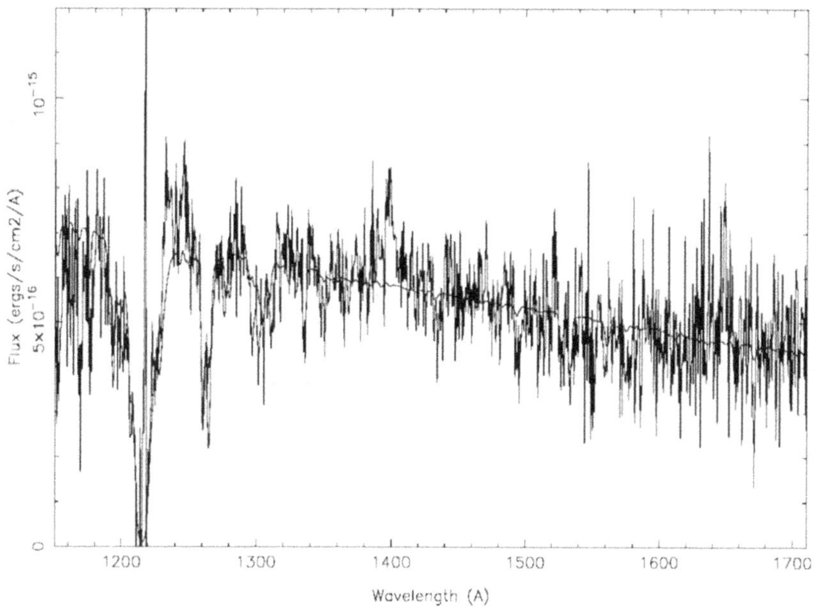

Figure 6.10. Flux distribution, flux versus wavelength, for the best-fitting hybrid composition, DBAZ, photosphere model with $\log g = 8$ (assumed), $T_{\text{eff}} = 17,000 \pm 1000$, He/H $= 10^3$, $Z = 0.05$, Vsini $= 400 \pm 50$ km s^{-1}, compared to the HST STIS spectrum of CP Eri. (Reproduced with permission from Sion et al. (2006); their figure 3.) © 2006. The American Astronomical Society. All rights reserved.

6.6 Masses of AM CVn Accreting WDs

The most important fundamental parameter of an accreting WD is its mass. While Zorotovic et al. (2011), Savoury et al. (2012), McAllister et al. (2019) and Pala et al. (2022) have determined the masses of 98 accreting WDs in H-rich CVs, relatively little is known about the masses of the helium-accreting WDs in AM CVn systems. The most robust determinations of WD masses in CVs are from deeply eclipsing systems for which very few were known among the AM CVn sample of Ramsay et al. (2018). However, recently a small fraction of the currently known sample of AM CVn systems were discovered to be deeply eclipsing. Campbell et al. (2015) reported that Gaia14aae is a deeply eclipsing system in which the accreting WD is totally eclipsed every 49.71 minutes. Prior to their work, only one other eclipsing AM CVn was known but Gaia14aae is the first one known in which the WD is totally eclipsed. These authors also reported that Gaia14aae underwent three outbursts over a 4 month time frame placing it at the long orbital period extreme of known outbursting systems. From the contact phases of the WD total eclipse, they found lower limits of $0.78\,M_\odot$ for the WD primary and $0.015\,M_\odot$ for the donor secondary. These values correspond to an inclination of 90°, a mass ratio of 0.019 and orbital separation of $0.41\,R_\odot$.

More recently, van Roestel et al. (2022) have found five new eclipsing systems resulting from a search for deep eclipses with the Zwicky Transient Facility (ZTF) using each system's Gaia parallax. Their orbital periods lie within the range 61.5 minutes to 35.4 minutes. The authors carried out phase-resolved spectroscopy and high speed photometry to confirm the AM CVn identifications and detected in all five systems, double-peaked line features due to helium and, in addition, they reported the detection of metals, two of which were not detected previously in any AM CVn system. These five new systems, together with the two previously known eclipsing AM CVns, YZ LMi and Gaia14aae, are shown in Table 6.2 that has been adapted from van Roestel et al. (2022). The masses and radii of the donors and their entropy offer the possibility of identifying the formation channel for each of the systems. Mindful of small number statistics, it is nonetheless interesting that the average mass of the WDs in the seven systems is $0.8M_\odot$, which is close to the mean mass of H-rich CV WD primaries of $0.83M_\odot$ (Pala et al. 2022).

It is clear that with an even greater number of eclipsing AM CVn systems in current and future surveys, deeper insights will emerge into the formation, structure and evolution of the very low mass, mostly substellar donor stars and the accreting WD primaries.

6.7 The Heating and Cooling of Accreting WDs in AM CVn Systems

Accreting WDs repeatedly explode as classical novae but also comprise the single degenerate channel for WD supernovae (Type Ia), the supernovae that provide the evidence that the expansion rate of the universe is accelerating and for the existence of dark energy. Since SNe Ia are hydrogen deficient and their light curves and spectra reveal remarkable uniformity, it is for this reason that attention focused on helium-rich objects devoid of hydrogen or with trace hydrogen. Many investigators turned to progenitors of Type Ia that are nearly pure helium or helium-rich but with carbon–oxygen cores. Of course, the mere existence of AM CVn binaries added great interest to these helium-accreting objects as possible progenitors for SNe Ia. Topics such as sub-Chandrasekhar mass supernovae, edge-lit detonations, as well as Chandrasekar-mass helium-rich degenerates with C–O cores became very active research areas for theorists and observers. However, all of these investigations of AM CVn binaries as SNe Ia progenitors will be covered in Chapter 7 on the broad topic "Accreting WDs as Progenitors of Supernovae". This chapter will be devoted to AM CVn binaries in the context of accretion physics, core re-heating, non-dynamical helium shell flashes, thermal structure and evolution.

Clearly, the accretion of helium and the declining rate of accretion as GWR drains angular momentum from the AM CVn binaries renders their evolution and their potential for thermonuclear outbursts to be distinctly different from the H-rich CVs. As their orbital periods increase, their accretion rates decline as we discussed earlier in this chapter. Hence, helium thermonuclear runaways are much less frequent than in the H-rich case. What also makes the AM CVn systems quite different from their H-rich counterparts is the behavior of the core in response to accretion. The question that naturally arises when following the evolution of a WD accreting from a binary companion, is whether or not the degenerate core could be re-heated during long-term accretion. To answer this question, the most relevant parameter is the timescale for heat conduction into the core. For AM CVn accreting WDs, core re-heating can indeed occur as discussed later in this chapter. However, before examining the heating and cooling of the helium-accreting WDs in response to accretion, it is important to first focus on the thermal structure and properties of the stellar source of the helium accretion, the Roche lobe-filling donor star.

Recalling the discussion in Section 6.4 of the three possible types of AM CVn donor stars emerging from the final common envelope, the core of the donor companion may be degenerate, in which case, the AM CVn binary will have a Roche lobe-filling WD donor, i.e., a double degenerate is formed. However, the binary could emerge from the 2nd common envelope with a non-degenerate or semi-degenerate helium star filling its Roche lobe in which case the WD is accreting from a very low-mass helium star. A third type of donor star identifiable if hydrogen lines are detected, can emerge from the common envelope if it had undergone significant nuclear evolution before reaching contact. The resulting system has an evolved donor with the binary evolving to short periods, becoming an AM CVn system. These three evolutionary channels are distinguishable based upon the N/C and N/O abundance ratios of the donor stars. Unfortunately, these low-mass donors are far too faint to analyze spectroscopically. However, the material accreting onto the WD will have the same composition as the donor itself so that the WD serves as a mass spectrometer for the composition of the donor. The donors emerging from all three possible channels will obey different mass–radius laws.

Nelemans et al. (2001), Deloye & Bildsten (2003) and Deloye et al. (2005) elucidated the properties of the He donor stars and their mass distribution, as likely AM CVn donors, using population evolution computations. They showed that the entropy of the donor at the onset of Roche lobe overflow determines the relationship between the time-averaged accretion rate, $\dot{M}$, and the orbital period P_{orb} while the donor star's mass–radius relation essentially determines the speed of evolution of the system. Deloye et al. (2005) and Bildsten et al. (2006) found that AM CVn systems with higher specific entropy donors, have hotter (i.e., higher T_{eff}) WDs than those in systems having a lower entropy (cooler) donor at a given value of the orbital period, P_{orb}. As discussed below, initially low entropy donors evolve more slowly with respect to P_{orb}. Therefore, the accreting WD with the low entropy donor will be cooler (appear older) at a given value of P_{orb}. As seen below, lower entropy donors have lower masses at a given value of P_{orb}, during the evolutionary stage when compressional heating is dominant, thus a lower time-averaged mass-transfer rate, $\dot{M}$, at a given value of P_{orb}, consequently leading to a lower luminosity. For the WD channel, Deloye et al. (2007) found that higher entropy donor stars have larger masses M_{d}, radii, R_{d}, at a given orbital period than zero temperature donors and hence a larger mass-transfer rate $\dot{M}$. They showed that among the AM CVn systems, variations in the donor star's specific entropy yields a range of mass–radius relations thus giving a range of possible donor masses and mass-transfer rates at a given P_{orb}.

Theoretical evolutionary models of accreting WDs in interacting binaries began with a series of seminal papers in the mid-1960s by Kippenhahn & Weigert (1967), Kippenhahn et al. (1967), and Giannone & Weigert (1967). Giannone & Weigart (1967) published a classic paper with the first detailed calculation of hydrogen-rich accretion onto a WD. Early quasi-static evolutionary sequences of accreting WDs were calculated by Paczynski & Zytkow (1978), Sion et al. (1979) and Iben (1982) while Fujimoto (1982a, 1982b) worked out the detailed theoretical framework for hydrogen shell flashes triggered by accretion of hydrogen. Among the earliest investigations of helium accretion onto a WD was carried out by Fujimoto & Sujimoto (1982) who investigated the evolution of accreting WDs from the onset of accretion through the helium shell flash. They studied the properties of the helium shell flashes in parallel with Fujimoto's (1982a, 1982b) generalized theory of shell flashes and with quasi-static model sequences with WD undergoing helium accretion at different rates and for different WD masses. Among their findings was that for a massive WD, if a helium layer mass is built up by H-fusion reactions, the energy of the helium shell flash can approach the supernova regime of energy. Of course, such a strong flash would have to be followed hydrodynamically. If the WD is hot, then a weaker helium shell flash ensues. What they found is that the strength of the helium shell flash is determined primarily by the accretion rate. If the accretion rate is rapid, then the He shell flashes are weaker and occur repeatedly whereas for lower rates of accretion, a much more energetic helium shell flash may lead to the explosive power of a supernova which could take place on a massive WD if the needed critical mass is reached on it.

Following up on the evolutionary studies of helium-accreting degenerates by Fujimoto & Sujimoto (1982), Limongi & Tornambe (1991) extended the exploration of helium-accreting degenerates using a different quasi-static stellar evolution code, the FRANEC Code (Chieffi & Straniero 1989). Their objectives were to (1) confirm the results of their quasi-static code with the helium-accreting evolutionary models of previous authors; and (2) explore new regions of parameter space not covered by previous investigations. The models were composed of 98% helium by mass, solar abundance of metals, and covered a wide range of WD masses ($0.4M_\odot$ to $1.0M_\odot$). They explored the response of a helium-rich WD with a C–O core to a range of helium accretion rates corresponding to the transfer of helium by Roche lobe overflow. Their goal was to follow the build-up of the helium layer mass sufficient to trigger a thermonuclear runaway which could lead to a the detonation of the highly degenerate C–O core. They found that the mass of the helium layer needed for a supernova explosion is relatively insensitive to the mass of the WD. In Figure 6.11, the results of their helium-accreting quasi-static evolutionary sequences with C–O cores are displayed in a plot of accretion rate in $M_\odot$ yr^{-1} and the total mass of the WD. The authors found that for accretion rates $\geqslant 5 \times 10^{-8} M_\odot$ yr^{-1}, weak and recurrent helium shell flashes occur but if the accretion rate is $\leqslant 10^{-10} M_\odot$ yr^{-1}, then the higher degree of degeneracy in the helium layer leads to the onset of a highly energetic explosion that must be followed hydrodynamically.

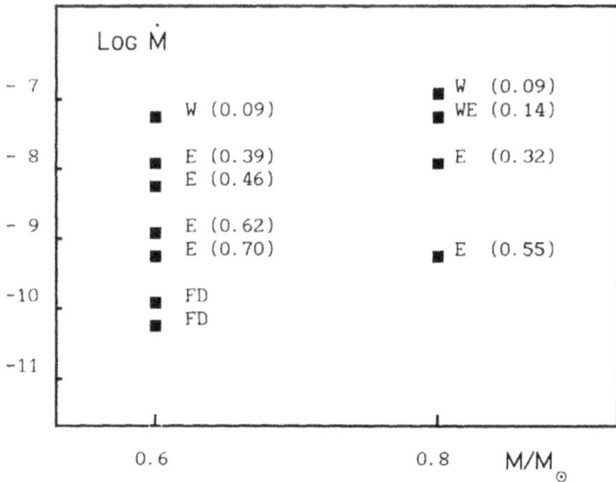

Figure 6.11. The fate of He accreting CO dwarfs versus the original mass of the dwarf (abscissa) and the accretion rate (ordinate). Numbers in parentheses give the helium layer mass accreted just before the onset of the helium thermonuclear runaway. E, W, and FD denote explosive, weak, and full degeneracy, respectively. (Reproduced with permission from Limongi & Tornambe (1991); their figure 4. © 1991. The American Astronomical Society. All rights reserved.)

Limongi & Tornambe (1991) computed evolutionary model sequences with helium accretion rates from $5 \times 10^{-11} M_\odot$ yr^{-1} to $1 \times 10^{-7} M_\odot$ yr^{-1}. Since they found that the critical mass for triggering a supernova level explosion appears relatively insensitive to the mass of the accreting star and that the accretion rate needed for the build-up differs little from system to system, then the explosions would appear spectroscopically homogeneous and have light curves that were uniform. Both of these observational characteristics are hallmarks of most SNe Ia.

Despite the above progress on understanding the build-up of a critical helium layer mass for He shell flashes and the leadup to a possible supernova level explosion, there remained many unanswered questions regarding the thermal evolution of an AM CVn binary and different stages that the binary evolves through as its orbital period keeps increasing and its accretion rate keeps declining. Yet the quasi-static evolutionary models of helium-accreting WDs, as discussed above in this chapter, focused upon the thermonuclear evolution of helium-accreting WDs (e.g., Fujimoto & Sugimoto 1982; Limongi & Tornambe 1991). There are no studies of the detailed thermal evolution of the accreted envelope and core of the WDs in AM CVn systems. Also, an important question to be addressed is the thermal interaction between the helium-accreting, non-degenerate He-rich envelope and the degenerate carbon–oxygen core. With the emergence of deeper insights into the thermal structure and properties of the low-mass donors (e.g., Nelemans et al. 2001; Deloye & Bildsten 2003; Deloye et al. 2005), the stage was set for deepening insights into detailed evolution of the accreting WDs using realistic WD envelopes and cores.

Bildsten et al. (2006) carried out quasi-static model sequences of accreting WDs that incorporated known parameters for their starting models' donor mass, WD mass and accretion rates that were established primarily from ground-based observational studies as well as population synthesis calculations. They chose a representative donor mass of $0.24 M_\odot$ but set the initial entropy of the donor in such a way that the orbital period of the system at the onset of mass transfer is greater than 30% and 90% of the donors in the population synthesis results from Deloye et al. (2005). This work established the variation of the time-averaged mass-transfer rate as a function of time. The WD models were then evolved quasi-statically using the model envelope methods that Townsley & Bildsten (2004) applied to H-rich CVs. Their calculation fixed the WD masses at $0.65 M_\odot$ and $1.05 M_\odot$. For both of these adopted WD masses, they used a constant accreted helium layer mass of $0.05 M_\odot$. The authors also justified their choice of keeping the WD mass and helium layer mass constant during the evolution. On the basis of their quasi-static evolutionary sequences, Bildsten et al. (2006) identified four stages in the evolution of an AM CVn WD primary that are displayed in Figure 6.12. The four evolutionary stages identified by Bildsten et al. (2006) are displayed corresponding to the variation of the time-averaged accretion rate, the core temperature and the luminosity of the accreting WDs for two values of the entropy of the donor star, a low entropy donor entropy and high entropy donor cases.

(1) The initial stage at formation is characterized by intense accretion at rates $\geqslant 10^{-7} M_\odot$ yr^{-1} on a timescale of $<10^5$ years leading to a helium thermonuclear runaway (helium nova?) followed by weaker helium thermonuclear shell flashes. This stage is so short that the core temperature, T_{core}, remains essentially unchanged. The amount of accreted helium that is ejected in this stage is poorly known and requires further modeling with codes like MESA. As P_{orb} increases to $\geqslant 10$ minutes, the accretion rate begins to decline more rapidly, lowering the temperature in the transition region between the core and accreted envelope, thus increasing the degree of degeneracy at helium ignition, resulting in a much more energetic He shell flash that presumably ejects a substantial amount, if not all, of the accreted envelope. As the orbital period continues to increase and the rate of accretion drops to below $2 \times 10^{-8} M_\odot$ yr^{-1}, then the accreting He-rich degenerate will likely not accrete an amount of helium needed to reach the helium ignition mass (proper pressure at the base of the accreted envelope) because the needed amount of accreted helium for ignition exceeds the remaining donor secondary star's mass.

(2) The second stage in the evolution of the accreting helium-rich degenerate is characterized by compressional heating resulting from the weight of the accumulating accreted helium. The surface temperature and surface luminosity is determined by this mode of heating. It depends only upon $\dot{M}$ and M_{wd}. During this stage, which lasts from $\sim 3 \times 10^5$ to 3×10^6 yr, released heat is diffusing upward toward the surface but also diffusing downward into the highly conductive degenerate core. It is in this stage that the core starts to undergo re-heating. Henyey & L'Ecuyer (1969) first showed that the heat conduction timescale for re-heating of an isothermal, non-convective core is essentially given by

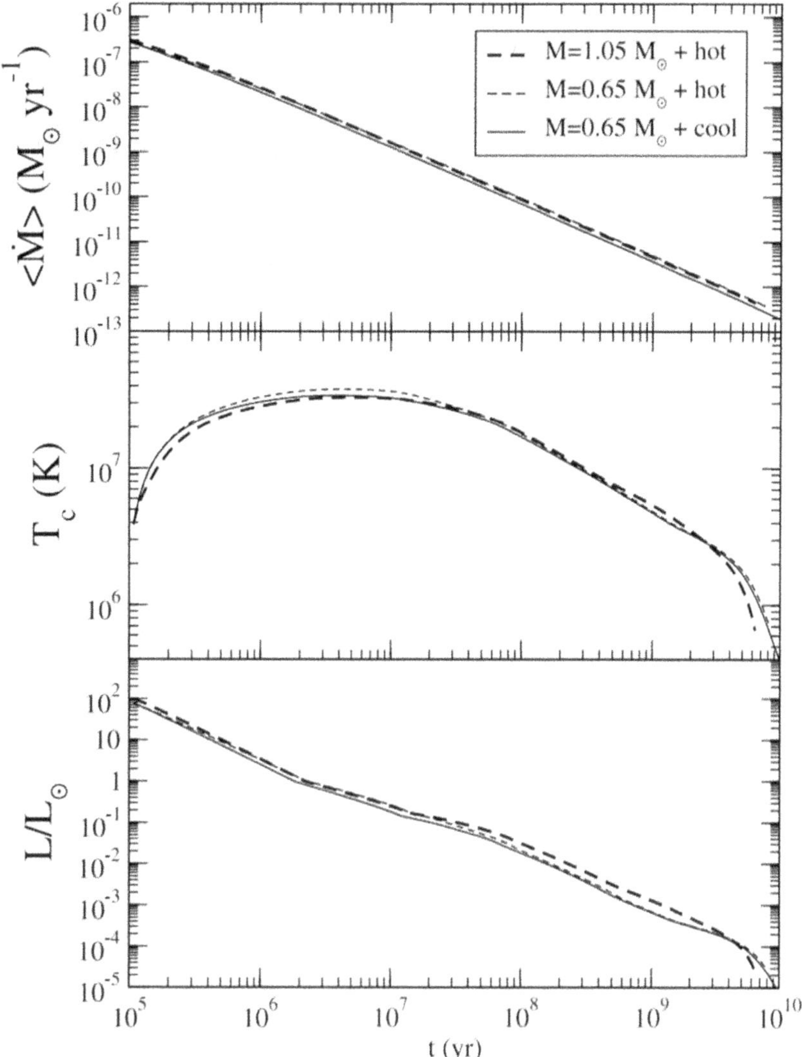

Figure 6.12. Time-averaged mass-transfer rate $\dot{M}$, WD core temperature, T_c, and WD luminosity, L, as a function of time since the onset of mass transfer for donors with fixed entropy. The thin (thick) lines show the evolution for a $0.65 M_\odot$ ($1.05 M_\odot$) WD, $0.05 M_\odot$ of which is a surface helium layer. For the lower mass we show two curves with low (solid line, cool) and high (dashed line, hot) donor specific entropy (from Deloye et al. 2005). There are four epochs: (1) (not shown) rapid accretion with helium shell flashes (2) T_c rises at early times when $\dot{M}$ becomes low enough that shell flashes are suppressed and the accreted helium layer can build up, (3) $\dot{M}$ falls enough that $T_{c,eq} = T_c$ and core heating ceases, (4) $\dot{M}$ becomes low enough that the cooling of the core dominates the outgoing luminosity. (Reproduced with permission from Bildsten et al. (2006), their Figure 1. © 2006. The American Astronomical Society. All rights reserved.)

$$\tau_{\mathrm{cond}} = 0.25 \left[\int_r^R \left(\frac{c_V}{\rho K} \right)^{0.5} \rho \, dr \right]^2$$

where r is the radius of the core–envelope interface, R is the total radius of the accreting WD, c_V is the specific heat at constant volume, ρ is the density and K is the thermal conductivity. As pointed out by Bildsten et al. (2006), the re-heating of the core of a helium-accreting AM CVn WD primary differs substantially from the re-heating of accreting H-rich degenerates with a C–O cores. During this phase, the WD core continues to re-heat as $\dot{M}$ continues its inexorable decline. The transition region between the core and envelope has a temperature, T_{tr}, which initially is hotter than T_{core} due to compression. However, gradually, T_{core} increases until it reaches essentially the same temperature as T_{tr}.

(3) The third stage represents a transition period when the core temperature T_c is gradually approaching an equilibrium value set by $\dot{M}$. This equilibrium state occurs when the radiated surface luminosity is balanced by the release of internal energy from below. This stage has a duration of $3 \times 10^6 - 10^8$ yr. As the time-averaged accretion rate continues declining, the compressional heating luminosity becomes lower than the thermal cooling luminosity emanating from the core. At this point, the evolution of the accreting WD proceeds independently of $\dot{M}$ and it cools like an isolated WD. Bildsten et al. (2006) estimated that the accretion rate at which this decoupling occurs is $1 - 3 \times 10^{-10} M_\odot$ yr.

(4) The last stage ($> 10^8$ yr) for the helium-accreting WD, measured from the time since decoupling from $\dot{M}$ is simply thermal cooling much like an isolated DB degenerate.

6.8 Conclusion

In the years ahead, the currently known sample of 56 AM CVn systems is expected to increase multifold from ever deeper ground-based surveys. The progress that has been made up to now has been truly remarkable given that as recently as 2005, less than two dozen helium-accreting CVs were known. A major leap forward will occur from an approved HST large GO project which includes 14 AM CVn systems with FUV spectra down to the Lyman Limit. Currently, the HST archive has FUV spectra for 8 AM CVn systems. When the HST Large GO COS observations commence, the 14 AM CVn systems combined with the 8 systems in the archive will comprise an FUV spectroscopic sample size of 22 systems. A full spectroscopic analysis of this sample will deliver accreting WD effective temperatures together with the scale factors of the best-fitting model atmospheres and accurate Gaia distances from which WD radii will be known. Using a WD mass–radius law, robust WD masses will be determined. A large body of photospheric chemical abundances and rotational velocities, Vsini, will be compiled for a sample size that is statistically significant. The WD surface abundances will potentially shed light on the formation of AM CVn systems. There is no question that existing theories and theoretical models of AM CVn systems will be subjected to ever more stringent tests.

References

Abbott, T. M. C., Robinson, E. L., Hill, G. J., & Haswell, C. A. 1992, ApJ, 399, 680

Bildsten, L., Townsley, D. M., Deloye, C. J., & Nelemans, G. 2006, ApJ, 640, 466

Bakowska, K., Marsh, T. R., Steeghs, D., Nelemans, G., & Groot, P. J. 2021, A&A, 645, 114

Campbell, H. C., Marsh, T. R., Fraser, M., et al. 2015, MNRAS, 452, 1060

Cannizzo, J. K., & Nelemans, G. 2015, ApJ, 803, 19

Chieffi, A., & Straniero, O. 1989, ApJS, 71, 47

Deloye, C. J., & Bildsten, L. 2003, ApJ, 598, 1217

Deloye, C. J., Bildsten, L., & Nelemans, G. 2005, ApJ, 624, 934

Deloye, C. J., Taam, R. E., Winisdoerffer, C., & Chabrier, G. 2007, MNRAS, 381, 525

Faulkner, J., Flannery, B., & Warner, B. 1972, ApJ, 175, L79

Fujimoto, M. Y., & Sujimoto, D. 1982, ApJ, 257, 291

Fujimoto, M. Y. 1982a, ApJ, 257, 752

Fujimoto, M. Y. 1982b, ApJ, 257, 767

Gänsicke, B T, Szkody, P, de Martino, D, et al. 2003, ApJ, 594, 443

Giannone, P., & Weigert, A. 1967, ZA, 67, 41G

Godon, P., & Sion, E. M. 2021, ApJ, 908, 173

Greenstein, J. L., & Mathews, M. 1957, ApJ, 126, 14

Groot, P. J., Nelemans, G., Steeghs, D., & Marsh, T. R. 2001, ApJ, 558, 123

Henyey, L., & L'Ecuyer, J. 1969, ApJ, 156, 549

Iben, I. 1982, ApJ, 259, 244

Iben, I., & Tutukov, S. 1991, ApJ, 370, 615

Kippenhahn, R., & Weigert, A. 1967, ZA, 65, 251

Kippenhahn, R., Kohl, K., & Weigert, A. 1967, ZA, 66, 58

Kotko, I., Lasota, J.-P., & Dubus, G. 2012, A&A, 544, A13

Krzeminski, W. 1962, PASP, 74, 66

Krzeminski, W., & Kraft, R. P. 1964, ApJ, 140, 921

Kupfer, T., Steeghs, D., Groot, P.J., et al. 2016, MNRAS, 457, 1828

Lambert, D. L., & Slovak, M. H. 1981, PASP, 93, 477

Levitan, D., Fulton, B. J., Groot, P. J., et al. 2011, ApJ, 739, 68

Levitan, D., Groot, P.J., Prince, T.A., et al. 2015, MNRAS, 446, 391

Limongi, M., & Tornambe, A. 1991, ApJ, 371, 317

Linnell, A. P., & Hubeny, I. 1996, ApJ, 471, 958

Lubow, S. H., & Shu, F. H. 1975, ApJ, 198, 383

Lubow, S. H., & Shu, F. H. 1976, ApJ, 207, L53

Marsh, T. R., Wood, J. H., Horne, K., & Lambert, D. 1995, MNRAS, 274, 452

McAllister, M., Littlefair, S. P., Parsons, S. G., et al. 2019, MNRAS, 486, 5535

Nasser, M. R., Solheim, J.-E., & Semionoff, D. A. 2001, A&A, 373, 222

Nelemans, G., Steeghs, D., & Groot, P. J. 2001, MNRAS, 326, 621

Nelemans, G., Yungelson, L. R., van der Sluys, M. V., & Tout, C. A. 2010, MNRAS, 401, 1347

Nyamai, M. M., Chomiuk, L., Ribeiro, V. A. R. M., et al. 2021, MNRAS, 501, 1394

Ostriker, J., & Hesser, J. 1968, ApJ, 153, L151

Paczynski, B. 1967, AcA, 17, 287

Pala, A. F., Gänsicke, B. T., Belloni, D., et al. 2022, MNRAS, 510, 6110

Paczynski, B., & Zytkow, A. N. 1978, ApJ, 222, 604

Podsiadlowski, Ph., Han, Z., & Rappaport, S. 2003, MNRAS, 340, 1214

Ramsay, G., Green, M. J., Marsh, T. R., et al. 2018, A&A, 620, 141

Roelofs, G. H. A., Groot, P. J., Nelemans, G., Marsh, T. R., & Steeghs, D. 2007, MNRAS, 379, 176

Ruiz, M. T., Rojo, P. M., Garay, G., & Maza, J. 2001, ApJ, 552, 679

Savoury, C. D. J., Littlefair, S. P., Marsh, T. R., et al. 2012, MNRAS, 422, 469

Sion, E. M., Acierno, M. J., & Tomczyk, S. 1979, ApJ, 230, 469

Sion, E. M., Solheim, J.-E., Szkody, P., Gaensicke, B., & Howell, S. B. 2006, ApJ, 636, L125

Sion, E. M., Linnell, A. P., Godon, P., & Ballouz, R.-L. 2011, ApJ, 741, 63

Smak, J. 1967, AcA, 17, 255

Smak, J. 1983, AcA, 33, 333

Solheim, J.-E. 2010, PASP, 122, 1133

Townsley, D., & Bildsten, L. 2004, ApJ, 600, 390

Tsugawa, M., & Osaki, Y. 1997, PASJ, 49, 75

Tutukov, S., & Yungelson, L. 1979, in IAU Symp. 83, 401

van Roestel, J., Kupfer, T., Green, M. J., et al. 2022, MNRAS, 512, 5440

Vila, S. C. 1971, ApJ, 168, 217

Wade, R. A., Eracleous, M., & Flohic, H. 2007, AJ, 134, 1740

Warner, B. 1995, Cataclysmic Variables (Cambridge: Cambridge Univ. Press)

Zorotovic, M., Schreiber, M., & Gänsicke, B. T. 2011, A&A, 536, 42

Accreting White Dwarfs
From exoplanetary probes to classical novae and Type 1a supernovae
Edward M Sion

Chapter 7

The Accreting White Dwarfs in Cataclysmic Variables

7.1 Introduction to Accreting White Dwarfs in Cataclysmic Variables

In Chapter 1, the structure, physical properties, and evolution of non-accreting isolated white dwarf (WD) stars, without any interactions with nearby stellar neighbors or close companions, undergo a simple cooling evolution along a line of constant radius, gradually leaking the thermal energy of the ions in their electron-degenerate cores into space while the atomic (and eventually molecular) opacity of their outer, non-degenerate, surface layers, only 60–100 km thick, regulates their rate of heat loss. This cooling rate, which also depends on core mass, determines the timescale of cooling evolution. This simple picture is complicated by a myriad of ongoing physical processes briefly discussed in Chapter 1, like gravitational and thermal diffusion, weak winds, radiative levitation, convective mixing, and very low rates of accretion from either debris disks or interstellar matter, all of which depend to some degree on the changing temperature regimes through which the WDs cool.

In stark contrast to the non-explosive, cooling evolution of isolated white dwarfs, the accreting WDs in cataclysmic binaries, whether non-magnetic and accreting via a disk or magnetic and accreting via an accretion column (polars) or accretion curtain, (intermediate polars), are heated and spun up by the accretion of mass and angular momentum. They undergo episodes of thermonuclear burning (simmering) and thermonuclear explosions. They are continually being injected with metal-rich gas, and in the long term, expand if their core masses are eroded by repeated classical nova explosions or shrink if their core masses increase due to continual accretion build-up. They can also undergo core-reheating which can reset their cooling timescales. In all of these objects moreover, cooling evolution takes on a very different meaning since their evolution proceeds with an external heat source (the boundary layer or accretion column) always shining downward upon them while undergoing the interplay of accretion with the same physical processes like gravitational and thermal diffusion, radiative levitation, and convective mixing, as single WDs.

7.1.1 Historical Context

It would be remiss for any monograph on accreting WDs to not provide an historical perspective on when, and by whom, the primary components in cataclysmic binaries were first identified as being WD stars. By the 17th century, novae and supernovae were being discovered and cataloged, but the mechanisms of their explosions and the kind of star that exploded was unknown until the work of Mestel (1952), which attributed the classical nova explosion to the ignition of hydrogen in electron-degenerate layers. Not long afterward, the bright dwarf novae, U Gem, SS Cygni, and AE Aquarii (originally classified a dwarf nova) were being extensively observed but the underlying causation of their variable optical brightness states was unknown. As reported in Warner (1995), statistical studies by Luyten & Hughes (1965) and Kraft & Luyten (1965) noted that post-novae have absolute magnitudes that cluster around $M_v \sim 4$ while dwarf novae in quiescence have absolute magnitudes that tend to cluster around $M_v \sim +7.5$, from which they reasoned that the hot primary components of CVs must be high gravity stars, either WDs or hot subdwarfs. Several years earlier however, the first mention that the hot blue primary components of dwarf novae were accreting WDs was in a classic paper by Crawford & Kraft (1956) on AE Aquarii.

By the 1950s, the basic structure of a CV was widely accepted: a compact binary (with an orbital period $\lesssim 1$ day) in which the primary, a WD star, accretes matter and angular momentum from the secondary star, a main sequence-like object, filling its Roche lobe. In non-magnetic systems, the matter is transferred, at continuous or sporadic rates, by means of an accretion disk around the WD. In magnetic systems, the WD magnetic field disrupts the formation of a disk and channels accreting gas onto the magnetic poles of the WD while in intermediate polars, the field strengths tend to be lower than in polars and an outer disk can form but its inner edge is truncated by magnetic pressure.

The long-term evolution of CVs is driven by the rate of the binary angular momentum loss (AML), which itself is a direct consequence of the mass transfer from the secondary to the WD. CV systems are believed to evolve from long binary period ($\sim$a day) to short period (fraction of an hour). At long orbital periods, AML is mostly driven by magnetic stellar winds of the donor star. As the donor star is eroded away by mass transfer above orbital periods of 3 hours, it eventually becomes fully convective. This jostles/disrupts the foot points of the magnetic field, shutting off the magnetic stellar wind, thus reducing AML. Mass transfer essentially stops, as the donor is detached from its Roche lobe and the system has entered the so-called CV period gap between orbital periods of 2 and 3 hours where very few CVs are found. However, now driven mainly by gravitational wave emission, AML continues to reduce the binary orbital period. At orbital periods ~ 2 hours, the secondary resumes contact with its Roche lobe and the CV once again undergoes mass transfer, but now below the period gap.

In order to fully understand the challenges of observing an exposed accreting WD in a CV, it is helpful to provide a brief introductory overview of CV evolution, the various subtypes of CVs, and the optimal times for each type of CV, that maximizes the possibility of spectroscopic detection of the underlying WD photosphere.

7.1.2 Overview of Cataclysmic Variable Outbursts, Evolution and CV Subtypes

The accreting WDs exhibit a variety of observed outbursts, of different amplitudes and durations. The dwarf novae (DNe, a type of CVs) release gravitational energy when a thermal-viscous instability in the accretion disk around the WD leads to rapid accretion at a high rate (the dwarf nova *outburst*, lasting days to weeks) until the disk has largely emptied and the system returns to *quiescence*, at which time the build-up of disk gas begins again (lasting weeks to months). If the accretion rate is high enough to nearly fully ionize the disk, then the disk instability is suppressed and stable accretion keeps the system in a *high state* or the high state is interrupted by rare and unpredictable *low states* of relatively short duration. In low optical brightness states of a CV (either low states of NLs or during the quiescence of dwarf novae), the WD is revealed in the ultraviolet (UV), as its emission greatly outshines the disk which is in a state of low mass accretion. At other times, DNe in outburst and NLs in high states are dominated in the UV by luminous accretion disk outshining the WD, and typically driving hot, fast wind outflow.

When the accreted layer on the WD reaches a critical pressure at its base (this happens on average every few thousand years), explosive thermonuclear runaway (TNR) shell burning is triggered: the (classical) nova explosion. If however disk accretion occurs at a very high rate (close to the Eddington Limit), novae explosions may be avoided if steady, stable shell burning occurs in equilibrium with the rate of accretion, because of the build-up of a helium buffer zone that prevents the all important CNO enhancements at the base of the accreted envelope needed to trigger a nova explosion. Steady H-burning in equilibrium with accretion is going on in the case of the supersoft X-ray binaries. Finally, if a WD in a CV is born massive ($M_{wd} > 0.8 M_\odot$) and grows in mass despite hundreds to thousands of nova explosions, then instantaneous collapse and thermonuclear detonation will occur if the CV WD with a C–O core reaches the Chandrasekhar limit: the Type Ia supernovae: SNe Ia.

The observed behavior of the different subtypes of CVs offers widely different possibilities for detecting the underlying, accreting WD, depending upon the level of optical brightness of each subtype. Radiation from the accretion disk or accretion column dominates the optical and far-UV (FUV) bandpasses during high optical brightness states and the WD is exposed only when the optical brightness is the lowest. Thus, the photometric behavior of each subtype must be examined in order to elucidate those times at which the white dwarfs' photospheric radiation is dominant. The optimal times for WD detection is described below for each CV subtype.

Z Camelopardalis systems are dwarf novae found above the CV period gap that undergo dwarf nova outbursts of typically two to four magnitudes but unpredictably enter a "standstill" state of intermediate brightness a few tenths of a magnitude below their brightness in outburst between outburst and quiescence, during which they exhibit very little variation in brightness. A handful of Z Cam stars reveal nova shells from past classical novae (Shara et al. 2017) including the prototype itself, Z Camelopardalis and AT Cancri. The light curve of Z Cam is displayed in Figure 7.1. During quiescence the WD is at least partially exposed and dominates the UV light. In outbursts and standstills, the disk's accretion light dominates. During a quiescence of a Z Cam system, the underlying accreting WD may be exposed, which offers the possibility of deriving its physical properties.

U Geminorum/SS Cygni dwarf novae are found above the period gap with orbital periods of 3–10 hours; they exhibit normal outbursts lasting a few days to a week with some normal outbursts longer in duration than others. The increase in brightness at outburst typically is 2–5 magnitudes. The observed recurrence time between outbursts ranges between a week to a few months. (see Figure 7.2: SS Cygni light curve). Since the donors above the gap tend to be: (1) more massive than the donor stars below the gap; (2) their rate of mass transfer is larger above the gap and; (3) the accretion disks are brighter and more spatially extended, the probability of unambiguously detecting the WD during quiescence is lower. The accretion disks in many of these systems are still contributing light in addition to the light of the WD during dwarf nova quiescence. This must be disentangled. CVs like U Gem, SS Aur, and TT Crt are notable exceptions where the contribution of accretion light is very small and the WDs are fully exposed.

SU Ursa Majoris dwarf novae are all found below the period gap, and exhibit two distinct types of outbursts, short (normal) outbursts and superoutbursts (for a review, see Warner 1995). The outbursts tend to be a bit brighter than normal outbursts and have durations of one to two weeks (Figure 7.3).

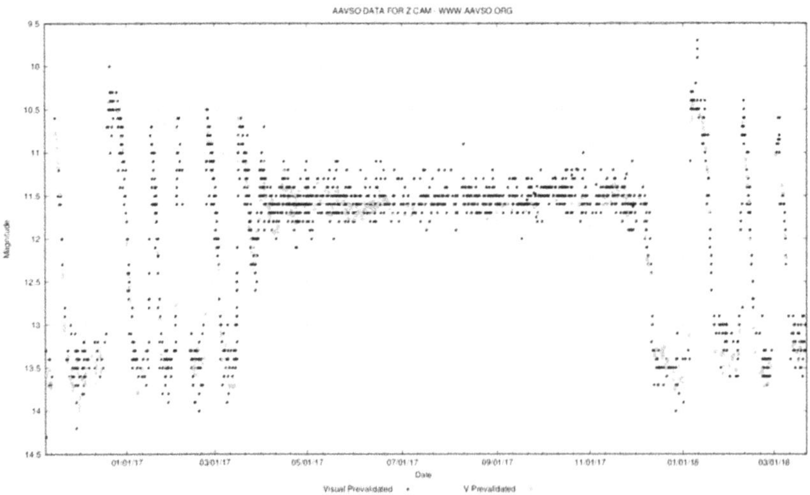

Figure 7.1. The AAVSO light curve of Z Camelopardalis itself. The only chance of unveiling the white dwarf is during the deep quiescences. During the outbursts and standstills the underlying white dwarf is not detectable unambiguously. (Reproduced with permission of AAVSO. https://www.aavso.org.)

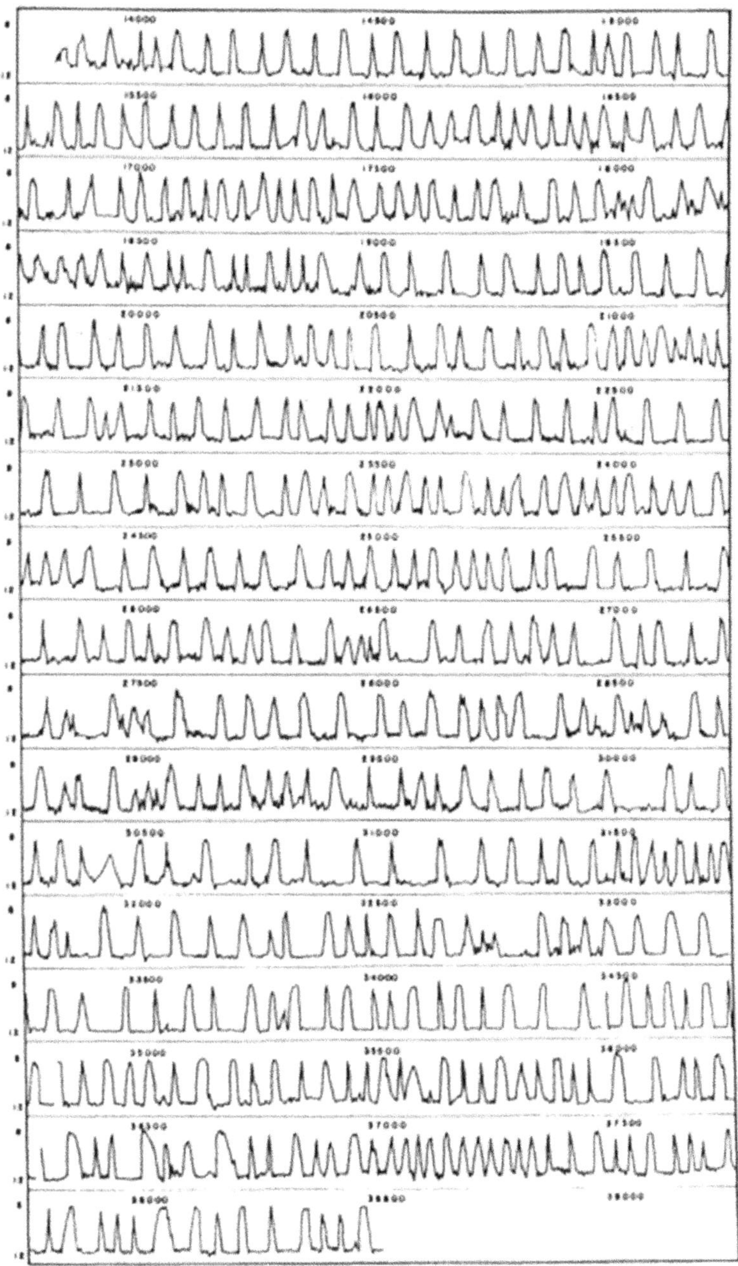

Figure 7.2. A long-term AAVSO light curve of the dwarf nova SS Cygni. (Reproduced with permission of AAVSO. https://www.aavso.org.)

UX Ursa Majoris-type Nova-like variables resemble classical novae just before the classical nova explosion or after the classical nova explosion. They are sometimes referred to as being dwarf novae stuck in permanent outburst. They all lie above the period gap, and for a reason yet unknown, are mostly clustered together between orbital periods of

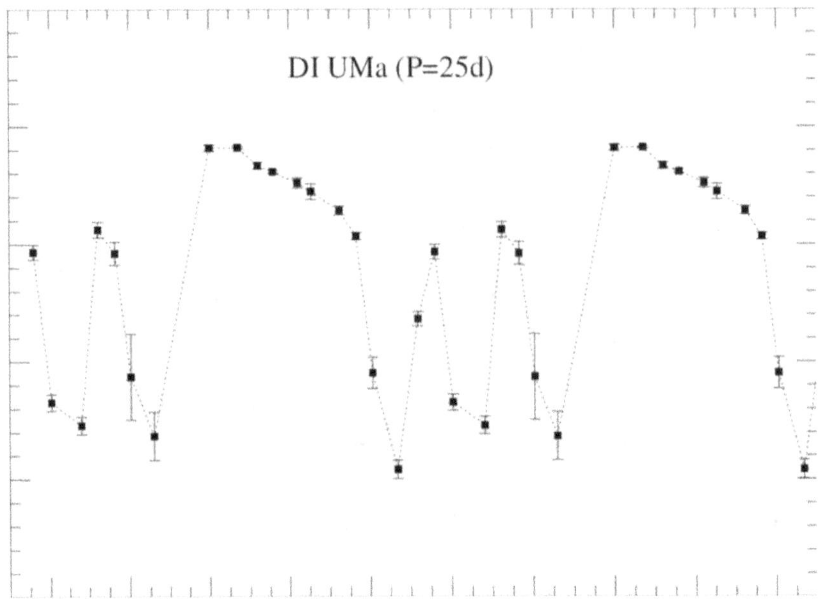

Figure 7.3. AAVSO light curve of an SU UMa type CV of the ER UMa subtype. Note the closely spaced normal outbursts and superoutbursts. The quiescences of this CV are so short that the chances of observing the accreting WD is remote. However, in the vast majority of SU UMa systems, the quiescences between outbursts and superoutbursts are much longer and the WD is essentially fully exposed for study. (Reproduced with permission of AAVSO. https://www.aavso.org.)

3 to 4 hours. They all have bright accretion disks that hardly vary in brightness but all show strong flickering, the signature of accretion onto a WD. The bright accretion disk always dominates the UV light in the system and the WD is hidden.

In Figure 7.3, the light curve of an ER UMa subtype of SU UMa-type dwarf novae, DI UMa, is displayed. These systems are characterized by having the most frequent outbursts and the shortest quiescent intervals between outbursts and superoutbursts. Thus the accretion disk continually dominates their light thus making it improbable that their underlying accreting white dwarfs can be exposed to spectroscopic analysis.

VY Sculptoris-type nova-like variables have the same spectral and photometric properties as the UX UMa stars but unpredictably enter deep low states of brightness (several magnitudes) in which accretion seems to greatly decline or even stop, the accretion disk becomes fainter or even disappears and the underlying accreting WD is exposed and very hot, at least in part, from prolonged accretion heating during the previous high state. In Figure 7.4, a sequence of IUE spectra reveal the spectral variations as the VY Scl nova-like descends downward in optical brightness from a high state (top spectrum) where the accretion disk and wind outflow dominate the UV spectrum to a low state (bottom-most spectrum) when the accretion disk has essentially faded or is absent and the bare, underlying hot white dwarf dominates the radiation in the UV (Hamilton & Sion 2006).

Magnetic Cataclysmic Variables—Polars and the intermediate polars (IPs) undergo magnetically controlled accretion but in polars the fields are so strong that an

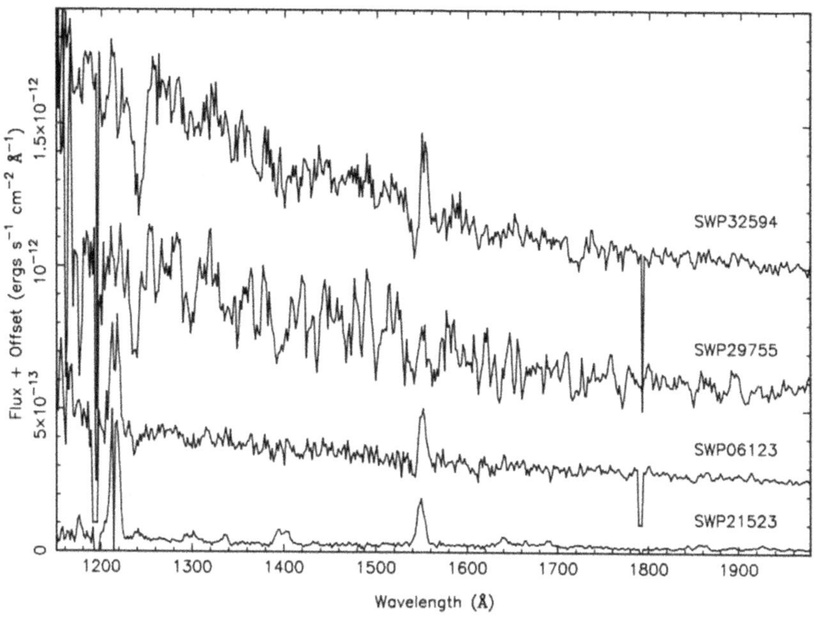

Figure 7.4. Four archival IUE spectra of VY Scl (Hamilton & Sion 2006) itself illustrating the different flux between a high state, intermediate declines and a deep low state when the underlying WD is nearly fully exposed for modeling and analysis. (Reproduced with permission from Hamilton & Sion (2008). © 2008. The Astronomical Society of the Pacific. All rights reserved.)

accretion disk cannot form because the magnetic pressure exceeds the ram pressure of gas flowing from the donor via L_1 (see Figure 7.6). The polars like AM Her itself have low states of optical brightness where accretion shuts off and the magnetic WD can be observed directly if the accretion column (usually a primary accretion cap and a secondary accretion cap) has cooled or gone away (see Figure 7.5). In contra-distinction to the polars, which have high magnetic field strengths in the range $B12$ to 10^9 Gauss, the IPs have lower field strengths of $B < 10^7$ Gauss in most cases and IPs do not have deep low states like many of the polars. Some polars appear to be in "permanent" low states like VV Puppis so the WD can be directly observed.

In polars, the donor gas is focused through the L_1 nozzle and quickly channeled along the magnetic field lines of force toward the white dwarf's magnetic poles at supersonic velocity, forming a standoff shock through which the gas temperature steeply rises and the infalling gas decelerates to roughly one-third of its pre-shock velocity (Aizu 1973). The post-shock gas is extremely hot (10 to 50 keV). This post-shock region cools down by two mechanisms, thermal bremsstrahlung and cyclotron radiation. Which of these two cooling mechanisms is dominant depends upon the magnetic field strength and the rate at which accretion is occurring. Thermal bremsstrahlung (mostly hard X-rays) by electrons in the Coulomb fields of ions emit a photon and cyclotron radiation, which is emitted in the near-infrared and optical domains. Which of these two mechanisms dominates the cooling is highly dependent on the field strength and the rate of accretion. In Figure 7.6, the accretion column,

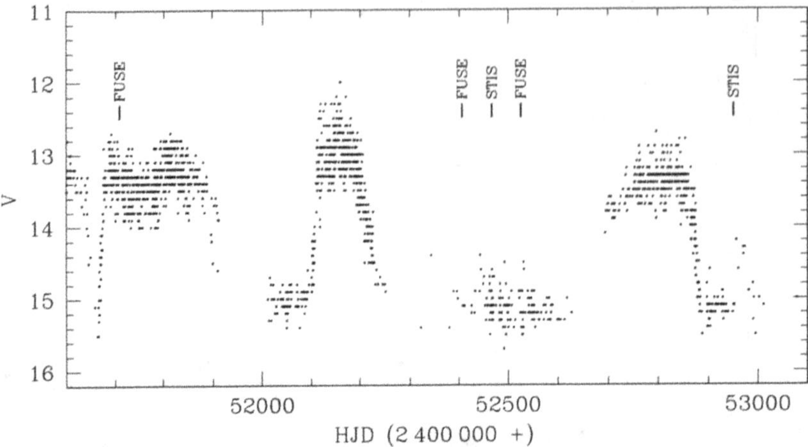

Figure 7.5. A long-term AAVSO light curve showing high states and low states of the polar AM Her. Note the placement of FUSE and HST STIS observations during a low state when accretion rate has declined or stopped and the WD photosphere is exposed (Gänsicke et al. 2006). Paradoxically, the accretion cap does not cool during a low state which likely implies a deeper penetration of heat into the envelope and thus a longer thermal timescale. (Reproduced with permission of AAVSO. https://www.aavso.org.)

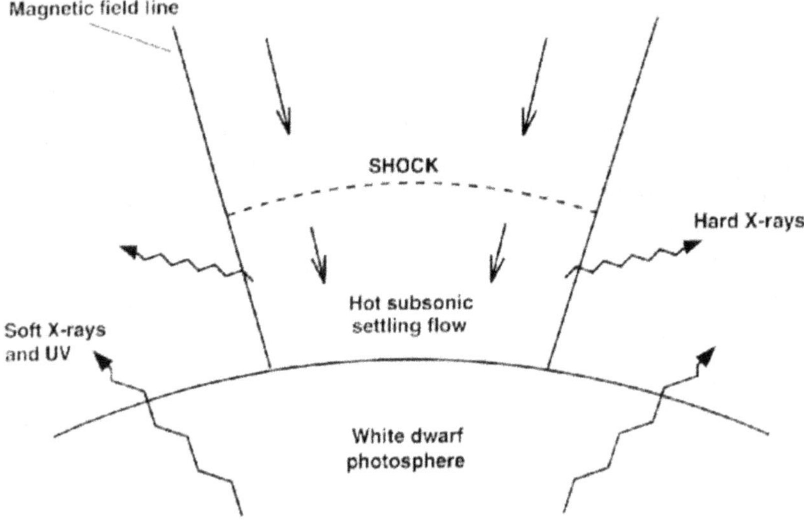

Figure 7.6. The accretion column, shock and post-shock settling flow are depicted as well as the hard and soft X-rays and UV emitted outwardly from the WD photosphere and accretion column. (Reproduced with permission from Sion (1999). © 1999. The Astronomical Society of the Pacific. All rights reserved.)

shock and post-shock settling flow are depicted as well as the hard and soft X-rays and UV emitted outwardly from the white dwarf photosphere and accretion column.

The IPs with their weaker magnetic fields can prevent the formation of an inner accretion disk but are insufficiently strong to prevent the formation of an outer disk.

In other words, the accreting white dwarfs in IPs are surrounded by a magnetically truncated accretion disk in which the gas is channeled by the field to thread itself along magnetic line of force onto the IP white dwarf at the magnetospheric radius, R_{m}, and the gas accretes in the form of accretion curtains (see Figure 7.7). Another distinguishing characteristic between polars and IPs is that the polars are synchronized with the spin period, P_{spin}, of the magnetic WD being the same as the orbital period of the system, P_{orb}. On the other hand, the IPs are asynchronous, typically with $P_{\mathrm{spin}} < P_{\mathrm{orb}}$ (Warner 1995; Mukai 2017). There are a relatively small number of polars in which the P_{spin} and P_{orb} are slightly different by typically only 1%; these systems are known as asynchronous polars (Mukai 2017).

While some polars enter low optical brightness states where the radiation associated with the accretion column and heated polar accretion caps weakens or disappears, allowing the WD photospheric radiation to be analyzed, the intermediate polars do not enter deep low states so the underlying accreting WD is not exposed because it is partially hidden by the emission from accretion curtains and the hot accretion caps. In rare circumstances, a favorable combination of the maximum versus minimum spin phase of the magnetic white dwarf and the orbital phase may allow the detection of the magnetic degenerate. For EX Hya (Belle et al. 2003) detected the magnetic white dwarf which has a surface temperature of

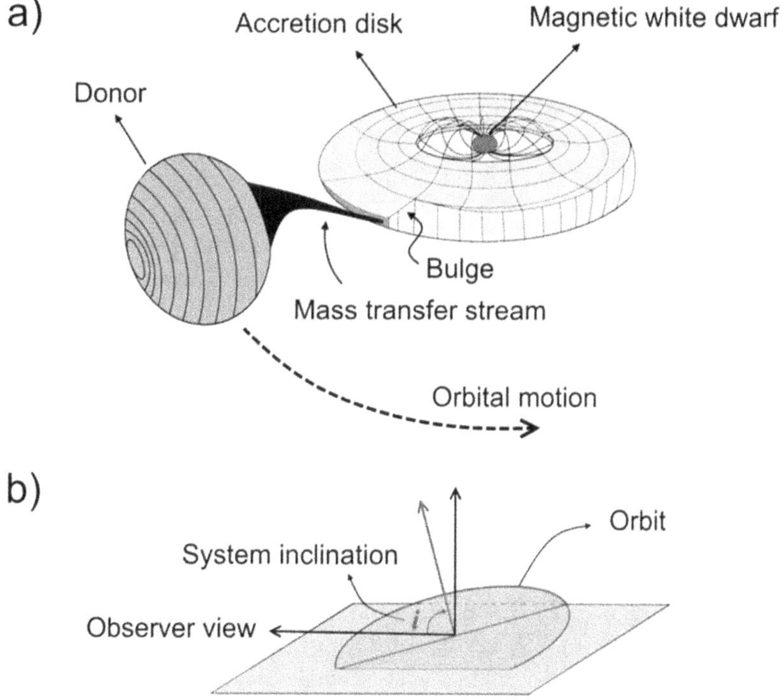

Figure 7.7. Depiction of an intermediate polar surrounded by a magnetically truncated accretion disk in which the gas is channeled by the field to thread itself along magnetic line of force onto the IP WD at the magnetospheric radius, R_{m}, accretes in the form of accretion curtains. (Reproduced with permission from Ghaderpour & Ghaderpour (2020). © 2020. The Astronomical Society of the Pacific. All rights reserved.)

23, 000 K $\pm$ 1000 K. A second intermediate polar with an exposed white dwarf is the system V709 Cas. The WD is a magnetic DA with a surface temperature of ~23,000 K, log g ~8 (Bonnet-Bidaud et al. 2001) and a stronger magnetic field than EX Hya.

7.1.3 Spectroscopic Detection of the Underlying CV White Dwarf

Since the WDs and accretion disks have their Planckian peaks in the FUV, they dominate the UV and the donor star is too cool to contribute. Therefore, the detection of the WDs in CVs had to await the advent of orbiting observatories: HST FOS, GHRS, STIS and COS spectroscopy together with FUSE FUV spectroscopy, IUE FUV spectroscopy, X-ray and EUV spectroscopy using HUT, Einstein, Orfeus, Chandra, XMM-Newton, EUVE, EXOSAT, ROSAT, and ASCA have led to a windfall in our knowledge of the underlying WDs in CVs and how they are affected by the accretion process. These space observatories have made it possible to detect numerous underlying WD accretors, the boundary layer between the accretion disk and the WD surface, the accretion disk itself and wind outflow in the wavelength domains where they emit most of their energy (between ~3 and 2000 Å). Thus, it is possible to determine many poorly known basic physical properties of these accreting degenerates both above and below the CV period gap. Among the newly determined physical properties to be sought are the all important accreting white dwarf masses, surface temperatures T_{eff}, mass accretion rates, gravitational redshift masses, gravity (log g), rotational velocities (Vsini), chemical abundances, the accretion energy budget, and how accretion and thermonuclear runaways can drastically alter the structure, evolution, and atmospheric chemistry of the accreting WD over time. Since the brightened accretion disk (with wind outflow) and the bright accretion columns dominate the FUV and optical spectra during high accretion states (dwarf nova outbursts and nova-like and Polar high states), the underlying white dwarf photospheres are exposed only during the quiescent state between dwarf nova outbursts, or during the low optical brightness states of nova-like variables, and the low accretion states of Polars when the emission lines due to cyclotron and thermal bremsstrahlung processes become very weak or vanish.

The earliest, unambiguous spectroscopic detections of the underlying accreting WDs in non-magnetic cataclysmic variables during the low optical brightness states of nova-like variables and the quiescence of dwarf novae were reported by Shafter for MV Lyrae during a very deep low state when they identified the accreting white dwarf with a surface temperature $\approx 50, 000$ K based upon IUE spectra. The underlying WDs in the two of the most famous dwarf novae, U Gem and VW Hydri, were detected with IUE spectra by Panek & Holm (1984) for U Gem and Mateo & Szkody (1984) for VW Hydri. In both of these objects, the Lyα profile and numerous absorption lines of metals were clearly detected in their photospheres during quiescence. Shafter et al. (1985) identified the underlying WD during the low state of another VY Sculptoris nova-like variable, TT Ari, based upon optical spectra that revealed the broad wings of the Balmer lines in the optical spectra were recognized as being the Stark-broadened wings of absorption features formed in the underlying white dwarf's photosphere. The optical spectrum of an accreting WD, in this case TT Ari during a deep low state, is displayed in Figure 7.8.

Among the magnetic CVs, the first reported spectroscopic detection of the underlying white dwarf in a polar was presented by Szkody et al. (1982), during a

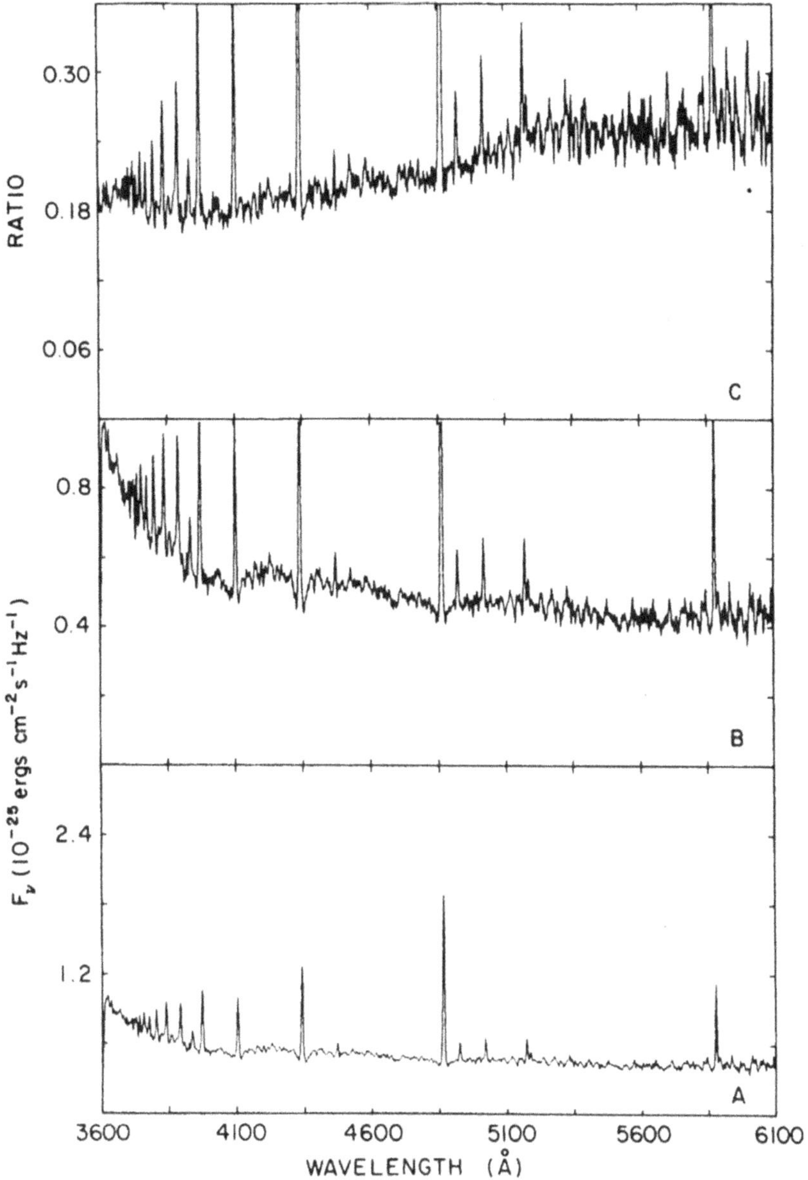

Figure 7.8. The VY Scl-type nova-like variable, TT Ari, captured during a deep low state of very low optical brightness revealing its underlying hot WD seen as pressure-broadened Balmer absorption line wings of the WD flanking central emission due to the accretion disk. (Reproduced with permission from Shafter et al. (1985). © 1985. The American Astronomical Society. All rights reserved.)

low state of AM Her. This finding was confirmed by Chiappetti et al. (1982) again with IUE spectra during a low state although the low state they observed was not as deep. Heise & Verbunt (1988) and Gänsicke et al. (1995) carried out UV spectroscopy of AM Herculis during its low optical brightness state.

7.2 CV White Dwarf Masses

Knowing the mass of an accreting WD in a CV is its most fundamental parameter and thus is of paramount importance. The accreting WDs are no less than the central engines of the observed outbursts, either as potential wells for the release of gravitational energy during accretion (dwarf novae, polars), or as the sites of explosive thermonuclear runaway (TNR) shell burning (classical novae), steady shell burning (supersoft X-ray binaries) or instantaneous collapse and total thermonuclear detonation if a WD reaches the Chandrasekhar limit (Type Ia supernovae SNe Ia). The cosmological importance of the WD supernovae (Type Ia supernovae) led to renewed efforts to determine the most precise masses by seeking the answer to a very basic question: can the mass of a CV WD grow with time despite the occurrence of hundreds to thousands of classical nova explosions during the lifetime of the compact binary? This question remains open. CV WDs are more massive (0.83 ± 0.03 Msun; Zorotovic et al. 2011) on average than their isolated WD counterparts (0.6 Msun), which itself remains poorly understood. If an accreting WD, born with a mass 0.8 Msun (or larger) with a carbon–oxygen core grows in mass due to accretion from the Roche lobe-filling donor, then it may reach the Chandrasekhar limit during its lifetime as a CV and undergo a Type Ia supernova explosion (SNe Ia), the standard candle providing the evidence that the expansion of the universe is accelerating, implying the presence of dark energy. Even a sub-Chandrasekhar white dwarf in a CV could undergo an SNe Ia event via a double detonation when the thick helium layer accumulating from many H shell flashes ignites, leading to the detonation of the underlying carbon–oxygen core. Both of these scenarios, along with the double degenerate pathway, will result in a SNe Ia. Therefore, the determination of precise white dwarf masses will help make it possible to constrain the single degenerate pathway and hence potentially the double degenerate pathway.

Since it is now reasonably well established that the CVs above the period gap evolve into the CVs lying below the period gap (Knigge et al. 2011), it is critical to determine the average CV WD mass above and below the CV period gap for comparison with mass distributions from theoretical evolutionary binary population synthesis models and the different angular momentum braking laws which drive mass transfer in CVs. Even the possibility of average WD mass differences between the different subclasses of CVs must be explored (e.g., U Gem/SS Cygni dwarf novae, Z Camelopardalis dwarf novae, SU UMa dwarf novae, classical (post-) novae, VY Scl/UX UMa Nova-like variables, ER UMa dwarf novae).

7.2.1 Radial Velocity Curves from Emission Line Velocities for White Dwarf Masses

CV WD masses can be determined by utilizing the emission lines detected and presumed to form in the hotter, inner accretion disk closest to the WD. By using relationships between the orbital period and the mass of the donor star, a WD mass is obtained and calibrated using double-lined spectroscopic binaries (Shafter 1983). The obstacles that arise in delivering a reliable WD mass by the radial velocity method are: (1) it is not clear that the emission line velocities manifest the true dynamical motion of the WD because velocities are determined from the different emission lines but these lines form in different parts of the accretion disk, (2) WD mass for a given system but derived by different investigators use different methods to determine the semi-amplitudes from the radial velocity curves and hence different velocity values for a given line feature. The WD masses determined from radial velocity curves generally lack the degree of precision and reliability needed for uncovering differences in CV WD masses above and below the period gap and allowing comparisons of different angular momentum braking laws.

7.2.2 The Photometric Eclipsing Binary CVs

This method delivers donor masses and radii leading to a precise value of the WD mass. As the donor eclipses the WD (primary eclipse), in very rapid succession (timescales of minutes) the WD and accretion disk and hotspot where the stream from L_1 impacts the outer disk rim are occulted, along with the gas stream trajectory known from the mass ratio. Examples of the eclipse model fits, adapted from McAllister et al. (2019) are displayed for 8 eclipsing CVs are displayed in Figure 7.9. The gray points represent the actual eclipse light curves, while the black points are the result of subtracting the Gaussian processes posterior mean (itself shown, A, by the red band covering the residuals below each plot). Also plotted are the separately color-coded components of the eclipse model: WD (purple), hotspot (red), accretion disk (yellow), and donor star (green).

From the high precision fast photometry of the primary eclipse, exact shape of the primary eclipse light curve, very precise masses are obtained from the eclipse structure (McAllister et al. 2019; Copperwheat et al. 2010; Savoury et al. 2011). Fortunately, this photometric method does not depend on detection of the light from the faint donor star, which is often undetectable and swamped by the radiation of the much more luminous accreting WD and accretion disk. What is critically needed for this method are high quality light curves of the eclipses which take place on timescales of minutes. The details of this method and the instrumental setups are described in Dhillon et al. (2007, 2014) and McAllister et al. (2019). The observations and analysis are carried out via collaborations of investigators at several universities in the United Kingdom. CV WD masses obtained by this method are considered the gold standard in the field of CVs. Only recently, masses for CV WDs have been derived with comparable precision by the spectroscopic method described below (e.g., Pala et al. 2022).

7.2.3 Gravitational Redshift Masses

If the UV spectrum reveals photospheric absorption lines of an exposed CV WD, then it is possible to measure a gravitational redshift. However, caution must be exercised to be certain the lines are forming in the photosphere. The white dwarf's

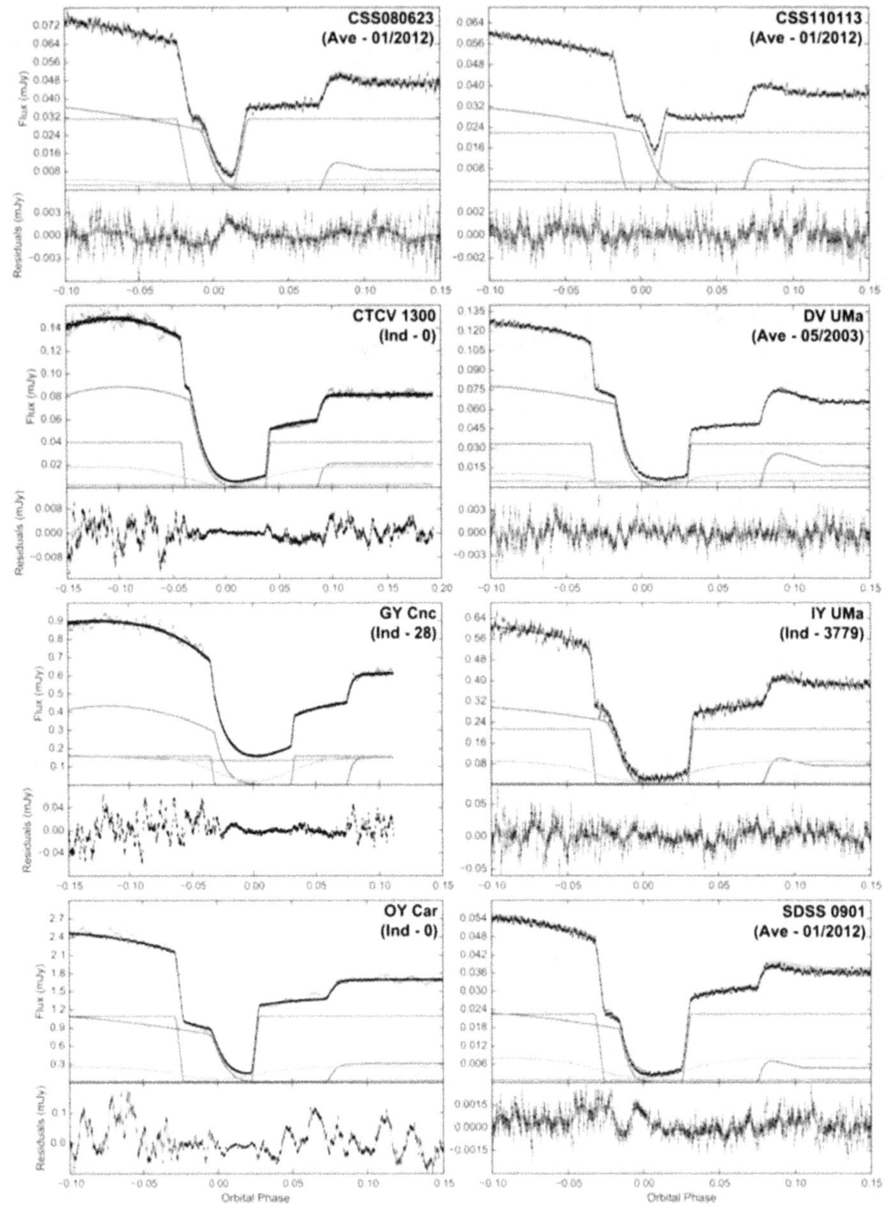

Figure 7.9. Eclipse model fits to g-band light curves of 15 CVs. The light curves are shown in gray points. The system name is displayed in the top-right corner of each plot, along with whether the eclipse is an individual (Ind) or average (Ave) eclipse. For individual eclipses the cycle number is shown, while for average eclipses the month and year of the eclipse is shown. (Reproduced with permission from McAllister et al. (2019). Copyright © 2019, Oxford University Press.)

lines should be in anti-phase with, for example, the optical emission line velocities of the donor star companion, which is typically a dwarf M star. A refined, accurate ephemeris is required. From the radial velocity measurements of the WD and the donor star, it is possible to infer the gravitational redshift mass of the WD. The systemic velocity of the system must also be known. The radial velocity curve of the measured WD radial velocities versus orbital phase is typically fitted with least squares sinusoidal functions of the form

$$v_{\text{obs}}(t) = \gamma_{\text{sys}} + \gamma_{\text{grav}} + K_{\text{wd}} \sin\left[2\pi(t - t_0)/P_{\text{orb}}\right], \tag{7.1}$$

P_{orb} is the orbital period of the binary, v_{abs} is the observed velocity of the photospheric absorption line, γ_{sys} is the systemic velocity, K_{wd} is the orbital velocity of the CV WD. CV WDs masses determined in this way have met with mixed success (e.g., Steeghs et al. 2007; Sion et al. 1994; Thorstensen et al. 1978; Vennes et al. 1991) due to a multitude of observational uncertainties (e.g., γ_{sys}).

7.2.4 Spectroscopic Modeling Methodology for CV White Dwarfs

The most reliable determinations of CV WD masses prior to Gaia came from the analysis of eclipse light curves of accreting WDs rather than from spectroscopic investigations based upon uncertain distances or based upon the use of disk emission line velocities as manifesting the true dynamical motion of the WD. Now however, it has become possible to determine CV WD masses from FUV spectroscopic observations of exposed WDs in CVs. This method has emerged as a means of determining precise CV WD masses (e.g., Pala et al. 2022; Godon & Sion 2021). If a best-fitting model photospheric spectral fit to the absorption lines and continuum of the exposed WD is achieved, then a degeneracy in the solution, namely T_{eff} is a function of $\log g$, must be removed. This is accomplished by scaling the model flux, F_{model}, to the observed flux distribution, F_{obs}. The normalization or "scale" factor, $S = 4\pi(R_{\text{wd}}^2/d^2)$.
  Thus,

$$F_{\text{obs}} = SF_{\text{model}} = 4\pi\left(R_{\text{wd}}^2/d^2\right)F_{\text{model}} \tag{7.2}$$

which yields S for the best-fitting model by using the Gaia (eDR3, DR3) parallax distance to solve for the WD radius R_{wd}. With R_{wd} known, one uses an appropriate non-zero temperature mass–radius relation, preferably from evolutionary models, to obtain the mass of the CV WD with an accuracy of ~5% (Pala et al. 2022; Godon & Sion 2021). Assuming that compressional heating alone determines the T_{eff} (Sion 1995; Townsley & Bildsten 2003), the accretion rate follows from

$$T_{\text{eff}} = 1.7 \times 10^4 \text{ K} \left(\langle \dot{M} \rangle/10^{-10}\right)^{0.25}[M_{\text{wd}}/0.9]. \tag{7.3}$$

Several dwarf novae, and a nova-like, out of many dozens with reliable CV WD masses delivered by the FUV spectroscopic modeling, are displayed in Figures 7.10, 7.11, 7.12, 7.13, and 7.14.

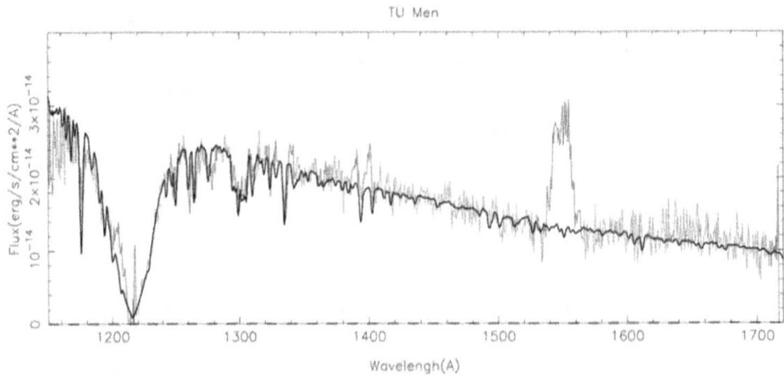

Figure 7.10. Modeling of the WD in TU Men. The HST STIS FUV spectrum of TU Men (in red) is fitted with a synthetic stellar spectrum (in black). The STIS spectrum has been dereddened assuming $E(B - V) = 0.08$. A first fit is carried out where the emission lines and absorption lines are masked to find the temperature and gravity; these regions are in blue. Once the best-fit temperature and gravity are found, the absorption lines are fitted to find the abundances and projected rotational velocity. The WD model has an effective surface gravity of log 8.25 ± 0.25 corresponding to a WD mass $M_{wd} = 0.77 + 0.13/ - 0.16$ by using the mass–radius grid for evolutionary models by Wood (1995), with an effective surface temperature of 27, 750 ± 1000 K. The model has a carbon to very high temperature abundance [C] = 0.2 ± 0.1, silicon abundance [Si] = 0.2 ± 0.1, and nitrogen abundance [N] = 20 ± 10, all in solar abundance units (Sun = 1). The projected stellar rotational velocity of the model is V_{rot}sini is 225 ± 75 km s^{-1}. (Reproduced with permission from Godon & Sion (2021). © 2021. The American Astronomical Society. All rights reserved.)

7.3 Masses of the Accreting Magnetic White Dwarfs in Polars and Intermediate Polars

There have been numerous attempts to determine reliable masses of the accreting magnetic WD primaries in polars and intermediate polars. From the time that Dexter et al. (1976) discovered the magnetism in AM Herculis (the "first" polar), the growth of the field both in the numbers of scientific papers per year and the numbers of known polars and intermediate polars has been virtually exponential. The reader is referred to an excellent review by Mukai (2017) for the exciting developments in X-ray emission studies of polars and IPs up to 2017.

A few years prior to the recognition of magnetic CVs as a subclass of CVs, a milestone seminal paper appeared by Aizu (1973) in which spherical accretion onto a WD, and the associated shock front from the radial accretion, as well as the structure and X-ray emission of a hot plasma was calculated. The gas that is flowing through the Lagrangian point L_1 has a ballistic trajectory until it is captured by the field at the magnetospheric boundary and threaded along the magnetic field lines with increasing gas density down to the white dwarf surface. The actual impact zones at the surface, i.e., the accretion caps, are at locations determined by the complex interaction between the channeled accreting gas and the magnetic field structure. This region will most probably be located where the ram pressure of the channeled flow and the magnetic pressure are the same. In a dipole field, there will typically be an upper accretion cap and lower accretion cap with the upper cap dominant.

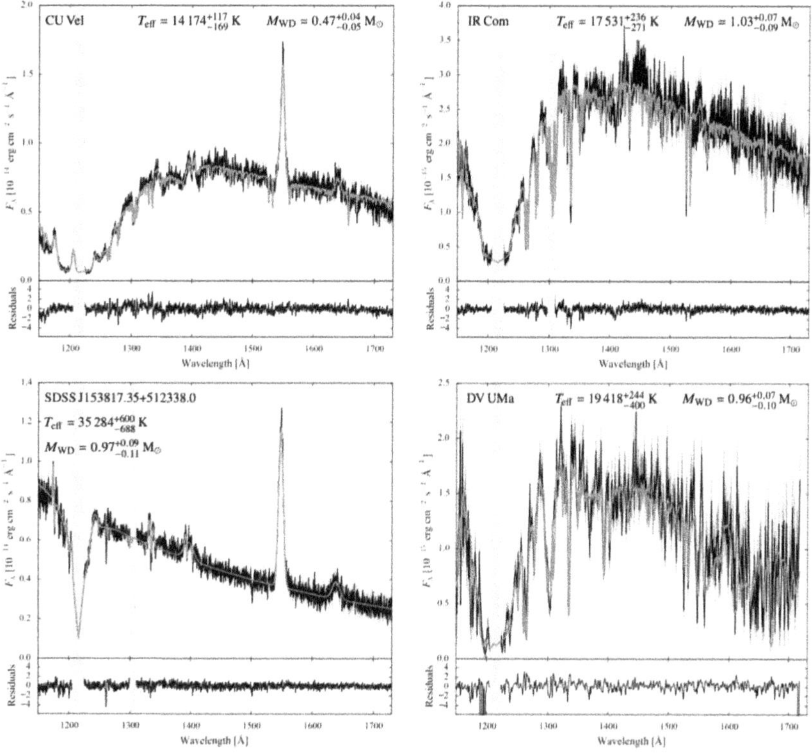

Figure 7.11. UV spectra (black) of sample CV WDs along with the best-fitting model (red), representative of cool (top-left panel), warm (top-right panel), hot (bottom-left panel), and eclipsing (bottom-right panel) systems. The best-fitting models are composed of the sum of a white dwarf synthetic atmosphere model, a second continuum emission component in the form of a blackbody, and the emission lines from the disk modeled with a Gaussian profile. In addition, the model for the eclipsing system DV UMa (bottom right) includes an absorption curtain component (see Section 3.2). The gray bands mask the geocoronal emission lines of Lyo (1216 Å…) and, whenever present, of O I (1302 Å…). (Reproduced with permission from Pala et al. (2022). Copyright © 2021, Oxford University Press.)

Aizu (1973) found that the temperature distribution, and the relative spectrum of thermal X-rays emitted by the hot plasma is determined only by the mass of the WD, while the density, the thickness of the hot plasma and luminosity depended on the accretion rate as well. Early attempts to determine the mass of the primary component in magnetic CVs utilized the radial velocity method (e.g., Mukai & Charles (1987)). Unfortunately, the resulting masses for a handful of systems were subject to uncertainties in the radial velocities of the very, cool, faint, secondary components, and large uncertainties in the orbital inclination which is known accurately only for eclipsing binaries.

Rothschild et al. (1981) were among the first to try to use the temperature of the accretion shock, T_{sh}, to find the WD mass. They found that for $M_{\mathrm{wd}} \approx 0.8$ Msun, the shock temperature as a function of mass was $T_{\mathrm{sh}} = 40.3 M_{\mathrm{wd}}^{(4/3)}$. The fundamental idea is that the maximum shock temperature, T_{max}, can be obtained by fitting the

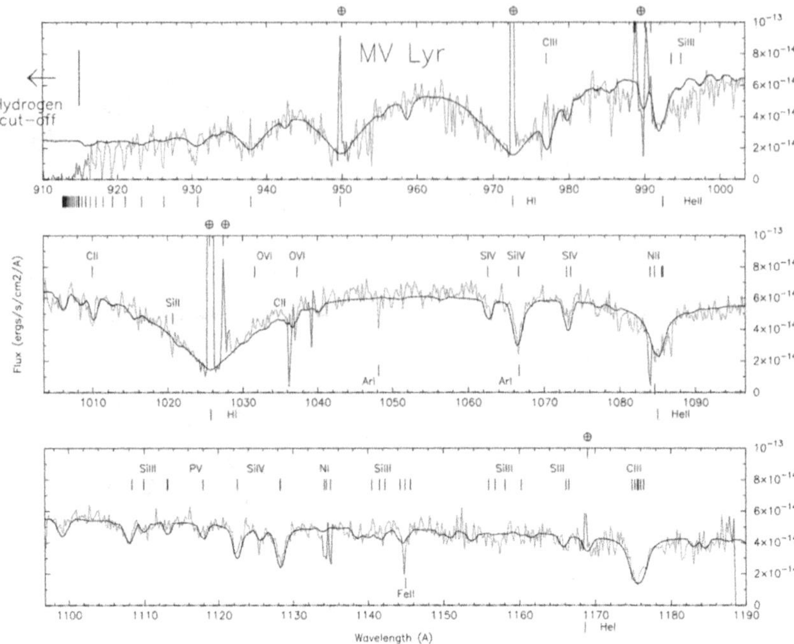

Figure 7.12. Best model fit to the FUSE spectrum of the exposed photosphere of the hot, accreting white dwarf in the VY Scl-type nova-like Variable MV Lyra during a deep low state when accretion has either greatly declined or is absent. $T_{\rm eff}$ = 45, 000 K ± 500 K, $M_{\rm wd}$ = 0.73 Msun, Vsini = 250 km s^{-1}, for solar composition. (Reproduced with permission from Godon et al. (2017). © 2017. The American Astronomical Society. All rights reserved.)

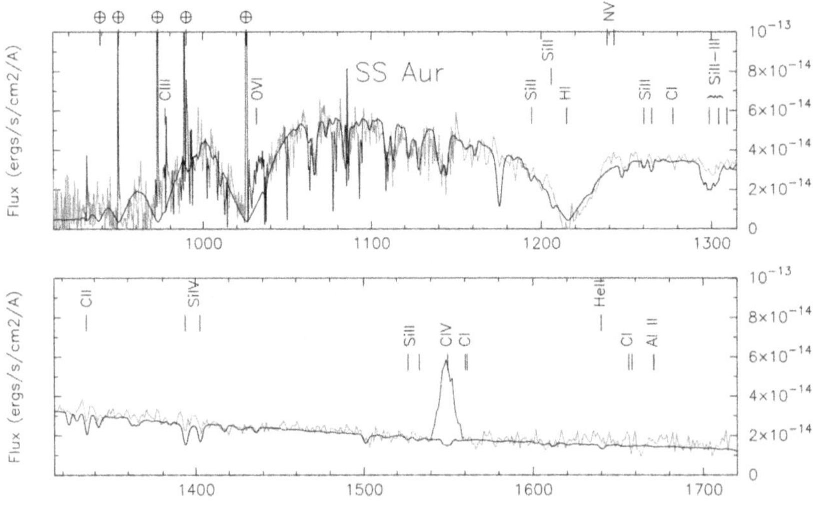

Figure 7.13. UV spectrum of the dwarf nova SS Aur. $T_{\rm eff}$ = 34, 000 K, log g = 8.93 V$_{\rm rot}$sini = 400 km s^{-1}. A best model fit to the combined FUSE + HST STIS spectra of the hot accreting, massive white dwarf in SS Aur exposed during dwarf nova quiescence when the accretion rate has greatly declined. $T_{\rm eff}$ = 34, 000 K, log = 8.93, Vsini = 400 ± 50 km s^{-1} for solar abundances. (Reproduced with permission from Godon et al. (2008). © 2008. The American Astronomical Society. All rights reserved.)

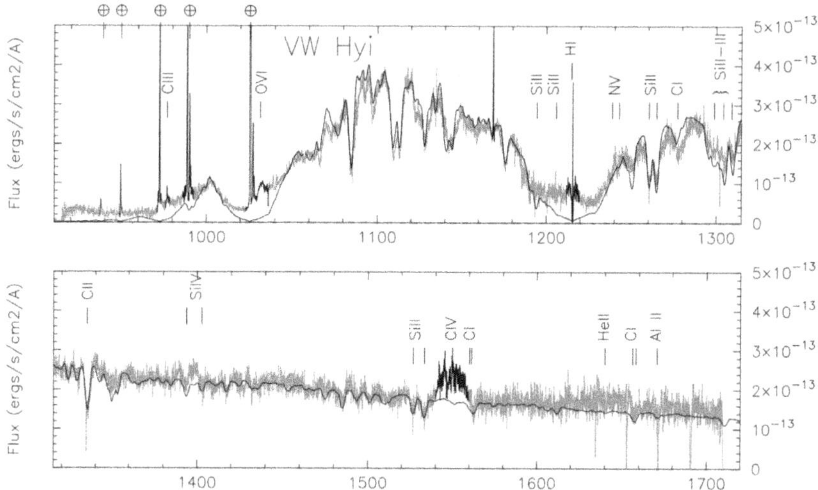

Figure 7.14. A best model fit to the combined FUSE + HST STIS spectra of the accreting white dwarf in the dwarf nova VW Hyi exposed during dwarf nova quiescence when the accretion rate has greatly declined. $T_{\rm eff} = 22,000$ K, $\log(g) = 8.0$, $V_{\rm rot} = 400 \pm 50$ km s^{-1}, chemical abundances [C] = 0.25, [Si] = 1.8, [N] = 3.0 × solar. (Reproduced with permission from Godon et al. (2008). © 2008. The American Astronomical Society. All rights reserved.)

X-ray continuum, which is assumed to be closely related to the depth of the white dwarf's gravitational potential well, and hence the WD mass.

With the advent of orbiting X-ray observatories (e.g., Chandra, Suzaku, Swift, RXTE, Ginga, ASCA, Integral, XMM-Newton, INTEGRAL, NUstar) the quest for reliable masses of polar and IP WDs relies upon X-ray spectroscopy. These X-ray spectroscopic methods are based upon on the fact that the temperature of the accretion shock, $T_{\rm sh}$, is sensitively dependent upon the mass and radius of the magnetic WD. Figure 7.15 displays the shock temperatures versus WD mass. This analysis involves model fits to the X-ray continuum and/or X-ray emission lines associated with the post-shock region (PSR). The geometry of radially infalling gas, magnetically entrained along field lines with the accretion flow onto the magnetic accretion cap of a polar, is seen. The radial infall is cool and supersonic forming a shock front where the infalling gas decelerates due to pressure forces, into a hot subsonic settling flow at the base of this accretion column, intensely heating the white dwarf surface at the magnetic accretion cap. The free-fall velocity is

$$v_{\rm ff} = (2GM_{\rm wd}/R_{\rm wd})^{0.5} \tag{7.4}$$

where $M_{\rm wd}$ is the WD mass, $R_{\rm wd}$ its radius, and G the universal gravitation constant. At the standoff shock, the gas is decelerated (by a factor of four), kinetic energy of the reduced infall velocity converted to heat, elevating the temperature, $T_{\rm sh}$, given by

$$kT_{\rm sh} = \frac{3}{8}[GM_{\rm wd}\mu m_{\rm H}/R_{\rm wd}] \tag{7.5}$$

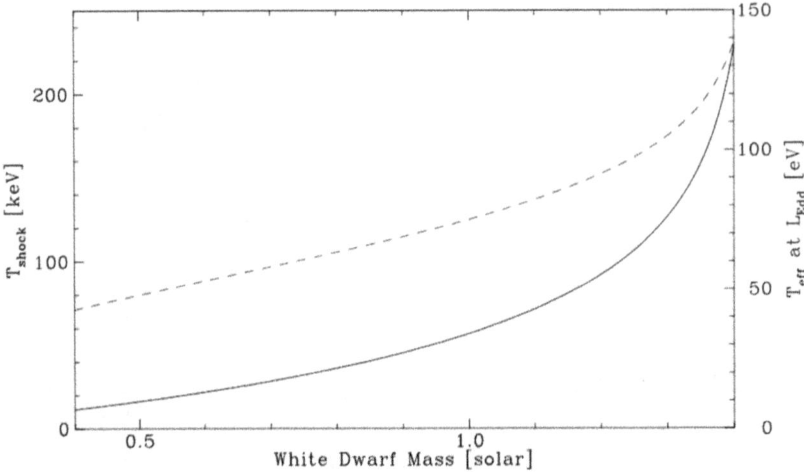

Figure 7.15. The shock temperatures as a function of the magnetic white dwarf mass, M_{wd}. (Reproduced with permission from Mukai (2017). © 2017. The Astronomical Society of the Pacific. All rights reserved.)

where T_{sh} is the shock temperature, k is Boltzmann's constant, m_H is the proton mass, and μ is the mean molecular weight of the infalling gas. Since a more massive magnetic degenerate has a smaller radius, a higher shock temperature results and hence harder X-rays are emitted. Early efforts to deliver the masses of accreting magnetic WDs in magnetic CVs by X-ray spectroscopic observations, were carried out by Rothschild et al. (1981), and Cropper et al. (1998) among others. Prior to 2000, X-ray measurement of WD masses in mCVs were limited to energies <20 keV (e.g., Cropper et al. (1998, 1999)). As a result, the early measurements of WD masses lacked the needed precision because the spectral cutoff >20 keV was missed yet must be covered for the most robust mass determinations. Without this needed coverage, the uncertainties were large.

Yuasa et al. (2010) estimated the masses of WDs in intermediate polars observed by the Suzaku satellite. They obtained wide-band spectra of 17 IPs with typical fitting errors of 0.1 to 0.2 Msun, in the range of X-ray energies 3–50 keV. They utilized the accretion column model of Suleimanov et al. (2005), with improvements to that model, and an optically-thin thermal emission code, simultaneously fitting the Fe line complex and the hard X-ray continuum of individual IP spectra. This led to greater accuracy than was previously achieved by X-ray spectroscopy of magnetic CVs. However, an examination of the intercomparison of their masses with those derived with other X-ray observatories in their Table 3 revealed, in some cases, large disagreement among the masses derived by different X-ray missions.

Since that time, there have been very important new advances in determining reliable masses of the accreting magnetic white dwarfs in intermediate polars using hard X-ray spectroscopy along with the enormous advantage afforded by Gaia distances. These X-ray spectroscopic methods rely on the fact that the temperature of the accretion shock, T_{sh}, is sensitively dependent upon the mass and radius of the magnetic white dwarf in that more massive white dwarfs have smaller radii, so the corresponding shock temperatures are higher. A higher shock temperature implies a hotter post-shock region, and hence a harder X-ray spectrum.

These analyses involve model fits to the X-ray continuum and/or X-ray emission lines associated with the post-shock region. The hard X-ray observations (largely but not exclusively by NuSTAR) of intermediate polars have yielded many dozens of reliable magnetic white dwarf masses in intermediate polars, as reported by Suleimanov et al. (2019); de Martino et al. (2020); Xu et al. (2019); Shaw et al. (2020) and references therein. As an example, in the work by Xu et al. (2019), use was made of the ratio of emission line fluxes of the Fe xxvi–Lyα and Fe xxv–He since their ratio, I7.0/I6.7, is very sensitive to the maximum shock temperature and hence the mass of the magnetic white dwarf primaries in IPs and Polars. Xu et al. (2019) have calibrated the theoretical I7.0/I6.7 relations, for which reliable empirical values of the shock temperature, T_{sh} and the ratio I7.0/I6.7 must be obtained. Their study used archival NuSTAR and Suzaku (10 keV–50 keV) X-ray spectral observations of 25 magnetic and non-magnetic CVs in the solar neighborhood. All of the WDs in their sample followed the theoretical I7.0 $- T_{\mathrm{max}}$ relation and the WD masses they derived from the I7.0/I6.7 ratio were in good agreement with dynamically measured WD masses. Their methodology has the advantages of small uncertainties in the measured I7.0/I6.7 ratio, and therefore less dependence on the continuum shape, which is subject to uncertainties associated with the shape of the X-ray continuum.

This large new sample of hard X-ray-derived magnetic WD masses in intermediate polars should lead to deepening insights into the formation, structure, magnetic accretion physics and evolution of magnetic CVs. Moreover, Kley (1991) obtained accurate mass accretion rates for many intermediate polars with the important conclusion, that the mass accretion rate in these objects does not depend on the orbital period. This has been further corroborated by Hamilton et al. (2011). A recent discovery of great importance to intermediate polars as well as polars, is the detection of rapid optical bursts from several polars that have been shown to be due to highly magnetically collimated thermonuclear bursts triggered by polar column accretion with energies reaching 10^{39} erg, about a millionth the energy of classical nova explosions (Scaringi et al. 2022a, 2022b). This finding confirms a theoretical prediction of stellar volcanoes in a theory paper by Shara (1982). It remains to be seen whether curtain accretion in intermediate polars could also trigger such explosions. The phenomenon appears to favor higher accretion rates and higher WD masses.

7.4 CV White Dwarf Temperatures

The photospheric temperatures of non-accreting WDs directly measure their cooling ages since their formation but for CV WDs, their photospheric temperatures are relics of their history of long-term average accretion and the effects of novae explosions. This makes their temperatures immensely important to understand their evolution (Sion 1991, 1999; Townsley & Bildsten 2003, 2004; Townsley & Gänsicke 2009; Pala et al. 2017, 2022). These temperatures have been derived from fitting their observed spectra during dwarf novae quiescence, nova-like low states and low states of polars with WD model atmospheres computed primarily with continually updated versions of the codes TLUSTY (Hubeney 1988) and SYNSPEC (Hubeney & Lanz 1995). In the simplest case of a dwarf nova below the period gap whose accretion rate during quiescence is extremely small and whose disk is in a cold state, single-temperature WD models have been successfully fit to the FUV and optical spectra (e.g., Szkody et al. 2002a; Gänsicke et al. 2005; Sion et al. 2004; Sion 1999; Godon et al. 2004). The best fits are determined with standard χ^2 minimization techniques, visual appearance of the fit to the continuum shape and profiles of absorption lines in the accreted atmosphere. In general, the best-fit is constrained by the minimum χ^2, the physical plausibility, fitting-derived parameters and the constraint that the combined optical model flux of the WD + disk yields an optical magnitude no brighter that the observed V-magnitude of the CV, since the disk and other cooler emitting components contribute flux in the optical. Prior to the Gaia data releases the largest uncertainties arose from the typically unknown CV WD mass (Lyα with its ground state electron is less sensitive to pressure broadening, hence gravity, than the Balmer lines) and of course errors in distance determinations and reddening values, $E(B - V)$.

Now with Gaia parallaxes, precision distances enable CV WD masses and surface temperatures to have essentially the same degree of accuracy as single isolated, non-accreting WD. The only remaining uncertainties are the reddening, unidentified FUV flux contributors other than the WD (e.g., 2nd components), temporal changes due to time variable accretion heating and cooling, variations imposed by the aspect of orbital phase, intrinsic differences in the different codes or versions of the same code employed by workers in the field and even the specific ways in which different workers eliminate emission line regions and artifacts in the fitting. Added to these difficulties are missing atomic physics, for example, unknown oscillator strengths, unknown opacity sources and instrumental flux calibration problems. There is also the problem of consistency between temperatures derived from spectral fitting in different wavelength domains, both in the FUV down to the Lyman limit (FUSE, HUT) and the optical, and in the FUV (HST, IUE) and the optical. In some cases where quiescent dwarf nova spectra are obtained by two observatories, the flux levels in overlapping wavelength regions match and spectral fitting has been carried out for the combined spectra yielding a broader wavelength coverage.

Experience has taught that temperature determinations of CV WDs approach, in successive approximations, an absolute value, but unlike the precision temperatures of single WDs, it is generally unrealistic to claim a unique value of surface temperature. Despite all of the above caveats, "reliable" surface temperatures have been derived (see Pala et al. 2017, 2022; Townsley & Gänsicke 2009; Godon & Sion 2021, 2022). The best existing effective temperature determinations with associated errors and mass determinations, all with Gaia distances, are given in Appendix A with orbital period and CV subtype for dwarf novae, nova-like variables in low states, and polars in low states.

The values and primary sources for T_{eff} in the Table 7.1, if different from Townsley & Gänsicke (2009) have been replaced by the Townsley & Gänsicke (2009) values.

Table 7.1. Temperatures of WDs in CVs

Object	Type	P_{orb}	T_{eff}	References
V485	DN	59	16,188 +313/-313	Godon & Sion (2023a)
GW Lib	DN	76.8	14,207 +56/ − 403	Pala et al. (2022)
BW Scl	DN	78.2	15,145 +556/ − 654	Pala et al. (2022)
LL And	DN	79.2	14,353 +1210/ − 743	Pala et al. (2022)
GZ Ceti	DN	79.7	14,430 +276/276	Godon & Sion (2023a)
V455 And	DN	81	10,500	Araujo-Betancor et al. (2005)
V386 Her	DN	81	14,500	Szkody et al. (2007)
WZ Sge	DN	81.6	13,190 +115/ − 105	Pala et al. (2022)
AL Com	DN	81.6	15,840 +1453/ − 1243	Pala et al. (2022)
SW UMa	DN	81.8	13,854 +189/ − 131	Pala et al. (2022)
SDSS1035	DN	82.1	11,475 +188/-188	Godon & Sion (2022)
V355 UMa	DN	82.5	12,500	Gänsicke et al. (2006)
HV Vir	DN	83.5	12,958 +257/ − 291	Pala et al. (2022)
WX Cet	DN	83.9	15,186 +1460/ − 1468	Pala et al. (2022)
T Leo	DN	84.7	16,000	Hamilton & Sion (2006)
EG Cnc	DN	86.4	12,295 +56/ − 57	Pala et al. (2022)
BC UMa	DN	90.2	14,378 +272/ − 327	Pala et al. (2022)
15	DN	90.8	14,453 +316/ − 366	Pala et al. (2022)
EK Tra	DN	91.6	18,800	Gänsicke et al. (2001)
CRT1538	DN	93.11	35,743 + 576/- 576	Godon & Sion (2022)
GY Cet	DN	98	14,500	Szkody et al. (2007)
IY UMa	DN	106.4	17,130 +/- 249	Godon & Sion (2022)
VW Hyi	DN	106.9	21,538 +235/-235	Godon & Sion (2023b)
CU Vel	DN	113.0	14,174 +117/ − 169	Pala et al. (2022)
QZ Ser	DN	119.7	15,893 +225/-225	Godon & Sion (2023a)
EF Peg	DN	123	16,644 +448/ − 570	Pala et al. (2022)
DV UMa	DN	123.6	19,325 +481/- 481	Godon & Sion (2022)
IR Com	DN	125.3	17,860 + 180/-180	Godon & Sion (2022)
TU Men	DN	168.8	27,750 +1000/-1000	Godon & Sion (2021)
MV Lyr	NL	191.0	47,000	Hoard et al. (2004)
DW UMa	NL	198.0	50,000	Knigge et al. (2000)
TT Ari	NL	198.0	39,000	Gänsicke et al. (1999)
UU Aql	DN	235.5	27,000	Sion et al. (2007)
WW Ceti	DN	253	26,000	Godon et al. (2006)
GY Cnc	DN	252.6	22,515 +292/- 292	Godon & Sion (2022)
U Gem	DN	254.7	31,000	Long et al. (1994, 1999, 2005); Sion et al. (1998)
HS2214	DN	258.2	27,643 +268/-268	Godon & Sion (2022)
BD Pav	DN	259.2	19,330 +312/- 312	Godon & Sion (2022)
SS Aur	DN	263.2	28,627 +190/ − 269	Pala et al. (2022)
RX And	DN	302.2	33,813 +869/-869	Sion et al. (2001a)

(*Continued*)

Table 7.1. (*Continued*)

Object	Type	P_{orb}	T_{eff}	References
TT Crt	DN	386.4	26,990 +557/−557	Godon & Sion (2021)
Z Cam	DN	417.4	57,000	Hartley et al. (2005)
HS0218	DN	428	17,980 +177/−177	Godon & Sion (2023a)
RU Peg	DN	539.4	70,000	Sion et al. (2004)
EY Cyg	DN	661.4	28,040 +1173/−1173	Godon & Sion (2023a)
V442 Cen	DN	662.4	31,032 +308/−308	Godon & Sion (2022)
BV Cen	DN	878.6	40,000	Sion et al. (2007)
SDSS2205	DN		15,000	Szkody et al. (2007)
V794 Aql	NL		>47,000	Godon et al. (2007)
EF Eri	Polar	81.0	9500	Beuermann et al. (2000)
DP Leo	Polar	89.8	13,500	Schenker et al. (2002)
V347 Pav	Polar	90.0	12,300	Araujo-Betancor et al. (2005
VV Pup	Polar	100.4	11,900	Townsley & Gänsicke (2009)
V834 Cen	Polar	101.5	14,300	Townsley & Gänsicke (2009)
MR Ser	Polar	113.5	14,200	Townsley & Gänsicke (2009)
BL Hyi	Polar	113.6	13,300	Araujo-Betancor et al. (2005)
ST LMi	Polar	113.9	10,800	Araujo-Betancor et al. (2005)
AR Uma	Polar	115.8	20,000	Gänsicke et al. (2001)
HU Aqr	Polar	125.0	14,000	Gänsicke et al. (1999)
QS Tel	Polar	139.9	17,500	Rosen et al. (2001)
AM Her	Polar	185.6	19,000	Townsley & Gänsicke (2009)
V1043 Cen	Polar	251.4	15,000	Gänsicke et al. (2000)
V895 Cen	Polar	285.9	14,800	Townsley & Gänsicke (2009)

7.5 Rotational Velocities of CV White Dwarfs

The WD rotation rates, Vsini, are required to achieve an understanding of CV evolution and spin-up of the CV WDs due to accretion with angular momentum. On theoretical grounds, it was long suspected that WDs in non-magnetic CVs should be rotating very rapidly and possibly near breakup due to the angular momentum transferred to the WD by disk accretion. In fact, rapid WD rotation was considered a viable explanation for the weak boundary layer luminosities that were at odds with standard disk accretion theory. Surprisingly however, the first rotation rates for white dwarfs in non-magnetic CVs measured with HST GHRS data (Sion 1999) were slow (<400 km s^{-1}). While this CV WD rotation is very rapid compared with the extremely slow rotation of single WDs (<40 km s^{-1}), it quickly became clear that rapid rotation of a CV WD could not account for the observed low or missing boundary layer luminosities.

In practice, the rotational velocities are derived from fitting the relatively sharp line profiles of ionized metals which form in the accreting WD photosphere, in those systems for which the photospheric origin of the features is unmistakable. The only difficulty arises when the rotation rate is rapid because then the chemical abundances of the ion species becomes tied in with the rotation rate determination. Thus, different rotation rates will be derived for different metal abundance values. Moreover, interstellar and circumstellar absorption as well as underlying emission and absorbing curtains (in high inclination systems) may complicate the rotation determination.

Up to now, there are measurements of over twenty H-rich CV WD rotational velocities, Vsini, or stringent upper limits on their rotation. These are listed in Table 7.2.

All of the velocities in Table 7.2 are well below 20% of the Keplerian velocity, far slower than predicted based upon the specific angular momentum transferred to the WD during long-term accretion. For example, quantitatively one can estimate the resulting spin up of a CV WD, expected over the lifetime of the CV accreting mass and angular momentum (Livio & Pringle 1998). For an estimated CV lifetime of, say 10^9 years, the ratio of the spin velocity of CV white dwarf Ω_{wd} to the Keplerian velocity Ω_K, is expressed as

$$\frac{\Omega_{wd}}{\Omega_K} = k^2 \left[1 - \left(\frac{M_{init}}{M_{wd}} \right)^{4/3} \right], \tag{7.6}$$

where k is the radius of gyration, Ω_K is the Keplerian velocity, $R_{wd} M_{wd}^{-1/3}$ and M_{init} is the initial mass. If $k^2 \sim 0.2$, then a one solar mass WD that accretes one-tenth of a solar mass should be spun-up to 45% of the Keplerian velocity, i.e., $\Omega_{wd} = 0.45\Omega_K$. Yet all of the dwarf nova WDs in Table 7.3 have rotational velocities well below this value. Nevertheless, their rotational velocities are faster by at least an order of magnitude or more than the slow rotational velocities (Vsini $\lesssim 30$ km s^{-1}), (Koester 2013) of isolated WDs. This provides clear evidence that *significant* angular momentum over time is being accreted along with the gas flow onto CV WDs.

The WD rotation, if near breakup would provide centrifugal pressure leading to an increased mass limit for a SN Ia (super-Chandrasekhar SNe Ia). Livio & Pringle (1998) examined the question of why CV WDs rotate slower than expected. They proposed a model in which the accreted angular momentum is removed during nova outbursts. They showed that the most likely source of the needed torque is a shear-generated magnetic field during the envelope expansion of the nova and that the efficiency of transport of angular momentum between the WD core and the extended nova envelope is rather low. Were this not the case, then the rotation rates of CV WDs would be even slower than what is observed.

While the sub-Keplerian rotational velocities of CV WDs are likely the result of nova explosions, Medvedev & Menou (2002) explored another possibility. Hot accretion flows during dwarf nova quiescence may have a role in limiting the rotational velocities of accreting WDs to sub-Keplerian values. The work of Medvedev & Narayan (2001) established that at large rotational velocities, dissipation in the hot flow is dominated by stellar braking in a hot settling flow (HSF) configuration whereas at low rotational velocities, the flow becomes a typical advection dominated accretion flow (ADAF). Although their work focused upon accreting neutron stars in X-ray binaries, the HSF solution should also apply

Table 7.2. Rotational Velocities of WDs in CVs

Object	Type	P_{orb} (minute)	Vsini (km s^{-1})	References
V485 Cen	DN	59	300 ± 50	Godon & Sion (2023a)
GW Lib	DN	76.8	40	Szkody et al. (2002b, 2012)
BW Scl	DN	78.2	122 ± 60	Godon & Sion (2023b)
LL And	DN	79.2	500 ± 50	Howell et al. (2002)
GZ Ceti	DN	79.7	500 ± 50	Godon & Sion (2023a)
V455 And	DN	81.1	630	Araujo-Betancor et al. (2005)
WZ Sge	DN	81.6	400/1200?	Sion (1995)
AL Com	DN	81.6	<800	Szkody et al. (2003)
SW UMa	DN	81.8	106 ± 50	Gänsicke et al. (2005)
SDSS1035	DN	82.1	150 ± 50	Godon & Sion (2022)
HV Vir	DN	83.5	400 ± 50	Szkody et al. (2002a)
WX Cet	DN	83.9	400 ± 50	Sion et al. (2003)
BC UMa	DN	90.2	238 ± 50	Gänsicke et al. (2005)
VY Aqr	DN	90.8	400 ± 50	Sion (2003)
EK Tra	DN	91.6	200 ± 50	Gänsicke et al. (2001)
CRTS1538	DN	93.1	400 ±100	Godon & Sion (2022)
IY UMa	DN	106.4	150 ± 50	Godon & Sion (2022)
VW Hyi	DN	106.9	500+/-100	Sion et al. (1995)
QZ Ser	DN	119.75	450 ± 50	Godon & Sion (2023a)
DV UMa	DN	123.6	150±50	Godon & Sion (2022)
EF Peg	DN	123	<300	Howell et al. (2002)
TU Men	DN	168.8	225 ±75	Godon & Sion (2021)
MV Lyr	NL	191.0	200 ± 50	Hoard et al. (2004)
DW UMa	NL	198.0	251 ± 66	Godon & Sion (2023b)
GY Cnc	DN	252.6	150 ± 59	Godon & Sion (2022)
WW Ceti	DN	253	600 ± 50	Godon et al. (2006)
U Gem	DN	254.7	150 ± 25	Long et al. (1994, 1999, 2005); Sion et al. (1998)
BD Pav	DN	258.19	200 ± 50	Godon & Sion (2022)
HS2214+28	DN	258.20	400 ± 50	Godon & Sion (2022)
SS Aur	DN	263.2	500 ± 50	Sion et al. (2004)
RX And	DN	302.2	500 ± 50	Sion et al. (2001a)
TT Crt	DN	386.42	400 ± 100	Godon & Sion (2022)
Z Cam	DN	417.4	330 ± 50	Hartley et al. (2005)
HS0218	DN	428	150 ± 50	Godon & Sion (2023a)
EY CygV	DN	661.4	—	Godon & Sion (2023a)
442 Cen	DN	662.4	> 250	Godon & Sion (2022)
RU Peg	DN	539.4	100 ± 50	Sion et al. (2004)

to accreting WD as well. They showed that optically-thin boundary layers of disk-accreting WDs can also be radially extended (of order a WD radius) and that energy advection is an important characteristic of their internal structure. Thus, Medvedev & Menou (2002) applied the model of Medvedev & Narayan (2001) to WDs in quiescent dwarf novae by constructing numerical solutions to the hydrodynamic equations for hot accretion in five well observed systems with relatively well-known dwarf novae system parameters. The five systems are WZ Sge, VW Hyi, SS Cyg, RX And and U GEM These solutions are displayed in Figure 7.16.

Medvedev & Menou (2002) posited that an accreting WD is spun down by hot accretion during dwarf nova quiescence and spun up during outburst, reaching a steady equilibrium between spin up and spin down and that that equilibrium occurs

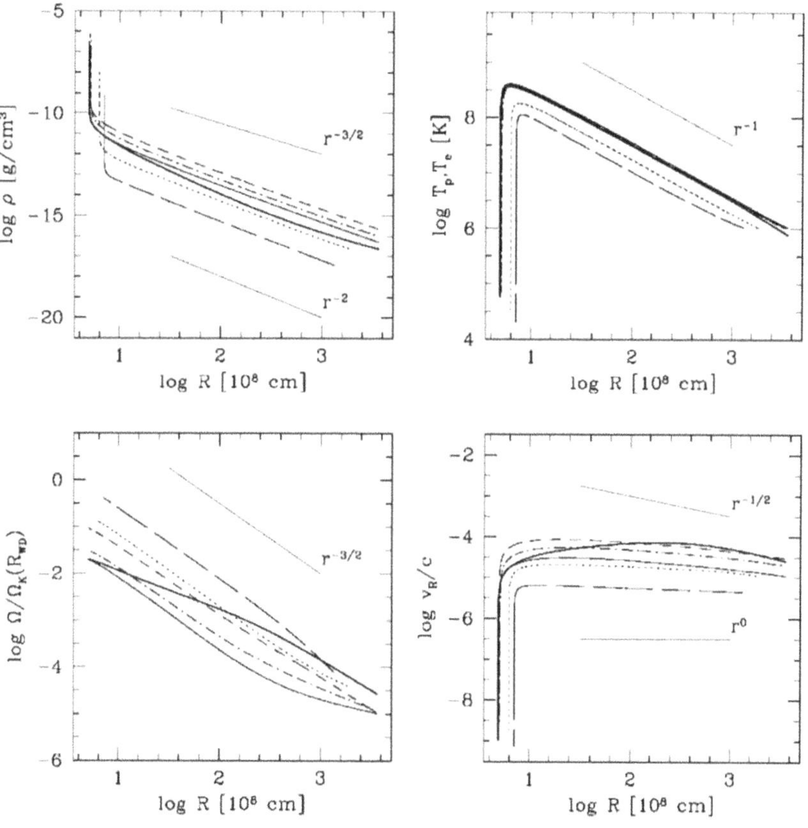

Figure 7.16. Radial profiles of density, r, proton and electron temperatures, T_p and T_e, respectively, angular rotation velocity, Omega, and radial velocity, V_r, for hot accretion with Gamma = 1.6 in the five systems listed in Table 7.1 (RX And = dot-dashed line, SS Cyg = short-dashed line, U Gem = thin solid line, VW Hyi = dotted line, and WZ Sge = long-dashed line). The density profiles for VW Hyi and WZ Sge have been scaled down by a factor 10 and 100, respectively. The thick solid line corresponds to the same model for U Gem, except that Gamma = 4/3 instead of 1.6. (Reproduced with permission from Medvedev & Menou (2002). © 2002. The American Astronomical Society. All rights reserved.)

at sub-Keplerian rotation velocities. Clearly, this possibility for explaining the relatively slow rotation of accreting WDs in dwarf novae merits further exploration.

It is clear that so far, with the relatively small number of velocities that there is no apparent difference in rotation between CV WDs above and below gap which might be expected from differences in system age and hence differences in accretion history. While nova explosions may spin down the rotation rates to well below the critical rotation, if the accretion rate is low and the WD is of average or moderate mass, then the recurrence time between novae is longer and an accreting WD could possibly have a faster rotation rate than higher mass accreting WDs. If nova explosions are the way to explain the slow rotation rates, the fact that nearly all the WDs in dwarf novae confirm the expectation that CVs do indeed cycle between dwarf nova states and classical nova systems. (Gänsicke et al. 2006). Moreover, depending upon how close the rotation velocities of massive CV WDs are to the critical velocity, SN Ia supernovae events could occur over a range of WD masses, thus possibly explaining some of the differences seen in SNe Type Ia explosions.

For these SNe Type Ia white dwarf supernovae therefore, the effect of stellar rotation velocity on the pre-supernova progenitor must be considered and its importance cannot be overstated. The effects of rotation could profoundly affect both the evolution of the SNe Type Ia progenitor evolution and the thermonuclear explosion of the WD. Piersanti et al. (2003) carried out a study of the effect of WD rotation on the evolution of accreting WDs with carbon–oxygen cores, using a one dimensional quasi-static evolution code. However, their investigation assumed rigid body rotation and neglected the angular momentum transport in their models. In effect, they assumed instantaneous angular momentum redistribution.

Yoon & Langer (2004) studied the effects of rotation on the evolution of accreting WDs with carbon–oxygen cores in which the accreting matter is carbon–oxygen rich gas. Their initial model was slowly rotating. Accretion of gas and angular momentum was assumed to be from a quasi-Keplerian accretion disk. With initial WD masses of 0.8, 0.9, and 1.0 Msun accreting at rates in the range of $3 - 10 \times 10^{-7}$ Msun yr^{-1}, they found that the model WDs developed differential rotation throughout, with the envelope and core both rotating at different rates. The angular momentum transport in the strongly degenerate core is dominated by dynamical shear instability while the angular momentum transport in the non-degenerate envelope is dominated by Eddington-Sweet circulation (i.e., related to Von Zeipel's Theorem), as well as Goldreich–Schubert–Fricke instability and secular shear instability. They established the pre-explosion angular momentum profile and found a wide range of WD explosion masses well above the 1.44 Msun Chandrasekhar limit of a non-rotating C–O WD.

Finally, all of the exposed CV WDs in the archival HST and archival FUSE data set should yield the most accurate stellar rotational velocities. This will permit a comparison of the distribution of WD rotational velocities both above and below the CV period gap. As the database of exposed CV WDs continues to grow, high quality HST COS spectra with sufficient resolution will inevitably enlarge the data set. Moreover, at the time of this writing, the first rotational velocities of the AM CVn helium accreting, exposed CV WDs will soon emerge from a large HST Guest Observer project in HST Cycle 29.

7.6 Chemical Abundances in CV WD photospheres

The chemical abundances of WDs in CVs have been derived largely from high quality HST and FUSE spectra. The observed metal lines are fitted with NLTE model atmospheres with rotationally broadened theoretical line profiles. For the most part, the S/N is just high enough to determine reliable individual chemical abundances. In Table 7.3, the chemical abundances determined to date for CV WDs are summarized. They present an overall picture of the abundance patterns characterizing the accreting WDs in CVs.

Table 7.3. Chemical Abundances of WDs in CVs

Object	Type	P_{orb}	Abundance	References
V485 Cen	DN	59	[C] 0.20+/-0.05 [N] 30+/-5 [O] 0.1+/-0.05	Godon & Sion (2023a)
GZ Ceti	DN	79.7	[Al] 10 +/-5 [Si] 2.0-5.0 [S] 0.1 – 1.0	Godon & Sion (2023a)
GWLib	DN		[Fe] 15 +/- 3	Szkody et al. (2002b)
			[C] 0.01 +/-0.005 [N] 15 +/-1 [O] -0.01	
			[Mg] 3 +/-1 [Al] 3 +/- 0.5 [Si] 1.0 +/-0.2	
			[P] -0.01 [Ca] 0.1 +/-0.05	
			Z ~0.1	
BW Scl	DN	78.2	[C] 1.0 +1.0/-0.5 [N] 5.0 ± 1.0	Godon & Sion (2023b)
			[O] 0.75 ± 0.25 [Mg] 3 ± 2 [Al] 5.0 ± 1.0	
			[Si] 0.3 ± 0.2 [P] $\leqslant$ 0.5	
			[S] 1.0 ± 0.3 [Fe] 3 ± 2	
LL And	DN	79.2	Z $\leqslant$~1	Howell et al. (2002)
WZ Sge	DN	81.6	[C] 0.5, [Fe] 0.1, [S] 0.1, [Si] 0.005	Sion (1995)
AL Com	DN	81.6	Z ~0.3	Szkody et al. (2003)
SW UMa	DN	81.8	[C] 0.025 ± 0.01 [N] 0.07 ± 0.02	Godon & Sion (2023b)
			[O] $\leqslant$ 0.01 [Al] 0.3 ± 0.05 [Si] 0.1 ± 0.05	
			[Ca] $\leqslant$ 0.1	
SDSS1 035	DN	82.1	[C] 0.03+/-0.02 Z ~0.3	Godon & Sion (2022)
HV Vir	DN	83.5		Szkody et al. (2002a)
WX Cet	DN	83.9	Z~0.1	Sion et al. (2003)
EG Cnc	DN	86.4	Z~0.3	Szkody et al. (2002a)
BC UMa	DN	90.2	[C] 0.20 ± 0.05 [Si] 0.3±0.1 [O] $\leqslant$ 0.01	Godon & Sion (2023b)
			[Al] 2.5 ± 0.5 [Si] 0.25 ± 0.05	
			[P] 0.3 ± 0.1 [S] 0.1 ± 0.1 [Ca] $\leqslant$ 0.1	
VY Aqr	DN	90.8	[Si] 0.05, [C] 1.0	Sion (2003)
EK Tra	DN	91.6	Z <1	Gänsicke et al. (2001)
CRTS1538	DN	93.1	[Si] 10^-4	Godon & Sion (2022)
IY UMa	DN	106.4	[P] < 0.01	Godon & Sion (2022)
VW Hyi	DN	106.9	[Si] 0.3, [C] 0.3, [N] 3, [O] 3, [Al] 2,	Sion et al. (1995)
			[P] 20, [Mn] 50	
QZ Ser	DN	119.75	[C] 0.05 +/0.02 [N] 15 +/-2	Godon & Sion (2023a)
			[O] 0.20+/-0.10 [Mg] 5 +/-1 [Al] 3 +/- 1	

(Continued)

Table 7.3. (*Continued*)

Object	Type	P_{orb}	Abundance	References
			[Si] 1 +/-0.3 [P] 0.3 +/-0.1	
			[S] 1.0 +/- 0.3 [Ca] 1.0+/-0.3 [Fe] 12 +/- 2	Howell et al. (1995)
EF Peg	DN	123	Z 0.1–0.3	
DVUMa	DN	123.63	[C] 0.1+0.05/-0.1 [Si] 0.3+/-0.1	Godon & Sion (2022)
IRComDN		125.3	[P]<0.01[S]<1.0[C] 1.0 +1.0/-0.05	Godon & Sion (2022)
			[Al] 5 +/-0.1 [Si]1+/-0.5	
			[Fe] 4+/-1.5	
TUMen	DN	168.77	[C]0.2+/-0.1, [N] 20+/-10,[Si] 0.12+/ -0.1	Godon & Sion (2021)
DW UMa	NL	198.0	[C] 0.06+/-0.04 [N]<0.1 [Si] 0.14+/-0.06	Godon & Sion (2023b)
GY Cnc	DN	252.6	[C] 0.01 +0.1/-0.005 [Si]0.1+0.10/-0.05	Godon & Sion (2022)
WW Ceti	DN	253	[P]<0.01 [S] 0.1+0.10/-0.05	Godon et al. (2006)
			[C] 0.1, [N[2, Si 0.3	
U Gem	DN	254.7	[C 0.3–0.35, [N] 35–41, [Si] 1.4–4, 6.6–10	Long et al. (2006)
BD Pav	DN	258.19	[C] 0.01+0.01/-0.005 [Si] 0.1 +0.10/-0.05	Long & Gilliland (1999)
HS2214	DN	258.20	[P] < 0.01 [S] < 1.0[Si] 0.2 +/- 0.1	Godon & Sion (2022)
SS Aur	DN	263.2	[C] 0.1, [Si] 0.1, [N] 2.0	Godon & Sion (2022)
				Sion et al. (2004)
RX And	DN	302.2	[C] 0.01 − 0.08 [N] 5 ± 2 [Si] 0.1 +0.1/−0.05	Godon & Sion (2023b)
TT Crt	DN	386.42	[C]0.01 +0.01/-0.005 [Si] 0.1 +0.1/-0.05[P]<0.01	Godon & Sion (2022)
HS0218		428	[S] 0.1	Godon & Sion (2023a)
RU Peg	DN	539.4	[C] 0.06+/-0.04 [N] 1.0 +/- 0.3 [Al] 1.0 +/- 0.3	Sion et al. (2004)
			[Si] 0.25 +/- 0.05 [Fe] 1.0 +/- 0.3	
			[C] 0.1, [Si] 0.1, [N] < 8	
EY Cyg	DN	661.4	[C] 0.05 +/-0.02 [Si] 0.03 +/- 0.02[C] < 0.001 [Si]	Godon & Sion (2023a)
V442 Cen	DN	662.4	< 0.01	Godon & Sion (2022)

The most telling characteristics of the abundances in Table 7.3 is their sub-solar metallicity values for some elements, suprasolar abundances in other elements, and variations in the mix of ion species from system to system. If the replacement time of the accreted atmosphere is shorter than the diffusion timescales of the accreted ions, then the surface abundances would tend to reflect the composition of the infalling matter from the donor star. There is little reason to expect the secondary mass donor stars to have sub-solar metallicity, with the exception of subsolar carbon from CNO processing. Rather, one expects that the effects of diffusion are, for the most part, responsible for these subsolar abundances. At temperatures below 20,000 K, radiative forces levitation of the ions is unimportant. However, on a case-by-case basis, one must be mindful that the reliability of the chemical abundances could be compromised by underlying emission, an upper disk atmosphere (curtain), circum-stellar absorption, interstellar absorption and variations of absorption features with orbital phase. The continually changing and unsettled atmosphere of a CV white dwarf where other forms of mixing and spreading of accreted matter are likely involved, also presents a challenging theoretical problem.

The two most reliable, straightforward cases are U Gem and VW Hyi where elevated N and depleted C are solidly established in their white dwarf atmospheres. Large enhancements above solar of N, Al, P, Si and Mn have been reported in VW Hyi (Sion et al. 1997, Sion et al. 2001), while large suprasolar metal enhancements were also found in U Gem by Long & Gilliland (1999) and Long et al. (2006). In a third well-studied system, WZ Sge, the photospheric origin of the absorption lines of metals seen in HST and FUSE spectra of the WD has been solidly established (Sion et al. 1997).

Indeed, Steeghs et al. (2007) showed that the rich spectrum of absorption lines due to metals seen in WZ Sge during a sequence of HST STIS spectra over five HST orbits, follow the orbital motion of the white dwarf and are gravitationally (Einstein) red-shifted.

It is also noteworthy that suprasolar Al abundances have been reported in three other systems shown in Table 7.3, BC UMa, BW Scl and SW UMa. Odd-numbered elements like Al, P and Mn can be enhanced via proton captures during CNO thermonuclear processing. The N abundance is elevated far above solar while the C abundance is depleted, also a result of CNO-processed gas in the white dwarf photosphere.

Large N/C ratios (discovered by their strong N V emission and weak or absent C IV emission) among 14 magnetic and non-magnetic systems were initially reported by Gänsicke et al. (2003) and references therein. These first recognized systems are AE Aqr, V1309 Ori, TX Col, MN Hya, and BY Cam while the non-magnetic CVs showing the N/C anomaly are BZ UMa, EY Cyg, RX J2329, CE315, GP Com and CH UMa.

The number of such CVs at the present time has substantially increased. Currently, the number of CV systems with large N/C abundances (hereafter the N/C anomaly), comprise 10 to 15% (or more) of all CVs. This N/C anomaly is strong evidence of CNO thermonuclear processing in a previously more massive secondary star presumably eaten away by mass transfer down to its CNO-processed core material.

Since these N/C values were inferred from emission lines presumably forming in the disk or accretion column, they should originate in the secondary. Indeed, if CNO-processed material from the stripped-down core of the secondary flows through the accretion disk or column onto the WD, then the surface abundances of the WD should manifest the same N/C anomaly which originated in the secondary donor's core.

Schenker et al. (2002) presented calculations suggesting that CV binaries originally consisted of a mass donor more massive than the current WD, resulting in a rapid phase of unstable thermal timescale mass transfer (hereafter TTMT). The early calculations by Schenker et al. (2002) and Podsiadlowski et al. (2001, 2003) show that systems with an initial donor mass of up to 2 Msun may survive the TTMT and subsequently evolve into a normal appearing CV. However, in this "normal" CV, the matter flowing over to the WD is CNO processed gas from the stripped core of the previously more massive donor. Schenker et al. (2002) predict that the secondaries in such systems should be hotter than expected for their orbital periods. That is exactly what one observes in systems having evolved donors that therefore have undergone significant nuclear evolution when they were initially more massive in the past. These donor secondaries are too hot and luminous for their orbital period to be considered unevolved main sequence star companions.

Evidence is also mounting that the CVs with the N/C anomaly may be far more populous than previously thought and may represent a significant evolutionary channel feeding into the general CV population above and below the period gap. El-Badry et al. (2021) have confirmed 21 such systems with evolved donors that are Roche lobe filling close binaries with 13 out of the 21 having ongoing mass transfer. Most of the secondaries are hotter than any previously known CV donors, with temperatures 4700 K < T_{eff} < 8000 K. Remarkably, all the secondaries that lie near or within the effective temperature >~7000 K, are detached from their Roche lobes. This happens to be where magnetic braking becomes inefficient due to the convective envelope of the donor vanishing. The secondaries have masses near 0.15 Msun and the WD companions have masses near 0.8 Msun. All of the objects in this sample will evolve to minimum orbital periods that match the orbital period range of the AM CVn stars. It seems likely that these evolved donor systems may represent an important link between the progenitors of the TTMT phase, SSXBs, evolved donor CVs, extremely low mass degenerates and AM CVn systems.

One of the most exciting aspects of TTMT is that it may allow the WD primary to grow in mass. During the relatively brief TTMT, supersoft X-ray binaries undergo stable, steady nuclear burning in equilibrium with their high rate of accretion. This builds up the mass of the WD accompanied by the rapid accumulation of helium ash at the core-envelope interface, thus blocking the enhanced CNO abundances needed to power the fast nova outburst. Thus, they can avoid classical nova explosions, therefore possibly avoiding a reduction of the WD mass over time. This scenario could be one of the most promising progenitor channels for producing Type Ia supernovae. Those CV systems that underwent TTMT and then appeared as normal CVs with the donor secondaries having undergone substantial nuclear evolution are sometimes called "failed" Type Ia supernova candidates.

The well-studied dwarf novae VW Hyi and U Gem are the two best examples of showing the accretion of CNO processed gas but at least five systems in Table 7.3 reveal the N/C anomaly from the abundance determinations of diagnostic photospheric absorption lines. Thus, these systems join over a dozen dwarf novae whose WDs reveal the N/C anomaly from WD photospheric absorption spectra. This distinction may be important because VW Hyi's WD has a strong C IV emission line in its HST spectra from either the disk, boundary layer or donor chromosphere/corona, while its photospheric absorption features yield an N/C ratio of 35 to 41 in its accreted atmosphere (Godon & Sion 2023b). The low mass M dwarf donor star in VW Hyi shows no evidence of having undergone significant nuclear evolution in its past. For VW Hydri's donor star, Hamilton et al. (2011) and Harrison (2016) confirmed that the abundance of carbon and the overall composition is solar in VW Hyi's donor star and that the donor star in U Gem is carbon deficient. At the very least, in VW Hyi, its donor having solar abundances might even suggest that the N/C anomaly could also originate in the WD itself in some cases, rather than the donor star, and subsequent contamination of the donor during the nova common envelope. For a few dwarf novae in Table 7.3, the donor secondary stars are normal cool dwarf companions with no evidence of prior nuclear evolution (e.g. BW Scl, VW Hyi, possibly SW UMa and RX And which is Z Cam object above the CV period gap).

The evidence is certainly compelling that the N/C anomaly is associated with evolved secondary stars that underwent substantial nuclear evolution. However, Nature may allow a second independent mechanism to cause the N/C anomaly in CVs. This second scenario involves the contamination of the donor star, by CNO-processed material from the nova shell and the common envelope when the bloated WD expands during the nova episode and engulfs the donor star in a common envelope. One can imagine that after hundreds or even thousands of classical nova cycles over long timescales, pollution of the donor star's atmosphere would result in the CNO-processed gas then being transferred back to the WD and being re-accreted.

Up until now, there has been no definitive theoretical calculations of the efficiency of the contamination process (Marks et al. 1999; Marks & Sarna 1999; Sion et al. 1998; Sion & Sparks 2014). Recently however, Sparks & Sion (2022) carried out analytical calculations of the contamination of the donor by the WD when it undergoes nova explosions. In this case, the N/C anomaly and suprasolar abundances of heavy elements like aluminum via the dominant proton captures that occur during hot CNO burning, could be pointing to contamination of the donor star via hot CNO processing by the nova explosion and accompanying expansion of the bloated WD to form a transient common envelope. The contamination of the donor by the common envelope leads to re-accretion of this material from the contaminated donor star to the WD (Sion & Sparks 2014, Sparks & Sion 2022). The companion star may be contaminated by CNO-processed material from the nova shell ejected by the WD during a classical nova explosion after hundreds or even thousands of classical nova cycles over long timescales.

Studies that considered only the secondary star's cross-section presented to ejected nova material to change the abundances in its convection zone, was a major objection to the nova-processed material affecting a change in abundances. However, numerous studies ignored the fact that during the thermonuclear runaway on a WD, a strong propagating shock wave not only ejects processed gas but also produces a large amount of non-ejected material that forms a common envelope. This nova-produced common envelope contains a large amount of non-solar material. Sparks & Sion (2022) showed that the secondary has the capacity and time to re-accrete enough of this material to acquire a significant non-solar convective region. This same envelope interacting with the binary will also produce a frictional angular momentum loss, that could be the consequential angular momentum loss needed for the CV problems involving the average CV WD mass compared to the average mass of single degenerates, the WD mass accretion rates, the period minimum, the orbital period distribution of CVs, as well as the space density of CV problems. Looking ahead, a much greater body of chemical abundances in the surface layers of CV WDs will be emerging from HST spectroscopy of an ever larger number of CV WDs, than the modest sample presented in Table 7.3.

7.7 CV WD Second Flux Components and Boundary Layer Structure: A Mystery

The surface temperature of a CV WD derived from model atmosphere fitting in the FUV may in some cases represent a kind of global average since the temperature distribution across the surface may be inhomogeneous due to the heating effects of accretion. It is well known that the synthetic spectral fitting to the FUV spectra of a number of systems during dwarf nova quiescence is significantly improved in both a statistical (numerical) sense and by visual inspection when a second component of flux is added to a model. Such improvements in the model fitting have been reported for both non-magnetic and magnetic CVs. In the disk accreting case, the nature of this second component flux source may tie in directly to four fundamental questions: (1) how and where does the accreted material actually enter into WD surface? (2) how much angular momentum is transported inward? (3) is the boundary layer structure related to the second component? and (4) is the second component due to flux from the region of the hot spot (shock front) where the gas stream from L_1 collides with the edge of the accretion disk around the WD?

In some cases, the second component, added to a WD photosphere, is favored by the fitting when single-temperature WD models and/or standard accretion disks cannot account for the spectral energy distribution. This second component is usually small (though it could be a disk or ring) and hot, with an emitting area usually smaller and hotter than a WD. Typically, it has the characteristic of being featureless and extends all the way down to the Lyman limit. There are a number of possible candidates including an optically thick inner disk region or ring, a second WD flux component with a higher temperature arising from a more heated region on the WD possibly rotating more rapidly than the cooler portions of the WD, or a hot rotating ring where the outer portion of the boundary layer (the UV boundary layer) interfaces with the inner-most accretion disk annulus and possibly heats the inner disk. Unfortunately, there are no clearly distinctive spectral features which could be used to identify this component. The continuum usually rises steeply toward shorter wavelengths but no absorption features are seen to uniquely characterize it. In some cases the FUV spectrum of a dwarf nova appears to be most successfully fit by an optically thick accretion disk model at the correct system distance but this appears implausible because the disk should be in a cold state with its inner region optically thin.

Striking examples are provided by two of the best observed and most extensively studied of all CVs, U Geminorum (Long et al. 1993, 1994, 2006; Sion et al. 1994, 1997) and VW Hyi (Long et al. 1996; Sion et al. 1997, 2001, 2004). In both systems, there has to be an unidentified second component contributing significantly in the FUV that probably arises in the boundary layer but may be a hot belt on the WD, may or may not be rotating rapidly, and is much hotter than the WD. In many cases, the second component contributes only a small fraction of the total FUV flux. For

U Gem, Froning et al. (2001) and Long et al. (2006) found that a two-temperature WD improves the fit over the best single-temperature WD but the physical reason behind this improved χ^2 is unclear. A FUSE study of Z Cam in quiescence by Hartley et al. (2005) reported a two-temperature solution in which a cooler WD with $T_{\mathrm{eff}} = 18,000$ K and a hotter "belt" with $T_{\mathrm{belt}} = 72,000$ K actually yields a statistically better fit than the best-fitting single-temperature model with $T_{\mathrm{eff}} = 57,000$ K. The two possible solutions are shown in Figure 7.17. These authors favor the simpler single-temperature solution. However, the issue must be resolved because it is essential to our understanding of CV evolution that we know the correct global surface temperatures since they are directly linked to the long-term mass accretion rates (Townsley & Bildsten 2003). Further evidence of superior two-temperature WD solutions to FUV spectra are reported by Long et al. (2005) for SS Cyg and Gänsicke et al. (2005) for WZ Sge-like systems. The model fits to the magnetic WDs in polars during low states are also significantly improved with the inclusion of a second component of flux (Araujo-Betancor et al. 2005), in this case the extra flux component being the hot accretion pole.

Sion et al. (2010) re-examined the case for detecting the WD in SS Cygni during quiescence. Their work employed SS Cygni's accurate HST FGS parallax

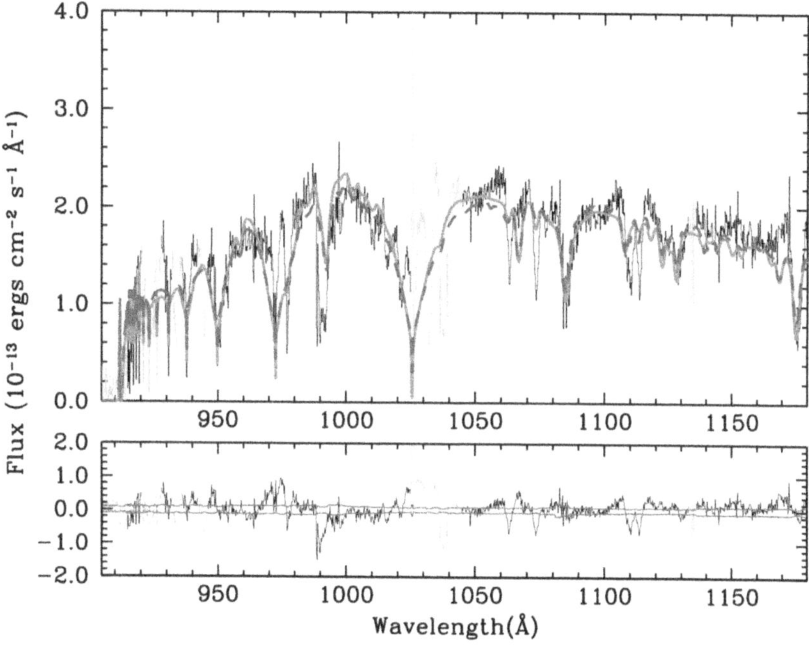

Figure 7.17. Solar abundance model WD fits to the mean FUSE spectrum of Z Cam in quiescence. In the upper panel the solid line plots the best statistical-fit single-temperature (57,000 K) model and the dashed line plots the best-fitting two-temperature (18,000 K + 72,000 K) model. The lower panel plots the difference spectrum of the observed and single-temperature model flux, with (1σ) errors in blue. The gray portions of the plotted spectrum were not considered in fitting the data. (Reproduced with permission from Hartley et al. (2005); their Figure 4. © 2005. The American Astronomical Society. All rights reserved.)

(giving $d = 166$ pc), with a newly determined mass for the accreting WD (Bitner et al. 2007) of $M_{wd} = 0.81$ Msun (lower than the previous, widely used 1.2 Msun) and reddening E_{B-V} values 0.04 (Verbunt 1987) and 0.07 (Bruch & Engel 1994) derived from the 2175Å absorption feature in the IUE LWP spectra. Their best-fit model solutions to the combined HST + FUSE spectral energy distribution, showed that the WD dominates the FUV flux and has a temperature T_{eff} of 45–55,000 K in quiescence, with $\log(g) = 8.3$ and a solar composition accreted atmosphere, depending on the value of the reddening. The presence of any second component, in this case an accretion disk, is a minor contributor to the FUV flux.

An intensive FUSE study of VW Hyi (Long et al. 2009) used a series of FUSE FUV spectroscopic observations to follow the cooling and spectroscopic behavior of the WD following a superoutburst. Their last observation in the time sequence reveals the WD has cooled from 24,000 K to 19,000 K with no evident second component present any longer. Their observation 12 has a temperature of 20,200 K, the same as the pre-outburst but does have a strong second component. A similar amount of cooling of the WD in VW Hyi was found by Sion et al. (1996) with HST observations following a normal outburst and a superoutburst. Long et al. 2009 explored the nature of the second component in VW Hydri and found it to be relatively featureless, showing a substantial modulation on the orbital (and just after outburst, the superhump) period. Surprisingly, they found that the second component is most likely associated with the hotspot where material from the secondary encounters the disk, rather than emission from the boundary layer region between the inner disk and WD. This second component fades about 10 days after the superoutburst. Long et al. 2009 also delineate a third component, detectable in the form of broad emission lines of C III (977), N III (991), and a combination of Ly α and O VI (1032), (1038), which appears to be accompanied by a flat continuum. These emission lines, which are probably associated with the accretion disk, remain relatively constant for the duration of the FUSE observations.

If there is an accretion belt as the second component in the FUV spectra of WDs during dwarf nova quiescence, then it is rapidly rotating, and any tell-tale absorption features would be washed out. However, past theoretical studies by Piro & Bildsten (2004a, 2004b) of the accretion onto the WD, find that at low accretion rates (like DN in quiescence) the amount of spreading is negligible and most of the dissipated energy is radiated back into the disk. When the accretion rate is high, the material spreads to latitudes high enough to be visible above the accretion disk. The timescale for spreading is very short, of the order of 50s, implying that the typical mass of the spreading region is much less than the typical "accretion belt" and therefore one is not likely to observe accretion belts in quiescence as "left over" from a dwarf nova outburst. This finding only adds to the quandary of identifying the source of the second component. It may not be an accretion belt on the WD surface. However, in an alternate picture of the boundary layer, Popham & Narayan (1993, 1995) predict that an optically thick structure may be present in the boundary layer. If this is correct, then optically thick rings could be present during dwarf nova quiescence. The two competing pictures of the boundary layer must be tested with multi-dimensional radiation hydrodynamic simulations of the accretion process.

For both the case of accretion rates determined for dwarf nova (nova-like) outbursts (high brightness states) and quiescence (low brightness states), it is essential to know from theory the predicted temperature difference between the polar and equatorial regions of the WD, where the accretion flow through the boundary layer preferentially heats the WD equatorial latitudes or produces a hot ring (e.g., "FUV" boundary layer). What is the surface area of the heated region? If it is spun up, what is the spin-down timescale? How rapidly does the accreted matter spread with latitude and depth on the WD? These questions must be addressed with simulations to shed further light on the physical nature of the second flux components. Radiation hydrodynamic simulations were carried out by Kley (1989, 1991) but did not include matter flowing into the WD which was treated as a solid boundary. Moreover, the computations were not run long enough to see any kind of steady state established. 2D hydrodynamic simulations begun in 2005 (Balsara & Fisker 2005) follow the matter into the WD for both optically thick and optically-thin boundary layer cases with the objective of following the evolution long enough to see a steady state behavior. In the first such 2D hydrodynamic evolutionary simulations following matter into the WD (Balsara et al. 2009) carried out numerical simulations for different values of the α viscosity parameter (α), corresponding to different mass accretion rates. In the high viscosity cases ($\alpha = 0.1$), the spreading boundary layer sets off a gravity wave in the surface matter. The accretion flow moves supersonically over the cusp making it susceptible to the rapid development of gravity waves and/or Kelvin–Helmholtz shearing instabilities. This BL is optically thick and extends more than 30° to either side of the disk plane after only 3/4 of a Keplerian rotation period (19 s). In the low viscosity cases ($\alpha = 0.001$), the spreading boundary layer does not set off gravity waves and it is optically thin.

Hertfelder et al. (2013) used 2D radiation hydrodynamics but with the slim disk approximation and thus could not address how the accreted gas from the boundary layer actually penetrates into the WD outer envelope. However, a more recent study by Hertfelder and Kley (2017) looked into the vertical structure of the boundary layer and how it depends on parameters such as stellar mass and rotation rate, as well as the mass-accretion rate. Their 2D simulation showed that if the WD is non-rotating, the boundary layer is characterized by a steep drop in angular velocity over a width of only 1% of the WD radius, a large depletion of boundary layer gas and, as expected, a very high temperature (~ 500,000 K) due to the strong shear. Depending on the value of the rotational velocity of the WD, the disk material travels up to the poles or is halted at a specific latitude. The rotational velocity of the WD also determines how much mixing with the stellar material occurs.

The highly important next step is to run longer 2D and 3D hydrodynamic simulations. As the computing power of the most massively parallel supercomputers continues to increase exponentially, the actual accretion of gas by the CV WD including numerically resolving the structure of the spatially thin boundary layer, must be followed, incorporating the most sophisticated microphysics and macro-physics of accretion and CV WD envelope models. The accretion flow into the WD envelope must be followed long enough for a steady state to be established. This is absolutely essential for needed insights into the physics of accretion within the accreting WD's envelope and the envelope-core, partially degenerate transition region.

Still another possibility for the source of the second component of continuum may be related to the formation of the high ionization emission lines observed during DN quiescence. If these broad emission lines are from the same source as the blue continuum, then a case may be made for the second component of flux to arise from the X-ray irradiation of the disk by the hot WD and boundary layer. Ko et al. (1996) made detailed, non-LTE calculations of the vertical structure and emergent spectrum of a temperature-inverted layer overlying the accretion disk in quiescent DN. They find a line-emitting accretion disk chromosphere is created by photo-ionization of the disk by an external X-ray/EUV source which could be the WD hot/boundary layer. Unfortunately, the accretion rate and disk mid-plane temperature they considered are inappropriately high for DN quiescence. More recently Kromer et al. (2007) have computed a time sequence of spectra of accretion disks as a dwarf nova transitions from emission to absorption during the rise to dwarf nova outburst. This work re-produces the emission lines of optically-thin accretion disks for the first time from first principles. They have modeled the emission and absorption lines of SS Cyg during its outburst and quiescence successfully but it remains unclear whether the same process gives a FUV continuum flux level in agreement with FUV observations of the second component. Obviously, with so many possibilities for the identity of the 2nd flux component, more theoretical work is required before the nature of the second component of flux is revealed.

7.8 Accreting Pulsating CV WDs

Among the important phenomena associated with accreting WDs in CVs is pulsa-tional instability. Isolated, non-accreting DA WDs known as the ZZ Ceti stars undergo non-radial g-mode pulsation when their surface temperatures lie within the ZZ Ceti instability strip, which is a downward extension of the Cepheid instability strip on the HR diagram. Asteroseismology has been a proven powerful technique to probe the interior structure of the envelope and even the core of an accreting, pulsating WD. The non-radial pulsation modes of a WD, whether accreting or non-accreting extend deep into the WD interior permitting constraints on the core and envelope mass, composition, and rotation rate (Fontaine & Brassard 2008). The complex interaction between accretion, non-radial g-mode pulsations, convection and rotation is a complex problem but not intractable (Fontaine & Brassard 2008). Indeed this methodology using non-radial pulsation modes has already delivered insights into the interior structure of the WD core, important constraints on the non-degenerate envelope, on core masses, and on the rotation rates of pulsating, accreting WDs.

When a dwarf nova outburst occurs, mass and angular momentum is accreted by the central WD, heating up a portion of the envelope followed by a cooling response occurring on the thermal timescale of the heated layer. Thus, a WD pulsator offers insights into WD cooling on timescales of years whereas the evolutionary core cooling occurs on vastly longer timescales of billions of years. Moreover, the effect of accretion heating and subsequent cooling can even shift the accreting pulsator into or out of the instability strip.

The primary way of identifying pulsators in CVs is to look for the hallmark optical spectroscopic signature of broad Balmer absorption wings, each with central emission cores. Among the most interesting systems are dwarf novae with very long quiescence intervals between outbursts like WZ Sge, GW Lib. The outbursts of these two systems are of such high amplitude that they were originally classified as classical novae. The superoutburst of WZ Sge was discussed in Section 7.10.1 of this chapter.

Currently, there are 18 known pulsating, accreting WDs in CVs, most of them from the Sloan Digital Sky Survey. These offer the exciting prospect of using the techniques of asteroseismology to probe the structure and physical properties of accreting WD pulsators. The CV systems, their orbital periods, pulsation periods and surface temperatures during quiescence are tabulated in Table 7.4.

In this section, two of the systems, GW Lib and V386 Ser, are highlighted as examples of the power of asteroseimology applied to these systems. Beginning with the first, pulsating accreting WD, GW Lib (Warner & van Zyl 1998), a complex pulsation spectrum was found (van Zyl et al. 2004). Specifically, GW Lib exhibits three main pulsation periods of 650 s, 370 s, and 230 s with variable amplitudes (van Zyl et al. 2004). An HST spectrum of GW Lib analyzed by Paula Szkody and her team (Szkody et al. 2002b) found that a single WD model atmosphere could not produce an acceptable fit to its quiescent spectrum. Therefore, Szkody et al. (2002b, 2012) tested a two-temperature model atmosphere with 63% of the photosphere being at 13,300 K and 37% at 17,100 K. This composite model worked in providing an acceptable fit. Both of these temperatures place GW Lib outside the ZZ Ceti instability strip for single ZZ Ceti pulsators. Szkody et al. (2002b) posited that the WD is massive, and theoretical pulsation modeling of the data was consistent with a massive WD pulsator. Curiously however, all other accreting pulsators are fit successfully with single-temperature model atmospheres. Theoretical work by Arras et al. (2006) showed that there could be two instability strips for pulsating accretors, one strip for lower gravity pulsating accretors with helium cores and a separate instability strip for hotter, higher gravity accreting degenerates with enhanced helium abundance from a presumed evolved donor secondary (Figure 7.18).

A superoutburst of GW Lib took place in 2007 April. During this most recent outburst, GW Lib rose from an apparent magnitude $V = 17$ to 8th magnitude. However, at 3–4 years after the 2007 outburst, the WD still had a surface temperature higher than its normal quiescent temperature and the non-radial oscillations that were undetected during the outburst, reappeared but with different periods (Szkody et al. 2012). Results on triplet spacing implied a long spin period for GW Lib while model atmosphere fitting revealed a short spin period (Szkody et al. 2012).

For GW Lib, the cooling response from its 2007 outburst, surprisingly, was found to be quite different from its cooling response from the 1983 outburst in that it was not a smooth decline (see Figure 7.19). First, the temperature from the UV spectra

Table 7.4. Known Pulsating Accreting WDs

Object	P_{orb} (minute)	P_{pulse} (s)	Outbursts	Temperature (K)	References
SDSS1507 + 52	66.6	500, 660, 1140	—	14,200	Uthas et al. (2011)
GW Lib	76.8	236, 376, 648	1983, 2007	16,000	Szkody et al. (2002b, 2012)
EQ Lyn	77.8	1192–1230	2006, 2012, 2019	15,100	Szkody (2022)
SDSS1457 + 51	77.9	582–642, 1200	—	—	Uthas et al. (2012)
BW Scl	78.2	618, 1242	2011	14,800	Gänsicke et al. (2005); Uthas et al. (2012)
V386 Ser	80.5	221, 305, 609, 175	2019	14,500	Szkody et al. (2007, 2021)
PQ And	80.6	634, 1263	1938, 1967, 1988, 2010	12,000	Patterson et al. (2005); Szkody et al. (2010)
LV Cnc	81.3	214, 260	—	13,500	Szkody (2022a, 2022b)
GY Ceti	81.5	335, 581, 595	1999, 2020	14,500	Szkody (2022a, 2022b)
V455 And	81	320–370	2007	10,500	Araujo-Betancor et al. (2005)
SDSS1339 + 48	83	642, 1065	2011	12,500	Gänsicke et al. (2006); Szkody et al. (2010)
VSDSS2205 + 11	82.8	330, 475, 575	2011	15,000	Szkody et al. (2007)
VSDSS0755 + 14	84.8	257–262	—	15,900	Szkody (2022)
SDSS0804 + 51	85.0	756, 256	2006, 2010	13,000	Pavlenko (2009); Pavlenko et al. (2012)
VSX0747 + 06	85.6	238, 684	—	—	Woudt & Warner (2011)
PP Boo	88.8	559	—	10,000	Nilsson et al. (2006); Szkody et al. (2010)
RXJ0232 − 37	95.3	267	2007	13,200	Szkody (2022)
REJ1255 + 26	119.5	668, 1236, 1344	1994	13,000 (opt)	Patterson & Thorstensen (2005)

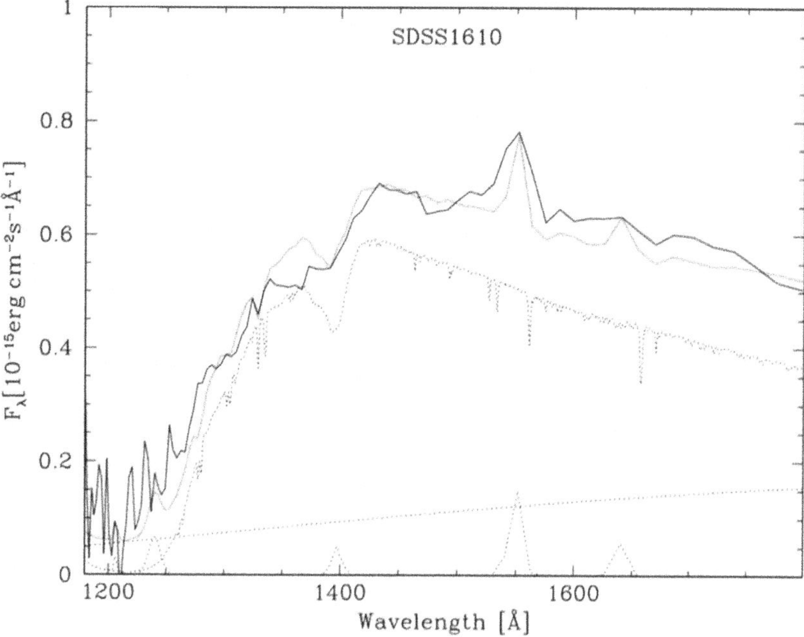

Figure 7.18. Best fit for the pulsating, accreting WD, SDSS J1610 with a 14,500 K WD at a distance of 480 pc and a 12,000 K blackbody fit. SDSS1610, 19th magnitude, P_{orb} = 81 minutes, pulsation periods 345 s, 607 s; amplitudes 7 mmag, 27 mmag, respectively by Woudt & Warner (2004). (Reproduced from Szkody et al. (2007). © 2007. The American Astronomical Society. All rights reserved.)

was not a smooth cooling transition from the 18,000 K determined from the observation 3 years after outburst to the 14,700 K measured at quiescence (Szkody et al. 2002b). Instead, there was an initial decrease to 16,000 K in 2011, and then an increase to the same average higher temperature near 17,000 K for the remaining three observations. Second, the pulsation spectrum was not a smooth change from a shorter period after outburst back to the 648, 376, and 236 s observed during quiescence. The pulsation spectrum contained a complex array of short periods (293 s in 2010 and 2011, 275 s in 2013 and 2017, and 370 s in 2015) as well as longer periods at 19 minutes and 4 hr that appeared for weeks at a time.

Following further systems after outburst will reveal information on the amount of heating due to the accreted mass and the impact on the pulsations. Further theoretical studies of pulsating, accreting WDs as well as FUV and ground-based monitoring over long timescales are essential to better understand the mode changes that occur at the base of the convection zone as the WD is heated followed by cooling in response to an outburst or superoutburst. An international campaign to resolve the modes of the pulsator V386 Ser during 11 days in 2007 May showed that the dominant pulsation mode at 609s is a triplet (Mukadam et al. 2010). The spacing of this triplet is a puzzle, implying either a very slow rotation of 4.8 days (contrary to results in Table 7.4) or differential rotation of the WD. Hubble COS spectra

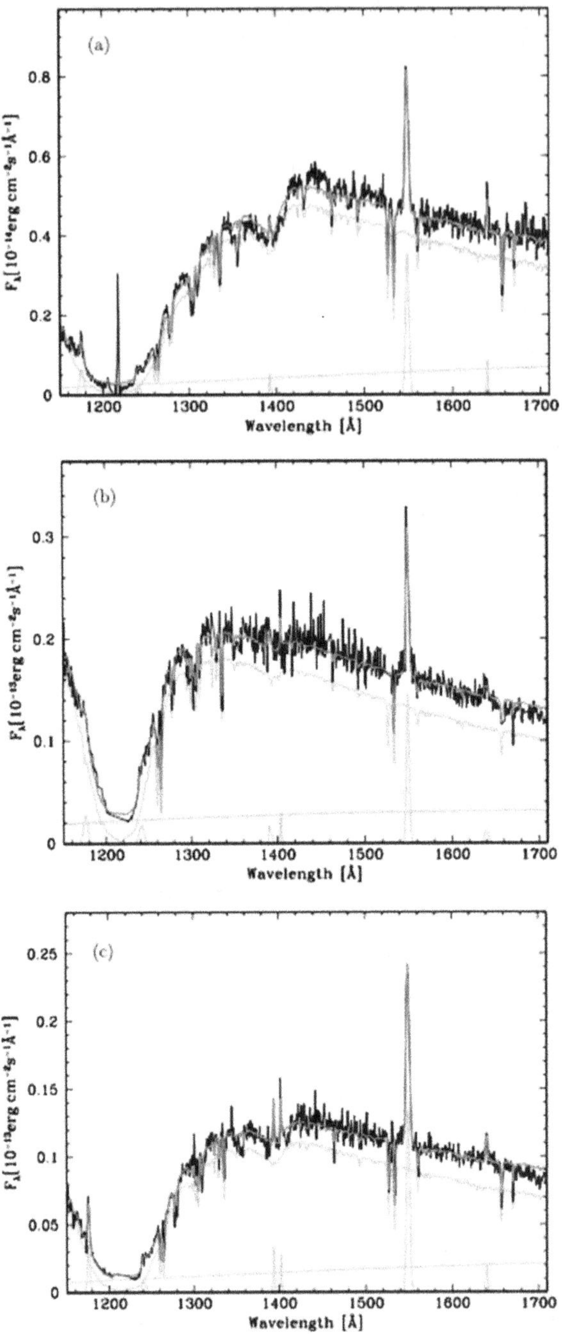

Figure 7.19. Three and four years after the 2007 April very large outburst of GW Lib, the fluxes and temperatures are still higher than quiescence ($T = 19$, 700 K and 17,300 K versus 16,000 K pre-outburst for a $\log g = 8.7$ and $d = 100$ pc). The K_{wd} of 7.6 ± 0.8 km s^{-1} determined from the C II 1463 absorption line, as well as the gravitational redshift implies a WD mass of 0.79 ± 0.08 Msun. The widths of the UV lines imply a WD rotation velocity of only Vsini = 40 km s^{-1}, a rotational velocity not unlike an isolated, non-accreting WD rather than an accreting, pulsating degenerate. (Reproduced with permission from Szkody et al. (2016). © 2016. The American Astronomical Society. All rights reserved.)

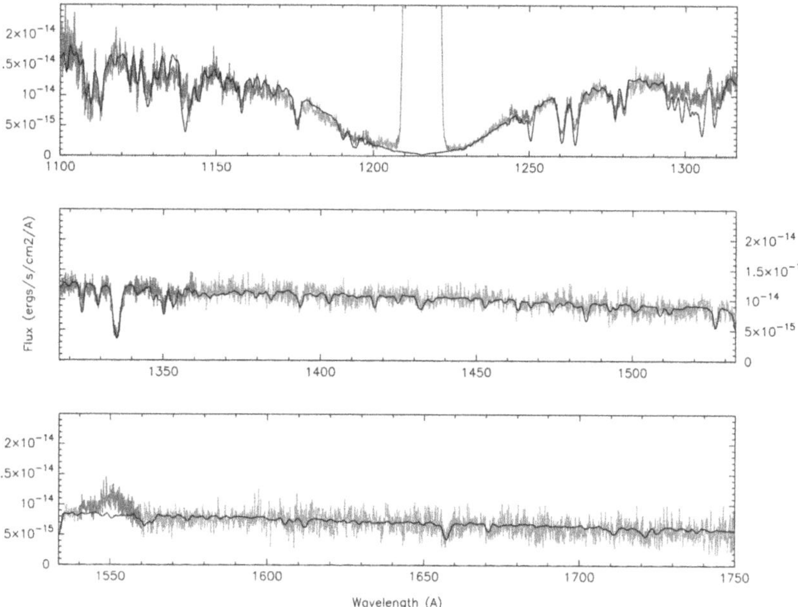

Figure 7.20. The first of two HST COS spectra, the 2019 August HST UV COS spectrum, obtained 7 months after outburst along with optical ground-based photometry The resulting spectral and pulsational analysis shows a cooling of the WD from 21,020 to 18,750 K (with a gravity log 8.1 along with the presence of strong non-radial g-mode pulsations evident in both UV and optical at a much shorter period after the 2019 outburst than at quiescence). (Reproduced with permission from Szkody et al. (2021), their Figure 6. © 2021. The American Astronomical Society. All rights reserved.)

obtained in 2019 August and 2020 February reveal the cooling of the WD between the two observations (see Figure 7.20; Figure 7.21) and the pulsational analyses showed strong pulsations in both the UV and optical with periods of 104.28 s (8/2019) and 174.56 s (2/2020).

7.9 A New Era in Research on Cataclysmic Variable Accreting WDs

Beginning with the launch of the HST in 1990, higher quality, higher resolution FUV spectra became available for the first time. While the IUE spacecraft with its much smaller telescope yielded a treasure trove of new knowledge on the properties of CV WDs (see Warner 1995; Sion 1991, 1999; Gänsicke et al. 2001; Urban & Sion 2006), the relatively small telescope aperture allowed the determination of temperatures of exposed CV WDs above and below the period gap, and a massive database of IUE spectra was built with acceptable signal to noise. However, IUE was restricted to relatively bright CV systems (dwarf novae, in outburst and quiescence), nova-like variables in continuous outburst (UX UMa nova-likes), nova-like variables in outburst with occasional deep low states (VY Scl nova-likes), and magnetic CVs (polars and intermediate polars). Moreover, there was insufficient resolution with IUE to extract rotational velocities, chemical abundances, gravitational redshifts, and reliable CV WD masses. The IUE era in

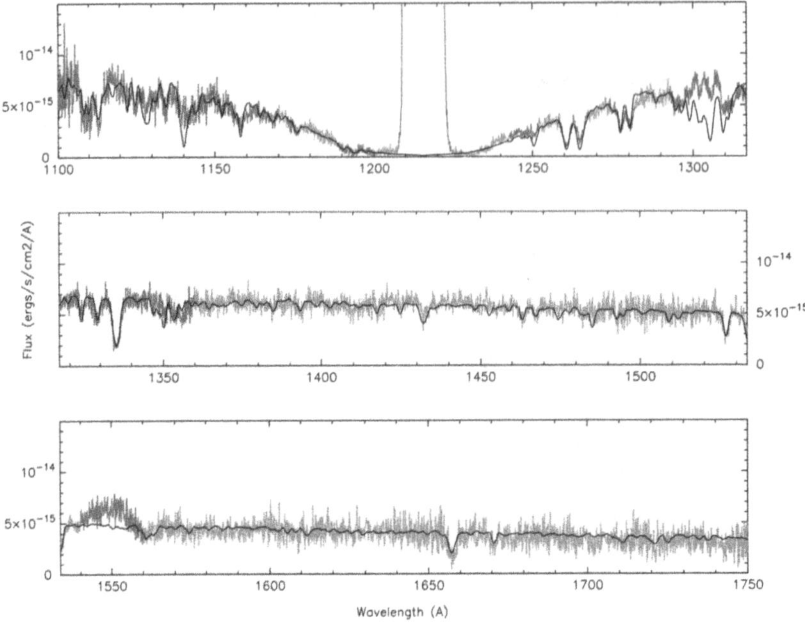

Figure 7.21. The 2020 February COS spectrum of V386 Ser (red line) is fit with a synthetic WD stellar photosphere spectrum (black line). A constant flux of 6×10^{-16} erg s^{-1} cm^{-2} A^{-1} has been subtracted from the spectrum. The WD model has a surface gravity of $\log(g) = 8.1$, surface temperature of 18,750 K, and scales to a Gaia distance of 239 pc. This model has solar composition and a projected rotational velocity of 200 km s^{-1}. (Reproduced with permission from Szkody et al. (2021); their Figure 9. © 2021. The American Astronomical Society. All rights reserved.)

studies of CVs and accreting WDs in CVs is nicely covered in reviews and UV spectroscopic atlases by La Dous (1991), Verbunt (1987), Sion (1999), IUE reviews and atlases of IUE spectra of CVs,

The first sample of robust CV WD masses was compiled in a benchmark paper by Zorotovic et al. (2011) who selected those CV systems in the Ritter and Kolb CV catalog with tabulated WD masses (116 out of 856 cataloged CVs) and dropped those CV WD masses that had no error estimates. Their final list consisted of 32 CV WDs deemed to have robust masses of which 22 of the systems were in eclipsing binaries. It was clear that no scientific investigations of CV WDs as potential SNe Ia progenitors could yield any definitive conclusion without precision WD masses. Insights into the physics of accretion onto CV WDs and their evolution also required robust masses. Zorotovic et al. (2011) determined a mean mass for their sample of 32 to be 0.82 ± 0.15 Msun and for the entire sample of CV WD masses with error estimates virtually the same at 0.83 ± 0.23. These robust CV WD masses are obviously higher than the mean mass of isolated field WDs, which is $M_{\rm wd} = 0.6 \pm 0.02$ Msun suggesting that perhaps the CV WDs gain mass as they age due to nova shell ejection masses being less than the amount of mass accreted during the quiescence between each nova explosion. This possibility was countered by the finding that the mean mass of CVs below the period gap was essentially the same as the mean mass of CV WDs above the period gap.

All of this changed when high quality HST spectra became available for large samples of CV WDs. For the first time thanks to HST, reliable temperatures, chemical abundances, rotation rates were derived, and finally by 2017, thanks to Gaia, a large body of reliable CV WD masses was derived. Beginning with guest observer (GO) programs in HST Cycle 2, CV WDs were observed including U Gem, VW Hyi and WZ Sge (Long et al. 1994; Sion et al. 1994, 1995a, 1995b, 1996). This was followed by more extensive CV WD projects using many dozens to over one hundred twenty HST orbits. A medium GO project (led by Paula Szkody), two HST snapshot projects (led by Boris Gaensicke) over two years, a large GO project (led by Boris Gaensicke) and most recently, a large GO project (led by Anna Pala) from which two landmark publications have emerged, the first paper being pre-Gaia by Pala et al. (2017), which yielded precise surface temperatures for a large sample of exposed, accreting WDs in non-magnetic CVs and several magnetic (polar) CVs with exposed WDs during low states (Araujo-Betancor et al. 2005) and the second paper (Pala et al. 2022) delivering the most accurate WD masses. These large CV WD samples were coupled with the highly anticipated Gaia mission data releases Gaia DR1, on 2016 September 14; Gaia DR2 data release on 2018 April 25, and DR3, on 2022 June 13. The impact of this large body of precision Gaia parallax measurements of CVs has been transformative. No longer were eclipsing CV binaries, with few exceptions, the only means of obtaining the most robust CV WD masses (Zorotovic et al. 2011; Pala et al. 2022). Precision temperatures were derived from scaling model atmosphere fits to HST spectra of CV WDs which together with Gaia distances led to WD radii and thus to the first derivation of a large body of precision, spectroscopically-determined CV WD masses (Pala et al. 2022). The impact of this paper on CV WD research is already immense (Pala et al. 2022; Godon & Sion 2021).

The comprehensive work by Pala et al. (2022) represents a culmination of the new science accomplished with the largest ever sample of CV WDs having Gaia parallaxes and precision CV white masses—a sample size of 89 CV WDs. The scientific highlights of this study, first and foremost, is that even with the largest sample size yet, there is no apparent difference between the average mass of CV WDs above the gap, $\langle 0.81 + 0.16/ - 0.20 \rangle$ and below the gap, $\langle 0.81 \pm 0.16 \rangle$. Since there is a growing consensus that CVs evolve from being above the gap (and younger) into older CVs below the gap, this result is indeed puzzling. Whether CV WDs gain more mass than they eject in nova explosions or lose more mass than they gain with time through nova explosions, one would expect a mass difference. However, there is currently a disparity between the number of precision CV WD masses above the gap (only 21 systems) versus below the gap (68 systems). Second, of the CV WDs in the sample, there are only 5 CVs with $M_{wd} < 0.5$ Msun, amounting to only $6\% \pm 4\%$ of all CV WDs. The overall average CV WD turns out to be $\langle 0.81 + 0.16/ - 0.20 \rangle$ Msun which notably, as in the past, is substantially higher than the average mass of isolated, non-accreting WDs (Zorotovic et al. 2011; Pala et al. 2022). Pala et al. (2022) found evidence that CV systems with orbital periods shorter than 3 hours, show an anti-correlation between the mass accretion rates and the mass of the WD, which implies the presence of an additional mechanism of angular momentum loss that is more efficient the lower the mass of the WD. Pala et al. (2022) also noted that CV systems above the period gap are hotter and accrete at a higher rate than CV systems below the period gap, thus confirming the same finding by Urban & Sion (2006) using the entire sample of exposed CV WDs with IUE archival spectra.

7.10 Structure, Evolution, Heating and Cooling of CV WDs

During a dwarf nova outburst, accretion onto the WD occurs at a high rate as much of the disk mass falls inward. After heating by irradiation and compression, the WD cools. This cooling in response to the outburst was one of several possible explanations for the observed decline of UV flux observed with IUE during dwarf nova quiescence (Verbunt 1987). Indeed, these declines appeared to contradict the standard disk instability theory for the outbursts, which predicts an increasing FUV flux during quiescence due to a gradual increase in the accretion rate. As more detections of the underlying WDs were made using the IUE and HST observations, Sion (1985, 1991) showed that WDs in CVs were hotter than field WDs of comparable age, and in Sion (1995) that compressional heating by the accreted material was the source of heating responsible for the measured surface temperatures, and observed FUV rate of flux decline. In the next section, the detailed physics of compressional heating in the context of accreting WDs is covered.

7.10.1 The Physics of Compressional Heating

To determine if these UV flux declines were due to cooling of the heated WD, theoretical simulations (for VW Hyi) were carried out by Pringle (1988) who considered the effect of the downward irradiation of the WD surface layers. The instantaneous release of gravitational energy when gas accretion onto the photosphere of a non-magnetic disk-accreting WD ($0.5GM_{\mathrm{wd}}\dot{M}/R_{\mathrm{wd}}$) is commonly referred to as accretion energy. However, the heating which occurs, mostly in the form of hard X-ray photons, does not penetrate very deeply beyond a few Compton depths. Any such heating is quickly released and the shallow depth to which it penetrates has a very short thermal timescale. This source of energy therefore could not account for the observed rate of FUV flux declines measured among dwarf novae in the aftermath of heating from a dwarf nova outburst, as was concluded by Pringle (1988).

Sion (1995) simulated compressional heating with time variable accretion by using a full, quasi-static, Henyey-type, evolutionary code with OPAL opacities and showed that compressional heating alone could account for the observed heating and subsequent cooling of the WDs before the next dwarf nova outburst occurs. Two of several such simulations are displayed in Figures 7.22 and 7.23. In Figure 7.22, a massive WD (1.2 Msun) undergoes one episode of accretion to determine the amount of heating from a single dwarf nova outburst. Accretion at 10^{-8} Msun was switched on for seven days to reveal the elevation of surface temperature and the subsequent rate of decline following a single outburst (Sion 1995). In Figure 7.23, another example of the evolution of an accreting WD was examined for a 0.8 Msun degenerate accreting at the rate of 10^{-8} Msun yr^{-1} for 7 days (the dwarf nova outburst) followed by 327 days of quiescence before the next

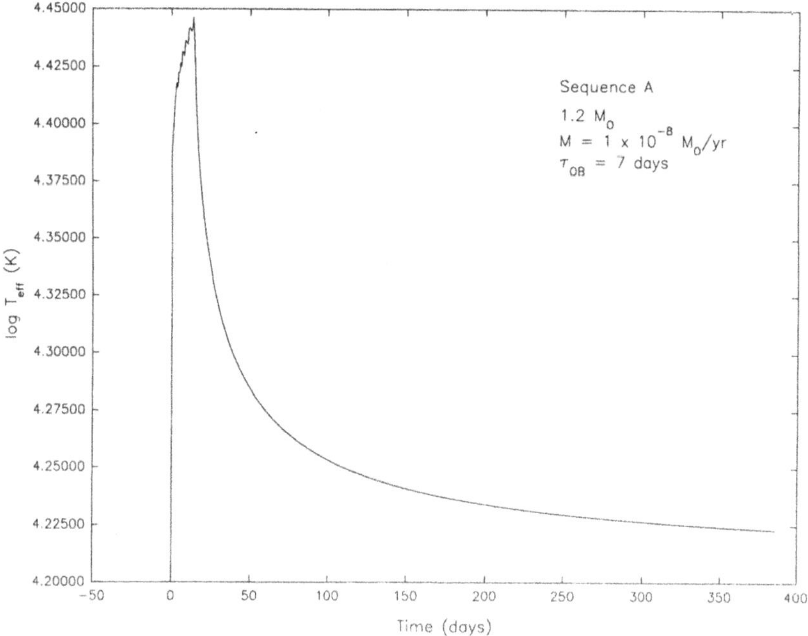

Figure 7.22. A massive CV WD (1.2 Msun) undergoes seven days in outburst accreting at the rate 1×10^{-8} Msun yr^{-1}, then undergoes cooling back to its surface temperature during dwarf nova quiescence. (Reproduced with permission from Sion (1995). © 1995. The American Astronomical Society. All rights reserved.)

outburst, and so on (Sion 1995). The evolution of the accreting WD was followed for 15 outburst-quiescence cycles. Sion (1995) and Godon & Sion (2003) ran further simulations of compressional heating with the inclusion of boundary layer irradiation, the WD rotation, and OPAL opacities. Independently, Piro et al. (2005) carried out a related study of the response to WD heating by the DN outburst, showing that long after the outburst, the fractional flux perturbation about the quiescent flux decays as a power law with time, not as an exponential decay. They confirmed the study by Sion (1995) that compressional heating alone was the source of heating (and subsequent cooling). They derived a simple fitting formula that yields estimates for both the quiescent flux and the accreted mass per unit area from which the WD mass can be derived if the accreted mass is known from observation.

One of the best observed cases of modeling the FUV flux decline following a dwarf nova outburst is WZ Sge which has the longest recurrence time of all dwarf novae and has extensive space-based and ground-based observational coverage. The flux decline was followed for several years after the 1978 outburst by Slevinsky et al. (1999), and after the 2001 July outburst that occurred 10 years early. Since a precision parallax was known (pre-Gaia), intense HST STIS and FUSE coverage deepened our insights. The observed light curve and model atmosphere-derived temperature decline of the accretion-heated WD were compared with the

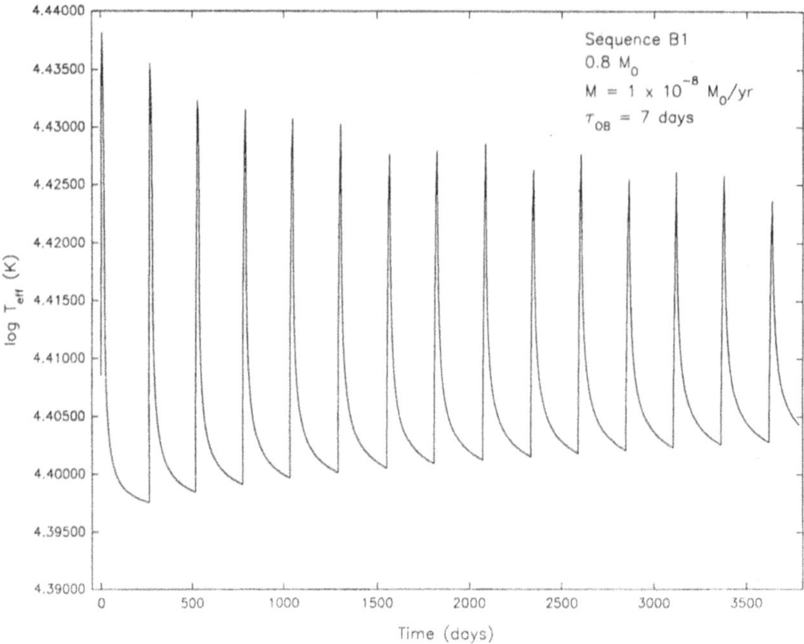

Figure 7.23. A CV WD with a mass of 0.8 Msun (the average mass of a CV WD) undergoes accretion at the rate 10^{-8} Msun for 7 days simulating the dwarf nova outburst followed by post-outburst cooling into quiescence, then another outburst after 327 days in quiescence, and so on. The evolution of the CV WD was followed for 15 outburst-quiescence cycles. (Reproduced with permission from Sion (1995). © 1995. The American Astronomical Society. All rights reserved.)

evolutionary simulations of Godon et al. (2004, 2006) displayed in Figure 7.24. These studies showed that compressional heating alone could account for the observed flux decline and thermal evolution of the WD but only if the WD is massive ($\sim 1.2 M_\odot$). This conclusion was confirmed independently by Piro et al. (2005). An example of their semi-analytic models of the heating and cooling is a comparison of their numerical results with the observed cooling of the WD in WZ Sge following the 2001 outburst and is displayed in Figure 7.25.

However, there are a few systems which have some coverage at different times during quiescence. Thus, changes in their temperatures during quiescence as a function of time since the last outburst have been studied. Unfortunately, detailed comparisons with observed flux declines are hampered by the lack of sufficient coverage of the light curve during dwarf nova quiescence. The response of the massive WD in U Gem to an outburst has been studied extensively with IUE, HUT, HST (FOS, GHRS, STIS), and FUSE. The temperature of the WD cools from 38,000 K shortly after the outburst down to 30,000 K in quiescence; however, as Long et al. (1994) pointed out, the UV flux decline is more gradual than the rate of cooling for a single-temperature WD at a fixed radius. The flux component responsible for this delayed cooling is discussed below.

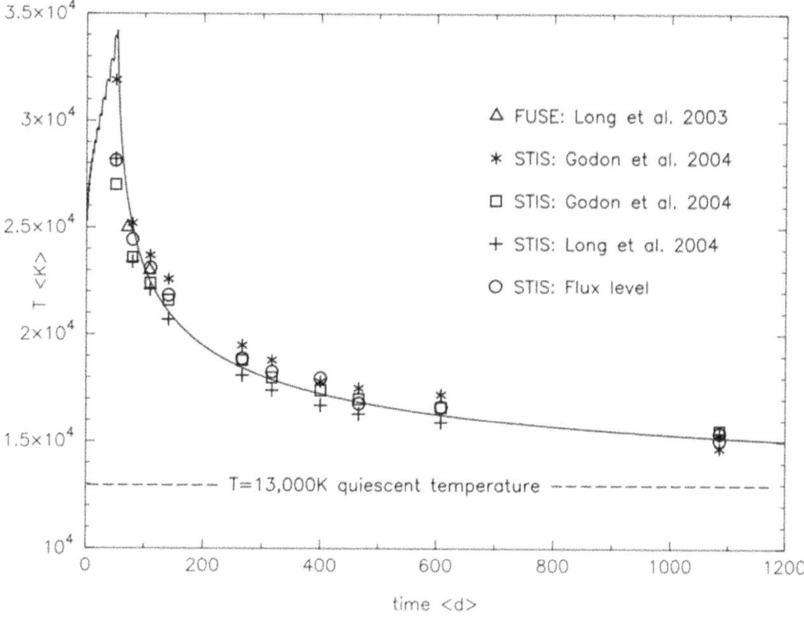

Figure 7.24. Modeling the heating and cooling of WZ Sge. The temperature (in Kelvins) of the WD is drawn as a function of time (in days) since the start of the outburst (2001 July 23). The solid line represents the compressional heating model with a 0.9 Msun WD (corresponding to $\log g = 0.85$ Msun) with an initial temperature of 13,000 K, accreting at a rate of 2.8×10^{-8} Msun yr^{-1} for 52 days. (Reproduced with permission from Godon et al. (2006). © 2006. The American Astronomical Society. All rights reserved.)

The response of the WD to heating was explored in an eclipse mapping analysis of light curves of HT Cas by Baptista (2012). His study revealed a very fast response of the underlying WD (which brightened by a factor of 2) to the increase in mass transfer rate, a simultaneous increase in the expansion rate of the accretion disk (which brightened by a factor of 3) and a relative amplitude of the high-frequency flickering which implies a high viscosity ($\alpha \sim 0.3 - 0.7$) for the quiescent disk and hence disagrees strongly with the key prediction of the disk instability model. This suggests that the outbursts of HT Cas may be due to bursts of enhanced mass transfer from the donor star. As a theoretical test of the observed fast response of the WD in HT Cas, Baptista (2012) carried out a simulation of the heating of the HT Cas WD by the dwarf nova outburst. For the basic parameters of the WD in HT Cas, he took $M_{wd} = 0.61 M_{\odot}$ ($\log(R_{wd}) = 8.962$), $V_{sini} = 200$ km s^{-1}, $T_{eff} = 14,000$ K (quiescence). To simulate the heating from irradiation and compression, Baptista (2012) took an accretion rate of $10^{-8} M_{\odot}$ yr^{-1} switched on for 2 days, followed by cooling. Baptista (2012) has displayed the theoretical simulation. The WD quickly reached a peak temperature of 18,600 K in response to the outburst. After ~10 days the WD cooled to its temperature in quiescence. This is in marked disagreement with the disc-instability model and implies that the outbursts of HT Cas could possibly be caused by bursts of enhanced mass-transfer rate from its donor star (Baptista 2012).

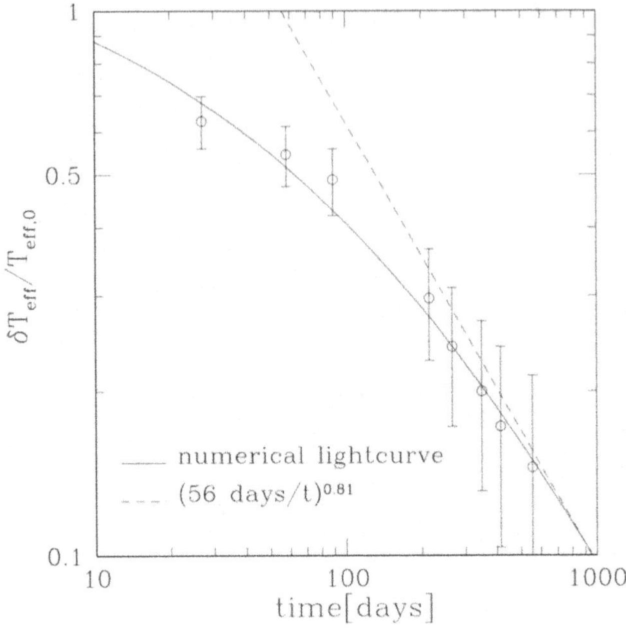

Figure 7.25. Fractional temperature during the 2001 July outburst of WZ Sge in comparison the cooling of a compressionally heated envelope. The circles show the temperature T_b from Godon et al. (2004), with the error bars of 1000 K. The solid curve is our numerical calculation with $T_{eff} = 14, 500$ K and constant $\dot{m} = 0.30$ g cm^2 s^{-1} for 52 days, giving Delta $y = 1.3 \times 10^6$ cm^{-2} corresponding to an average accretion rate of $\dot{M}10^{-8}$ Msun yr^{-1} for an $M_{wd} = 1.1$ Msun, $R_{wd} = 5 \times 10^8$ cm WD. The analytic formula given in equation (20) of Piro et al. (2005) is for $t_{late} = 56$ days and is shown by the dashed line. (Reproduced with permission from Piro et al. (2005). © 2005. The American Astronomical Society. All rights reserved.)

7.10.2 Accreting WDs: Heating and Cooling in Response to Long-Term Accretion

At the outset of this section, it is helpful to define often confused terms between "steady state" accreting models and steady nuclear burning models. Gravitational potential energy will be converted into internal heat in response to overall structural changes brought on by accretion but ultimately, a steady state is approached in the sense that following the first or second thermonuclear pulse, the accreting model "forgets" the effect of the initial conditions and the characteristics of successive nuclear pulses become increasingly identical (for a given WD mass and accretion rate) and more and more independent of the initial conditions. However, "steady state" nuclear burning refers exclusively to an equilibrium balance between the rate at which matter is being accreted and the rate at which hydrogen is being destroyed by nuclear burning.

The instantaneous release of gravitational energy when gas accretion onto the photosphere of a non-magnetic disk-accreting WD ($0.5GM_{wd}\dot{M}/R_{wd}$) is commonly referred to as accretion energy. However, as discussed earlier in this chapter, the heating which occurs, primarily due to hard X-ray photons, has a very short thermal timescale due to the shallow depth to which it penetrates (Pringle 1998). Any such heating is released too quickly to account for the observed FUV flux declines following a DNe outburst.

Fundamentally, compressional heating is a consequence of the heat stored in the envelope of the accreting WD due to the weight of the accreted material. In its simplest form, the luminosity and the surface temperature of an accreting WD due to compressional heating can be expressed approximately as

$$L_{\rm s} \approx \beta \left[\left(\frac{GM_{\rm wd}\dot{M}}{R_{\rm cold}} \right) - \left(\frac{GM_{\rm wd}\dot{M}}{R_{\rm wd}} \right) \right] - L_\nu \tag{7.7}$$

where $R_{\rm cold}$ is the Hamada–Salpeter radius of a zero-temperature WD, and beta is the fraction of the total accretion energy stored in the WD. In a classic paper by Icko Iben (1982), the value of beta turned out to be surprisingly high, amounting to between 0.15 to 0.25 of the total accretion energy, for spherical accretion, a result that could only be known from evolutionary accreting WD model sequences.

The surface temperature of the accreting WD follows from the Stefan–Boltzmann law

$$T_{\rm eff} = [L_{\rm s}/(4\pi R_{\rm wd}^2\sigma)]^{0.25}. \tag{7.8}$$

For a WD envelope undergoing accretion and with no nuclear energy release, energy conservation for a given mass element may be written

$$\frac{dL_r}{dM_r} = \epsilon_{\rm g} \tag{7.9}$$

where M_r is the mass internal to a shell of radius r and L_r is the luminosity crossing that shell. From the 1st law of thermodynamics, the gravitational energy release can be expressed as

$$\begin{aligned}
\epsilon_{\rm g} &= \frac{-du}{dt} - P\frac{dv}{dt} \\
&= \frac{-du}{dt} + \frac{P}{\rho^2}\frac{d\rho}{dt} \\
&= -c_{\rm p}\frac{dT}{dt} + \frac{\delta}{\rho}\frac{dP}{dt} \quad \text{and} \\
\delta &= -\left(\frac{d\ln\rho}{d\ln T} \right)_{\rm p}.
\end{aligned} \tag{7.10}$$

Since the compressional heating is essentially adiabatic, both for the envelope and for the core, the time derivative of the internal energy can be written

$$\frac{du}{dt} \approx \frac{P}{\rho^2}\left(\frac{d\rho}{dt} \right). \tag{7.11}$$

Therefore, from Equation (7.10), since $dP/dt < 0$, then $dT/dt < 0$.

Thus, the work done against the pressure is absorbed as internal energy.

Now, the time-averaged surface luminosity of an accreting WD is given by:

$$\langle L_{\rm s} \rangle = \langle L_g^{\rm acc} \rangle + \langle L_{\rm nuc} \rangle + \langle L_{\rm shear} \rangle + L_{\rm rad} + L_{\rm cool} - L_\nu \tag{7.12}$$

where $\langle L_g^{\rm acc} \rangle$ is the time-averaged rate of gravitational potential energy release due to structure changes and compressional heating as the envelope mass, $M_{\rm env}$, increases due to accretion, $L_{\rm rad}$ is the accretion luminosity radiated away immediately at the surface, $\langle L_{\rm cool} \rangle$ is the thermal cooling luminosity, $\langle L_{\rm shear} \rangle$ is the luminosity generated by shear mixing of the tangentially accreted gas into the envelope which may happen in accreting WDs but needs to be explored with 2D and 3D hydrodynamic simulations, $\langle L_{\rm nuc} \rangle$ is the cycle-averaged nuclear luminosity, and L_ν is the neutrino luminosity.

An insightful examination of the microphysics of heating and cooling of an accreting WD in the long term was carried out by Townsley & Bildsten (2004) who considered the detailed physics of its thermal structure and evolution. They made use of a thin layer approximation, properties of radiative zero envelopes, justified the use of static envelope models for both the effect of accretion on the thermal structure of the accreted non-degenerate envelope and under what conditions the core of the accreting WD is also heated and responds to continued accretion. The exceptional importance of this paper makes it worthwhile to follow their derivation of one of the three key equations (the conservation of energy), the others being heat transport and hydrostatic equilibrium, that Townsley & Bildsten (2004) used to numerically integrate down to the core–envelope boundary and beyond and derive the thermal structure of their "static" model envelope and ascertain the relationship between the core luminosity and the surface luminosity.

For an accreting WD, internal energy is released in the atmosphere by the continual sinking of accreted gas within the gravitational potential well of the WD, its mass/$\log g$. This freshly accreted gas as it moves downward at some velocity v_r displaces already accreted gas layers pushing it downward and adiabatically compressing the layers below. This leads to compression of the envelope by gravitational energy release as well as by any thermonuclear fusion energy released from slow, stable burning at the bottom of the accreted envelope as the mass of the accreted envelope increases with time. This internal energy release modifies the L–$T_{\rm c}$ relation, giving an accreting WD a higher luminosity than an isolated, non-accreting WD with the same core temperature, $T_{\rm c}$.

In the case of a non-accreting isolated WDs, the relation between the luminosity L and the core temperature, $T_{\rm core}$, depends upon the WD mass, $M_{\rm wd}$ (i.e., $\log g$) and the chemical composition of the non-degenerate envelope whose opacity is the thermostat regulating the rate of cooling of WDs. On the other hand, for an accreting WD, its surface luminosity, L, versus the core temperature $T_{\rm c}$ relation depends on three parameters, namely the mass of the WD, $M_{\rm wd}$, the time-averaged rate of accretion $\langle \dot{M} \rangle$, and the mass of the accreted envelope, $M_{\rm env}$.

They showed that for a fixed $M_{\rm wd}$ ($\log g$), the local heat equation is given by

$$T \frac{DS}{Dt} = T \frac{\partial S}{\partial t} + T v_r \frac{\partial S}{\partial r} = -\frac{dL_{\rm r}}{dM_{\rm r}} + \epsilon_{\rm nuc} \tag{7.13}$$

where the first term on the left-hand side is the time dependent rate of change of the specific entropy and the second term is the advection of specific entropy. On the right-hand side, the first term is the loss of energy by flux divergence and the second term is the energy gain due to thermonuclear fusion. Now $v_r = -\langle \dot{M} \rangle / 4\pi\, r^2 \rho$ is the sinking speed of accreted gas motion which forces previously existing gas downward in the layer, where S is the specific entropy.

The time derivative of the entropy, $\partial S/\partial t$, depends on the current thermal structure of the non-degenerate envelope and $\langle \dot{M} \rangle$. If the specific entropy remains constant, then $\partial S/\partial t = 0$ and the envelope is unchanged. If $\langle \dot{M} \rangle$ is higher but with an unchanged thermal structure in the envelope, then a local build-up of heat results in making $\partial S/\partial t > 0$, which means heat will build-up rapidly leading to a thermonuclear runaway, because $|v_r|$ increases (due to the larger $\dot{M}$) but with no change in the flux divergence dL_r/dM_r. On the other hand, if $\langle \dot{M} \rangle$ becomes smaller, then the envelope would cool.

Accordingly, Townsley and Bildsten (2004) converted to pressure coordinates and defined the accretion rate per unit WD surface area, $\dot{m} = \langle \dot{M} \rangle / 4\pi r^2$ to translate measurements of CV WD values into measurements of accretion rate per unit WD surface area averaged over the thermal time of the radiative envelope (~ 1000 yr).

With no (or negligible) nuclear energy release into the envelope, Equation (7.13) with $\epsilon_{\mathrm{nuc}} = 0$ becomes

$$T v_r \frac{dS}{dr} = g\dot{m}T\frac{dS}{dP} = g\dot{m}\, c_p \frac{T}{P}\left[\left(\frac{d\ln T}{d\ln P}\right) - \nabla_{\mathrm{ad}}\right] \tag{7.14}$$

where the specific heat at constant pressure $c_{\mathrm{p}} = T(\partial S/\partial t)_{\mathrm{p}}$ and the adiabatic gradient $\nabla_{\mathrm{ad}} = (\partial \ln T/\partial \ln P)_{\mathrm{s}}$.

The advection of specific entropy is expressed by how much the actual temperature gradient differs from the adiabatic temperature gradient. In the case of a static envelope $dS/dt = 0$, then the gradient of the luminosity becomes, in pressure coordinates,

$$\frac{dL}{dP} = \langle \dot{M} \rangle c_p \left[\frac{dT}{dP} - \left(\frac{\partial T}{\partial P}\right)_{\mathrm{s}}\right] - \frac{4\pi r^2 \epsilon_{\mathrm{nuc}}}{g}. \tag{7.15}$$

Equation (7.15) expresses the conservation of energy which Townsley & Bildsten (2004) integrate inward along with the equations for heat transport and hydrostatic equilibrium to construct a "static" model envelope with the radius R_{wd} of the WD kept fixed at the Hamada–Salpeter zero temperature mass–radius relation, and no nuclear energy generation. With the three structure equations integrated inward, then the thermal structure of the envelope will be determined and hence the values of L_{s} and L_{core} that depend on the same four parameters, namely the core temperature T_{c}, the WD mass M_{wd}, the time-averaged accretion rate $\langle \dot{M} \rangle$, and the mass of the accreted envelope, M_{acc}. After some simplifying assumptions, Townsley & Bildsten (2004) kept the radius r fixed and assumed no nuclear release $\epsilon_{\mathrm{nuc}} = 0$. With those assumptions, Equation (7.15) for the gradient of the luminosity can be integrated analytically to yield an important relation between the surface luminosity L, time-averaged accretion rate, $\langle \dot{M} \rangle$ and core temperature, T_{c}.

$$L \approx 3\langle \dot{M}\rangle kT_{\rm c}/\mu m_{\rm p} \tag{7.16}$$

where μ is the mean molecular weight of the envelope of the accreted material, k is the Boltzmann constant and $m_{\rm p}$ is the proton mass. Thus, the authors established a useful relation between the surface luminosity and core temperature, as long as the radius remains constant and $\epsilon_{\rm nuc} = 0$.

They also justified their use of a static envelope by showing that

$$\left| \partial T/\partial t \right| \ll T/\tau_{\rm acc} \left| \nabla_{\rm ad} - \frac{d\ln T}{d\ln P} \right| \tag{7.17}$$

where $\tau_{\rm acc} = P/g\dot{m}$ and $M_{\rm acc}$ increases with time as $\langle \dot{M}\rangle \Delta t$.

To do this, the authors use finite differencing, set $\Delta t = \tau_{\rm acc}$ and compared the thermal timescale of the accreted envelope, $\Delta M\, c_{\rm p}T/L$, and the time to accrete the thin layer under study, viz. $\tau_{\rm acc} = \Delta M/\langle \dot{M}\rangle$ where ΔM is the mass of the thin layer, L is the surface luminosity, and T is the temperature of the thin layer sustained by compressional heating from deeper layers. The ratio of the thermal timescale to the accretion timescale for the layer leads to the condition $\Delta T/T \ll 2/5$ where ΔT is the temperature change at the base of the envelope during the accumulation of accreted gas. Since the increase in the envelope mass takes place at constant core temperature $T_{\rm c}$, $\Delta T/T \ll 2/5$ is always satisfied. Therefore the approximation of adopting a static envelope by the authors is superbly justified.

Townsley and Bildsten (2004) show that $\tau_{\rm th}/\tau_{\rm acc} \sim T/T_{\rm c}$, which essentially means that when $T \ll T_{\rm c}$, the accreted gas is cooling much faster than it is being accreted. Hence, there is no effect on the thermal structure of the deeper layers in the envelope. However, the condition $T \ll T_{\rm c}$ breaks down and is no longer valid during a dwarf nova outburst or superoutburst where the thermal structure can be significantly altered including the deeper layers and possibly the upper core at the core–envelope boundary. However, the depth to which the envelope has been heated has a Kelvin time of order the decline timescale back to dwarf nova quiescence, regardless of whether it is a normal outburst or a superoutburst.

The authors also examine the effect of sedimentation of helium ions in the H-dominated background gas. In particular, He ions are undergoing thermal and gravitational diffusion. Unfortunately, there is considerable uncertainty about the diffusion coefficients all the way back to the work of Fontaine & Michaud (1979), Muchmore (1984) and Paquette et al. (1986). See for example Koester (2013) and Heinenon et al. (2020). Nevertheless, mindful of the uncertainties, it is worthwhile to apply new improvements and updated calculations of diffusion. Townsley & Bildsten (2004) compare the downward drift velocity, $v_{\rm drift}$, of the helium ions under a macroscopic force $-2m_{\rm p}g$, with the downward velocity of the accreting gas, $v_{\rm acc} = \langle \dot{M}\rangle/4\pi R_{\rm wd}^2\rho$ where $v_{\rm drift} = 2m_{\rm p}gD/kT$ where D is the thermal diffusion coefficient, $m_{\rm p}$ is the proton mass, g is the surface gravity, k is the Boltzmann constant and T is the temperature. They show that the effect of energy release due to Helium sedimentation for the lowest accretion rates comprises a non-negligible fraction of the energy released by compressional heating. They also showed with numerical estimates that for a surface luminosity of $L = 10^{-3}$ Lsun, the CNO cycle during the long interval of say, 10^6 years between two successive nova explosions, the temperature of the outer 10^{-3} of the core mass can be elevated by $\sim 10\%$ or about 10^6 K.

In Figure 7.26, the modeling of Townsley & Bildsten (2004) illustrates the relationship between the surface luminosity, L_s and the core temperature, T_c is for a 0.6 Msun WD accreting over a range of accretion rates 10^{-11}, $10^{-10}.4$, 10^{-10} and $10^{-9.4}$ Msun yr^{-1} and a fixed layer mass of 5×10^{-5} Msun. As expected, the surface luminosity steeply increases, the higher the accretion rate and the core temperature T_c. Shown for comparison is the L_s versus T_c relation if $\langle \dot{M} \rangle = 0$ and the surface luminosity for a cooling, non-accreting DA WD with hydrogen layer mass, M_H of 6×10^{-5} Msun and a helium layer mass of 5×10^{-3} Msun.

From their examination of the surface luminosity L, and envelope thermal structure versus T_{core}, their models reveal that for low values of $\langle \dot{M} \rangle$ and high values of the core temperature T_c, radiation from the hot core flows upward into the bottom of the accreted envelope layers heating it, and the hot core dominates the surface luminosity. However, at a given core temperature T_c below which energy release due to accretion becomes important, the accretion rate $\dot{M}$ is the controlling parameter. For higher values of the accretion rate $\dot{M}$, the energy from compressional heating is

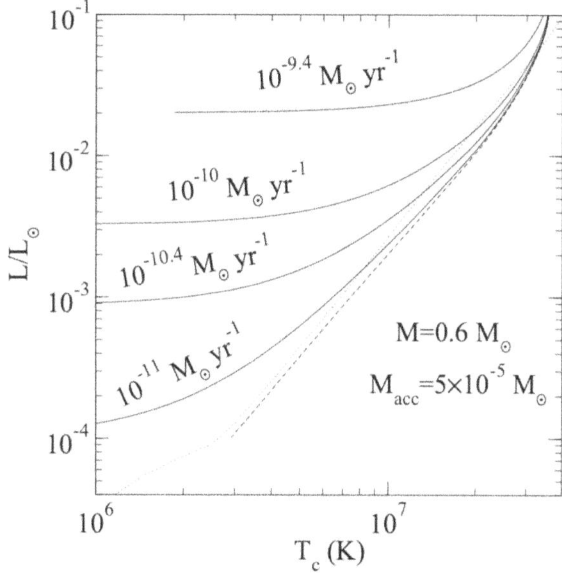

Figure 7.26. Exiting surface luminosity L as a function of core temperature T_c for a 0.6 Msun WD. The accumulated H/He layer has a mass $M_{acc} = 5 \times 10^{-5}$ Msun. Various accretion rates are shown as labeled solid lines with $\langle \dot{M} \rangle = 10^{-11}$, $10^{-10.4}$, 10^{-10}, and $10^{-9.4}$ Msun yr^{-1} The dashed line has $\langle \dot{M} \rangle = 0$, and the dotted line is the luminosity–core temperature, $L - T_c$, relation of an isolated cooling DA WD with pure hydrogen layer of mass $M_H = 6 \times 10^{-5}$ Msun and pure helium layer mass $M_{He} = 6 \times 10^{-3}$ Msun. (Reproduced with permission from Townsley & Bildsten (2004). © 2004. The American Astronomical Society. All rights reserved.)

released faster and thus the luminosity L is dominated by accretion at a higher value of T_{core}. However, at the minimum values of overall luminosity L, for a given accretion rate, the accreted envelope determines its own temperature and the surface luminosity L is decoupled from the core temperature T_{c}. There is a temperature inversion when integrating downward with the temperatures increasing with depth as expected but once the edge of the core is reached, the temperature gradient flattens out due to the efficiency of electron conduction and the temperature then decreases into the core. Most importantly, the crossover behavior in T_{c} between core heating at low values of T_{c} and core cooling at high values of T_{c} is revealing. At low values of $\langle \dot{M} \rangle$, any long-term compression of the WD core due to mass gain, has little impact on the envelope. By the same token at low values of $\langle \dot{M} \rangle$, long-term expansion of the WD core due to mass loss has little impact. Core heating occurs at high values of the mass of the accreted layer, M_{acc}, and core cooling at low values of M_{acc}. What is $T_{\mathrm{c,eq}}$ over the classical nova cycle?

In Figure 7.27 from Sion (2002) and Figure 7.28 from Pala et al. (2022), the surface temperatures of exposed CV WDs of different subtypes are displayed versus orbital period P_{orb}. The overall distribution function of CV WD temperatures reveals a close clustering of CV WD surface temperatures below the lower boundary

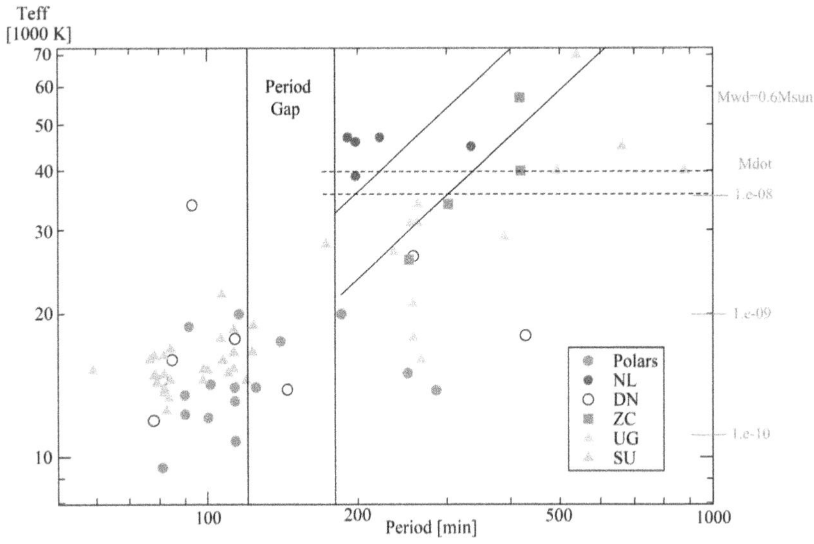

Figure 7.27. Effective WD temperature as a function of the orbital period. The references for the individual temperatures can be found in Sion et al. (2008), Townsley & Bildsten (2009), and Araujo-Betancor et al. (2005) and references therein. The traditional magnetic braking above the period gap (Howell et al. 2001) is shown between the parallel diagonal solid lines. On the right-hand side are the time-averaged accretion rates corresponding to the temperature scale on the left-hand side of the diagram based upon the T_{eff} versus $\langle \dot{M} \rangle$ relation of Townsley & Bildsten (2003) for a 0.6 Msun WD. Shown for comparison between the dotted lines is the effective temperature evolution (the horizontal dashed lines) during the long-term evolutionary path of a 0.8 solar mass WD (with an initial core temperature of 30 million degrees K) which has undergone 1000 nova outburst cycles accreting at the long-term, time-averaged rate of 10^{-8} Msun yr^{-1}, shown with permission by Dr. Irit Idan.

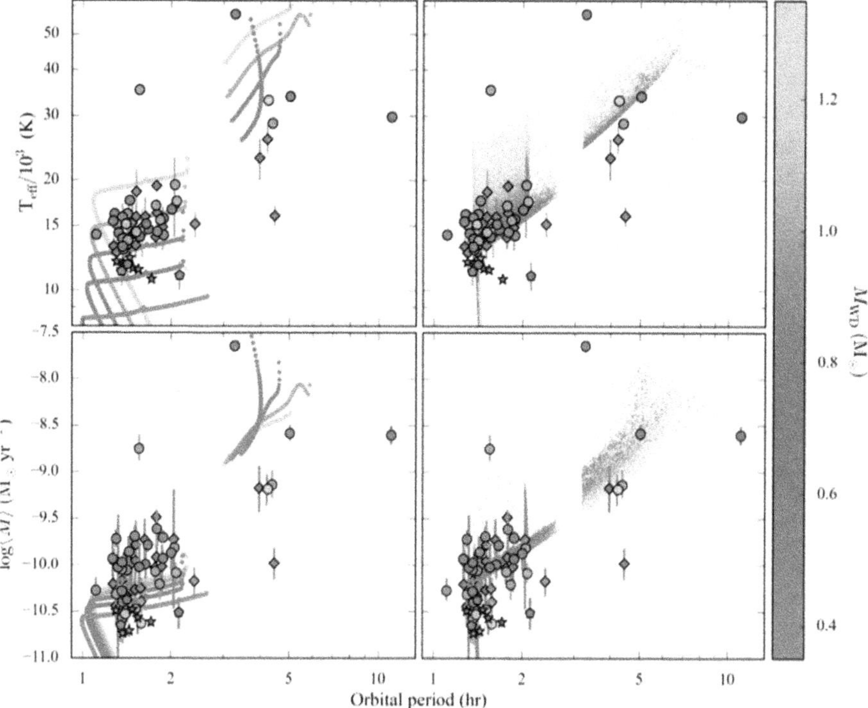

Figure 7.28. Effective temperatures (top) and mass accretion rates (bottom) as a function of the orbital period, for the systems in our HST sample (circles for pre-bounce and stars for period bounce CVs) and those from the literature (diamonds for pre-bounce and pentagons for period bounce CVs). The inset shows a closeup of the period bounce systems. (Reproduced with permission from Pala et al. (2022); their Figure 14. Copyright © 2021, Oxford University Press.)

($P_{orb} < / = 2$ hours) of the CV period gap, which is almost entirely populated by SU UMa-type dwarf novae (with a few other objects "known" to be period bouncers) while above the upper boundary of the period gap ($P_{orb} \approx 3$ hours) the CV WD surface temperatures are widely dispersed at a given orbital period. Sion et al. (2003, 2008) addressed the close clustering of SU UMa and WZ Sge dwarf novae with a relatively small dispersion in T_{eff} centered roughly at ~15,000 K. These authors commented that 15,000 K is roughly the temperature expected from compressional heating when mass transfer is driven by gravitational wave emission at accretion rates in the range of 10^{-10} to 10^{-11} Msun yr^{-1}. This author is unaware any other published comments on a physical interpretation of the very different distributions of CV WD T_{eff} above versus below the CV period gap. It seems to this author that the insightful modeling by Townsley & Bildsten (2004) could possibly explain this difference. It could possibly be due to the equilibrium temperature at which the heating of the core and the cooling of the core over classical novae limit cycle evolution is reached, i.e., the equilibrium core temperature, $T_{c,eq}$, where heating and cooling of the core balance each other. This equilibrium core temperature, $T_{c,eq}$, may possibly have been achieved by the older CV systems below the gap, while above the gap, CV systems (mostly U Gem systems, Z Cam systems, the UX UMa and VY Scl nova-likes, and post-novae) being younger, have not had sufficient time to reach $T_{c,eq}$. Hence, this may contribute in part to the much larger dispersion of CV WD temperatures at a given orbital period, above the period gap compared with the tight concentration of CV WD Temperatures below the period gap.

The accreted envelope mass versus ignition temperature (at the envelope base) was also explored quantitatively by Townsley & Bildsten (2004). They compared the results of their calculations with two highly cited monographs on the semi-analytical theory of hydrogen shell flashes, flash strength, recurrence times and steady nuclear burning at the accretion supply rate (Fujimoto 1982a, 1982b). We recall here that the most important parameter characterizing the strength of a hydrogen shell flash is the "proper pressure" defined by

$$P_{sh} = \frac{GM}{r_1^2}\left(\delta \frac{M_i}{4\pi r_1^2}\right) \tag{7.18}$$

where r_1 is the radial distance of the nuclear burning shell from the stellar center, P_{sh} is simply the product of the column mass above the burning shell (i.e., the weight of the layer) and the surface gravity of the accreting WD. The results of Townsley & Bildsten are within a few percent of Fujimoto's results with the small differences due to their inclusion of ^{3}He fusion and OPAL opacities.

7.11 Classical Novae, Recurrent Novae, Symbiotic Variables: The Single Degenerate Pathway to Supernovae Type Ia?

From the perspective of research advances on the structure and evolution of accreting WDs in CVs, the rate at which new discoveries are occurring is rapidly accelerating. First, the determination of unprecedentedly precise fundamental parameters such as CV WD masses, their surface temperatures and accreted atmospheric chemistry (thanks to Gaia parallaxes), offer totally new insights into their structure and evolution (Pala et al. 2022). The so-called single degenerate channel leading to Type Ia supernovae can be tested. With the precision masses of the accreting degenerates, an answer to the fundamental question of whether or not more massive CV WDs can reach the Chandrasekhar limit (1.44 Msun for a carbon–oxygen core) and thus be viable candidates for being progenitors of Type Ia supernovae, may emerge. Do their masses increase or decrease over evolutionary timescales in consequence of hundreds to thousands of nova explosions during typical CV lifetimes of 10^9 years? While it is now established that younger CVs are predominant above the period gap and that they do indeed evolve across the CV period gap (between orbital periods of two and three hours during which accretion has largely shut off), into older systems below the period gap, this fundamental question remains unsettled (e.g., Hillman et al. 2020, Starrfield et al. 2020a, 2020b). If CV WDs accrete more matter than they eject with each nova, then their masses should be increasing. This has been predicted to happen for massive WDs, based upon hydrodynamic evolutionary simulations through thousands of nova explosions (Hillman et al. 2020).

All three of these object classes could potentially be major channels into the cosmologically critical population of Type Ia supernovae because all three types of objects contain accreting white dwarfs. If the white dwarfs in these systems grow in mass via accretion to reach the Chandrasekhar Limit of 1.44 $M_\odot$ for a carbon-oxygen core, then in principle, a Type Ia supernovae explosion should result (Whelan & Iben, 1973).

While a massive WD accreting wind outflow from a red giant binary companion in a symbiotic variable could trigger a Type Ia supernove explosion (single degenerate channel), the symbiotic systems could also contribute to the double degenerate merger channel. Since the lifetime of symbiotics is only 10^5 to 10^6 years, the accreting white dwarf may eventually merge with the core of the red giant during a common envelope phase in which the combined mass of the two WDs exceeds the Chandrasekhar Limit of 1.44 $M_\odot$ (e.g. Di Stefano 2010).

Briefly, the observed properties of Type Ia supernovae illustrate the challenges to identifying the progenitors as well as to theoretical modeling. The spectra exhibit no spectral features of either hydrogen or helium, and the accreting white dwarf is totally obliterated in the explosion leaving neither a neutron star (pulsar) or black hole. During the late stages of SNeIa explosions, the light curve is powered by the decay chain of the radionuclides $Ni^{56} --> Co^{56} --> Fe^{56}$. The peak luminosity depends upon the amount of Ni^{56} that has been nucleosynthesized. The reader is referred to the reviews by Hillebrandt & Niemeyer (2000), Leibundgut (2000, 2001), and a more recent review by José (2017).

The single degenerate pathway is principally focused upon cataclysmic variables because of their large population, their more secure accreting white dwarf parameters like their masses, M_{wd}, surface temperature, T_{eff} , and accretion rates. We start by mapping out for the reader, the regions in the parameter space of envelope mass, white dwarf mass and accretion rate where accreting white dwarfs undergo thermonuclear runaway instabilities. This is provided by an important plot from Nomoto et al. (2007), displayed in Figure 7.29 and is a followup to an earlier similar plot by Fujimoto (1982).

7.11.1 The Classical Nova Explosion

The classical novae undergo thermonuclear explosions that release $\sim 10^{45}$ ergs and reach maximum luminosities between 10^{39} and 10^{40} erg s^{-1}, with ejection velocities reaching many thousands of km s^{-1} and ejected masses of $10^{-3.5}$ to 10^{-7} $M_\odot$ depending on the mass of the accreting white dwarf at the time of the explosion. Higher mass white dwarfs require less accreted mass prior to ignition of the thermonuclear runaway (TNR). This explosive hot CNO burning of hydrogen into helium can reach temperatures as high as 300–400 million Kelvins. At these temperatures, proton captures on nuclei occur outside of the CNO bi-cycle and contribute to the nucleosynthetic yield of the explosion. The reader is referred to a comprehensive review of the thermonuclear runaways and the classical nova explosions by Starrfield, Iliadis & Hix (2016) and the monograph on Stellar Explosions by Jordi José (2017).

The classical nova explosion itself is a deflagration, i.e.subsonic. which adds substantial complexity to the hydrodynamic evolution of the outburst. The region in the accreting white dwarf of the greatest importance is the transition region lying

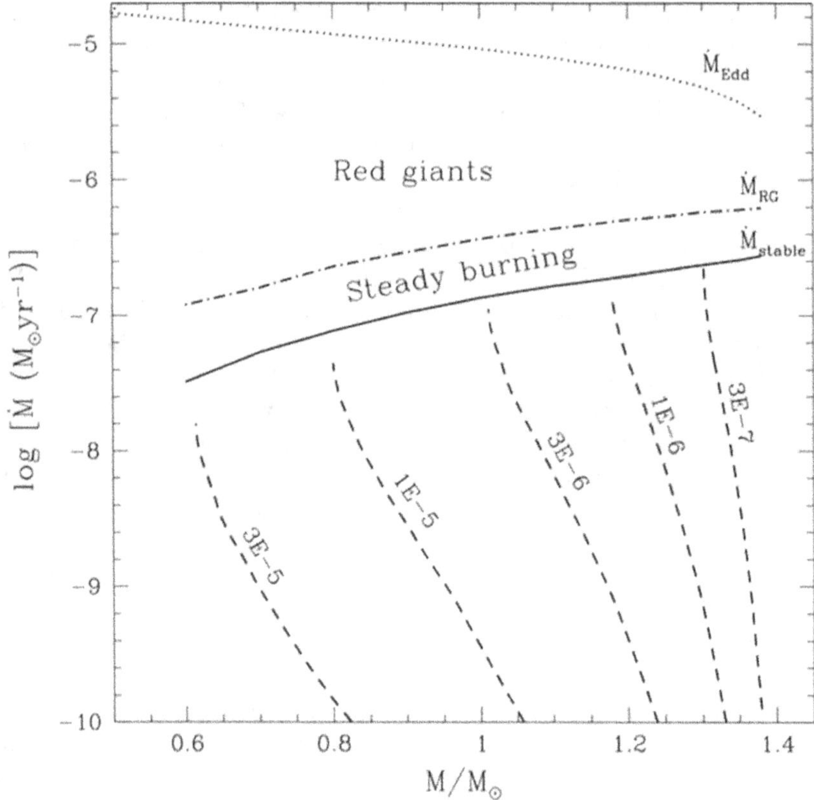

Figure 7.29. Properties of H-burning shells in accreting white dwarfs, shown in the plane of white dwarf mass M_{wd} vs. accretion rate $\dot{M}$. If the accretion rate is lower than $\dot{M}_{\mathrm{stable}}$ (solid line), then H-burning shells are thermally unstable. Dashed lines trace the loci of the envelope mass M_{env}. For a given M_{env} and $\dot{M}$, the envelope masses of these steady state models are smaller than the envelope masses of ignition model envelopes. In the area between the solid ($\dot{M}_{\mathrm{stable}}$) and dash-dotted ($\dot{M}_{\mathrm{RG}}$) lines, the H-burning shell burns steadily and the star is located around the "knee" or the horizontal branch on a steady state white dwarf model locus. Above the dash-dotted line for $\dot{M}_{\mathrm{RG}}$, the stellar envelope has expanded to red giant size and a strong wind outflow occurs. The dotted line indicates the Eddington accretion rate $\dot{M}_{\mathrm{edd}}$ as a function of white dwarf mass. Since thermonuclear runaways power classical nova explosions, a summary of classical novae properties will prove helpful for the upcoming discussion of various quasi-static and hydrodynamic evolutionary sequences. Reproduced from Nomoto et al. (2007). Copyright American Astronomical Society. All rights reserved.

between the degenerate core and the partially degenerate bottom-most layers of the accreted, compressed non-degenerate envelope. The actual mixing that occurs in the core-envelope interface remains vexing. This is due to poorly understood effects of turbulence, the mixing of core material with accreted matter, hydrodynamic instabilities in the presence of envelope convection and other processes (José 2017). All of these physical processes are operating virtually simultaneously during the explosion.

The nova mass ejection itself is not well understood and could involve shocks as well as the intense radiation from the explosion itself. What fraction of the ejected material fails to achieve escape and falls back into the white dwarf? The properties of the common envelope inevitably formed during the expansion and bloating of the accreting white dwarf following the explosion, remain poorly known. The effect of the binary itself on the ejection of mass is also poorly understood.These uncertainties are underscored in the review article by Chomiuk, Metzger & Shen (2021). Mindful of the many uncertainties regarding how much mass is ejected compared with the amount of mass accreted by the white dwarf since its last nova, the results of simulations are briefly discussed in the following section.

7.11.2 History of Quasi-Static and Hydrodynamic Studies of Classical Novae Explosions

Before examining the latest evidence for or against CV white dwarf growth to the Chandrasekhar limit, it is illuminating to follow the historical path of studies, up to the present, by groups around the world. The underlying mechanism for nova explosions was first proposed to be thermonuclear fusion in an electron degenerate gas by Schatzman (1950) and Mestel (1952) although Schatzman mistakenly identified the fusion process as He^3 burning. When actual evolutionary studies of unstable thermonuclear burning, triggered by accretion, were carried out (Giannone & Weigert 1967, Saslaw 1968, and Rose 1968), they shed considerable light on accretion physics and unstable thermonuclear burning resulting from the ignition in a degenerate gas as well as well as runaway nuclear burning due to the Schwarzschild & Harm (1965) thin shell thermal instability. The barrier faced by all of these studies however was that they used only quasi-static evolution codes (e.g. Henyey-type; Kippenhahn-type) and could not follow the evolution of a thermonuclear runaway (TNR) through to the peak of the explosion, dynamical ejection and subsequent evolution because their time steps could not be shorter than minutes due to numerical difficulties. Moreover, if the time step is too large, physical instabilities can be missed.

A milestone paper by Starrfield (1971) set the stage for the first hydrodynamic studies of the nova explosion. Although using a semi-implicit, time-dependent quasi-static code, it was designed to take timesteps as short as minutes. This was done by adopting the entropy as the independent variable. The input physics and micro-physics was the most sophisticated available at the time. For the first time, the peak of the hydrogen TNR was reached and helium burning in the outer layers of the He core commenced.The nuclear burning produced more than 10^{45} ergs.

This important work was quickly followed by the first hydrodynamic evolutionary study of the nova outburst (Starrfield, Truran, & Sparks 1972) using a fully implicit Lagrangian hydrodynamic code. Their code incorporated a nuclear reaction network. The amount of mass ejected and the kinetic energies were in agreement with observations. This paper also showed for the first time that the strong enhancement of CNO nuclei is essential to power the classical nova explosion and the ejection of mass.

A definitive answer to whether or not an accreting white dwarf gains mass or loses mass after each nova explosion remains elusive. It is obvious however that the evolution of cataclysmic variables has to be followed hydrodynamically through more than one outburst. In this author's view, the break-through came with the first multi-cycle hydrodynamic study in 1986, published by Dina Prialnik (1986), a Ph.D student of Giora Shaviv, at the University of Tel Aviv. This was the first evolutionary hydrodynamic sequence through more than one nova explosion. This seminal paper led to extensive, increasingly sophisticated hydrodynamic evolutionary studies through hundreds, even thousands of nova explosions, with the establishment of a grid of evolutionary, multi-cycle nova models covering a wide range of parameter space (Prialnik & Kovetz 1995). This was soon followed by even more refined grids of multi-cycle evolutionary sequences by Yaron et al. (2005), Epelstain et al. (2007), and Hillman et al. (2016) with careful attention to whether the accreting white dwarf gained mass after each nova ejection or lost more mass than it accreted.

Once the theoretical parameters were available from multi-cycle accreting models over a wide range of white dwarf mass, accretion rate, and core temperature, covering thousands of nova explosions, direct comparisons with observations could be carried out. Shara et al. (2018) conducted such a comparison with the database of observed light curves (Strope et al. 2010, Schaefer et al. 2010) of known classical novae. From their light curve fitting, Shara et al. (2018) determined that the mean mass (frequency-averaged mean mass) of 82 Galactic classical novae is 1.06 (1.13) $M_\odot$, while the mean mass of 10 RNe is 1.31 $M_\odot$. These white dwarf masses, and the observed nova outburst amplitude and decline time distributions allowed Shara et al. (2018) to also determine the long-term mass accretion rate distribution of classical novae. Remarkably, that value is unexpectedly low, 1.25×10^{-9} $M_\odot$ yr^{-1}. The result of this work is displayed in Figure 7.30.

7.11.3 Single Degenerate Pathway to Type Ia Supernovae

The central question of whether white dwarfs in cataclysmic variables are a viable pathway to Type Ia supernovae remains unanswered, as discussed earlier in this chapter. Two observational results possibly support the gain of mass by accreting white dwarfs in CVs after evolution through thousands of nova explosions:(1) the average mass of CV white dwarfs (0.83 +/- 0.03 $M_\odot$) is significantly larger than the average mass of essentially non-accreting field degenerates (Pala et al. 2022) and; (2) the average mass of white dwarfs in Novae (Shara et al. 2018) is even larger, with respect to the average mass of field degenerates, though this is likely a selection effect (see below). On the other hand, there is no statistical difference betwen the average mass of CV white dwarfs above the period gap and those below the period gap. Both average masses are 0.81 (Pala et al. 2022). This is contrary to expectations since CVs above the gap almost certainly evolve into CVs below the gap. They are not two distinct CV populations.

If one appeals to hydrodynamic evolutionary studies of multi-cycle accreting white dwarf models through hundreds to thousands of nova explosions, the results are mixed. Prialnik & Kovetz (1995) found that, surprisingly, recurrent novae can occur on lower mass white dwarfs (~1 $M_\odot$) if they accrete at a rate $\dot{M}$ greater than or equal to 10^{-7} $M_\odot$ yr^{-1} and could lead to a SNeIa. For accretion rates $\dot{M} \sim 10^{-9}$ $M_\odot$ yr^{-1}, they find that the white dwarf mass always decreases with time and thus cannot be progenitors of SNeIa.

In still another, long term multi-cycle study however, Yaron et al. (2005) find that the accreting white dwarf gains more mass than it ejects at the highest accretion rates

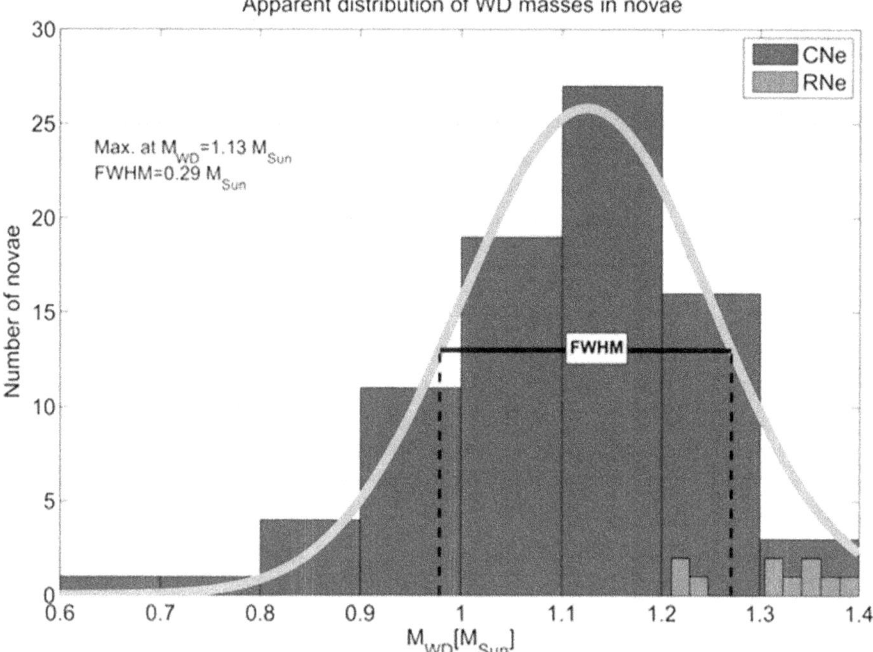

Figure 7.30. The average nova WD mass is 1.13 $M_\odot$ and the average accretion rate 1.25×10^{-9} $M_\odot$ yr^{-1}. The average mass agrees exceptionally well the mean nova WD mass from the theoretical estimate obtained by Ritter et al. (1991), who predicted the mean WD mass, weighted by nova frequency (that is, the value determined directly by observations), to be found between 1.04 and 1.24 $M_\odot$. Reproduced from Shara et al. (2018). Copyright American Astronomical Society. All rights reserved.

of 10^{-7} $M_\odot$ yr^{-1}, for all nova white dwarf masses and all three different adopted core temperatures, $T_c = 10$, 30, and 50 million K.

Epelstain et al. (2007) report that their equilibrium core temperatures of 0.65 and 1.0 $M_\odot$ increase only very slowly and on a very long timescale which disagrees with the results of the Townsley & Bildsten (2004) analytical results, discussed earlier in this chapter. The WD mass decreases steadily for the cases considered, but not by much. The mass of the donor star, by contrast, decreases considerably: by 0.16 $M_\odot$ for the 0.65 $M_\odot$ model, and by 0.15 $M_\odot$ for the 1.0 $M_\odot$ model. Still more numerical hydrodynamic simulations have shown that such systems (massive WDs and high rates of accretion) eject less mass than has been accreted during each nova (Starrfield et al. 2012; Hillman et al. 2015), thus allowing the white dwarf to grow and hence be a possible SNe Ia progenitor.

The prospects for progenitorship of SNeIa began to brighten even more when Hillman et al. (2016) launched a study of the fate of the steadily accumulating helium layer built up over the long intervals of hydrogen-rich shell flashes. In effect they assessed the accretion rate of helium at the base of the H-rich accretion layer and examined what happens when helium flashes eventually are triggered. They found that the helium layer has a net mass increase as the helium flashes are followed.

During the first several helium flashes, the mass loss is a substantial fraction of the helium layer mass accumulated since the last previous He-flash. Surprisingly however, the fractional loss of helium layer mass through mass ejection started declining for subsequent helium shell flashes and eventually no mass was lost with each helium shell flash. This behaviour constituted a steady state evolution of the helium layer with no mass loss due to helium runaways. It is noteworthy that this Hillman et al. (2016) result disagrees with the diagram displayed in Figure 7.29 from Nomoto et al. (2007). Helium runaways are also discussed in Chapter 6 on AM CVn systems, in this book. Hillman et al. (2016) revealed a broad range in parameter space for growth of the CV white dwarf mass to eventually reach the Chandrasekhar Limit and a SNeIa. All of the foregoing hydrodynamic evolutionary sequences yielded valuable results and a surprisingly wide parameter space in M_{wd}, $\dot{M}$ and T_c for cataclysmic variable white dwarfs to reach the Chandrasekhar Limit. However, they all assumed a constant mass transfer rate and did not explcitly include the effects of the system's evolution on the red dwarf donor itself. This would cause mass transfer variations that had not yet been included in the long-term evolutionary models. For the first time, the Israeli hydrodynamic code of Hillman et al. was modified to follow the evolution of the red dwarf donor and the accreting white dwarf in tandem. The response of the donor red dwarf to its loss of mass, to irradiation by the hot white dwarf, and to changes in the orbital period and orbit size from continued angular momentum losses, completely changed the results of the multi-cycle evolutionary sequences dramatically (Hillman et al. 2020; Hillman 2021). They found that all of the accreting white dwarfs up to $\sim$ 1.25 $M_\odot$ with red dwarf companions up to 0.7 $M_\odot$, *will not* grow in mass. Thus, only accreting white dwarfs with masses greater than 1.25 $M_\odot$ and donor masses greater than 0.7 $M_\odot$ could possibly be progenitors of SNe Ia. If confirmed, these results leave only the recurrent novae with massive white dwarfs and high accretion rates like RS Oph (Mikolajewska & Shara 2018), as well as symbiotic variables (Patat et al. 2007) as the remaining single degenerate channel candidates to explode as SNeIa.

On the other hand, a long-term study involving the hydrodynamic code NOVA (Starrfield et al. 1972) combined with MESA multi-cycle hydrodynamic evolutionary studies by Starrfield et al. (2020) explored a wide range of mass accretion rates onto the WD as well a broad range of initial white dwarf masses. They found that nuclear burning occurs leading to TNRs in every evolutionary sequence with all of the white dwarfs growing in mass or when the radius reaches a red giant radius of 10^{12} cm. These authors remove mass from the white dwarf when the explosion and subsequent expansion becomes optically thin when the density has dropped in these zones to 10^{-13} g cm^{-3}. An example of the multi-cycle hydrodynamic evolutionary models computed with MESA (Starrfield 2014) is displayed in Figure 7.31 where H-rich nova TNR cycles are displayed for a 0.7 $M_\odot$ accreting white dwarf undergoing accretion at four different mass accretion rates.

Finally, we are left with the open question: what is the overall amount of mass ejected in nova explosions? This remains uncertain and a discrepancy still exists between theory and observation. Yet this uncertainty directly impacts the question of whether the white dwarf mass increases or decreases after each nova explosion. The amount of mass lost during each nova depends on the mechanism operating to remove the mass and the rate of growth in WD mass also varies accordingly and is

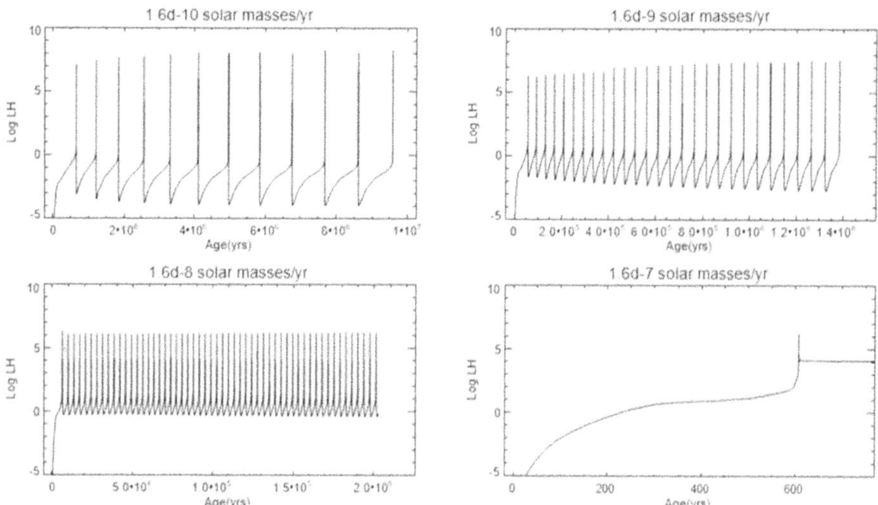

Figure 7.31. caption: log of the hydrogen luminosity as a function of time for ac- cretion onto a 0.7 $M_\odot$ white dwarf accreting at four different mass accretion rates identified at the top of each panel. Note that as the mass accretion rate increases, the time between outbursts decreases, however, the peak luminosity remains approximately the same. The sequence with the highest (lower right hand side) accretes until the flash occurs and then grows rapidly to a red giant radius of 10^{12} cm and the evolution is ended. Nonetheless, the surface luminosity for this sequence exceeds the values normally measured for CVs. Reproduced from Starrfield (2014). CC BY 3.0.

highly model-dependent among groups that carry out multi-nova cycle 1D evolutionary sequences using widely different algorithms. The nova mass ejection itself is not well understood and could involve shocks as well as the intense radiation from the explosion itself. What fraction of the ejected material fails to achieve escape and falls back into the white dwarf? The properties of the common envelope inevitably formed during the expansion and bloating of the accreting white dwarf following the explosion, remains poorly known. These uncertainties are underscored in the review article by Chomiuk, Metzger and Shen (2021). Even the effect of the binary motion itself on the amount of mass ejected must be further explored (Shen and Quataert 2022). More work in this area is definitely warranted. The amount of mass ejected in each nova explosion remains highly uncertain and groups that carry out multi-nova cycle hydrodynamic evolutionary simulations employ different criteria to determine how much mass is removed due to the explosion. Thus, at this point, in view of the remaining uncertainties, it seems premature to claim that we know the answer to whether the mass of an accreting CV WD increases or decreases in the long term.

References

Aizu, K. 1973, PThPh, 49, 104

Araujo-Betancor, S., et al. 2005, ApJ, 622, 589

Arras, P, Townsley, D M, & Bildsten, L 2006, ApJ, 643, L119

Balsara, D., & Fisker, J. 2005, ApJ, 635, L69

Balsara, D., Fisker, J., Godon, P., & Sion, E. M. 2009, ApJ, 702, 1536

Baptista, R. 2012, Mem. S.A. It., 83, 530

Beuermann, K., Wheatley, P., Ramsay, G., Euchner, F., & Gänsicke, B. T. 2000, A&A, 354, L49

Belle, K., Howell, S., Sion, E., Long, K., & Szkody, P. 2003, ApJ, 587, 373

Bitner, M. A., Robinson, E. L., & Behr, B. B. 2007, ApJ, 662, 564

Bonnet-Bidaud, J. M., Mouchet, M., de Martino, D., Matt, G., & Motch, C. 2001, A&A, 374, 1003

Bruch, A., & Engel, A. 1994, A&A, 104, 79

Chiappetti, L., Maraschi, L., Tanzi, E. G., & Treves, A. 1982, ApJ, 258, 236

Chomiuk, L, Metzger, B D, & Shen, K J 2021, ARA&A, 59, 391

Copperwheat, C. M., Marsh, T. R., Dhillon, V. S., et al. 2010, MNRAS, 402, 1824

Crawford, J. A., & Kraft, R. P. 1956, AJ, 123, 64

Cropper, M., Ramsay, G., & Wu, K. 1998, MNRAS, 293, 222

Cropper, M., Wu, K., Ramsay, G., & Kocabiyik, A. 1999, MNRAS, 306, 684

de Martino, D., Bernardini, F., Mukai, K., Falanga, M., & Masetti, N. 2020, AdSpR, 66, 1209

Dexter, L. H., Gottlieb, E. W., & Liller, W. 1976, B.A.A.S, 8, 511

Dhillon, V. S., Marsh, T. R., Stevenson, M. J., et al. 2007, MNRAS, 378, 825

Dhillon, V. S., Marsh, T. R., Atkinson, D. C., et al. 2014, MNRAS, 444, 4009

Di Stefano, R. 2010, ApJ, 719, 474

El-Badry, K., Rix, H-W., Quataert, E., Kupfer, T., & Shen, K. J. 2021, MNRAS, 508, 4106

Epelstain, N, Yaron, O, Kovetz, A, & Prialnik, D 2007, MNRAS, 374, 1499

Fontaine, G., & Brassard, P. 2008, PASP, 120, 1043

Fontaine, G., & Michaud, G. 1979,231, 846

Froning, C., Long, K. S., Drew, J., Knigge, C., & Proga, D. 2001, ApJ, 562, 963

Fujimoto, M Y 1982, ApJ, 257, 767

Fujimoto, M. Y. 1982a, ApJ, 257, 752

Fujimoto, M. Y. 1982b, ApJ, 257, 767

Gänsicke, B. T., de Martino, D., Marsh, T. R., Haswell, C. A., Knigge, C., Long Knox, S., Shore,, & Steven, N. 2006, Ap&SS, 306, 177

Gänsicke, B. T., Sion, E. M., Beuermann, K., et al. 1999, A&A, 347, 178

Gänsicke, B. T., Beuermann, K., de Martino, D., & Thomas, H. C. 2000, A&A, 354, 605

Gänsicke, B. T., Szkody, P., Sion, E. M., et al. 2001, A&A, 374, 656

Gänsicke, B. T., Szkody, P., de Martino, D., et al. 2003, ApJ, 594, 443

Gänsicke, B. T., Szkody, P., Howell, S. B., & Sion, E. M. 2005, ApJ, 629, 451

Gänsicke, B. T., Rodriguez-Gil, P., Marsh, T. R., et al. 2006, MNRAS, 365, 969

Gänsicke, B. T., Beuermann, K., & de Martino, D. 1995, A&A, 303, 127

Gänsicke, B. T., et al. 2006, ApJ, 639, 1039

Gianonne, P, & Weigert, A 1967, Z.Astrophys, 67, 41

Ghaderpour, E., & Ghaderpour, S. 2020, PASP, 132, 114504

Godon, P., Seward, L., Sion, E. M., & Szkody, P. 2006, AJ, 131, 2634

Godon, P., Sion, E. M., Balman, S., & Blair, W. P. 2017, ApJ, 846, 52

Godon, P., Sion, E. M., Barrett, P., & Szkody, P. 2007, ApJ, 656, 1092

Godon, P., Sion, E. M., Barrett, P. E., et al. 2008, ApJ, 679, 1447

Godon, P., & Sion, E. M. 2021, ApJ, 908, 173

Godon, P., & Sion, E. M. 2022, ApJ, 928, 26

Godon, P, & Sion, E M 2022, ApJ, 928, 168

Godon, P., & Sion, E. M. 2003, ApJ, 586, 427

Godon, P, & Sion, E M 2023a, ApJ, accepted

Godon, P., & Sion, E. M. 2023b, ApJ, submitted

Godon, P., Sion, E. M., Cheng, F., et al. 2006, ApJ, 642, 1018

Godon, P., Sion, E. M., Cheng, F., Gänsicke, B. T., Howell, S., Knigge, C., Sparks, W. M., & Starrfield, S. 2004, ApJ, 602, 336

Godon, P., Sion, E. M., Cheng, F., Long, K. S., Gänsicke, B. T., & Szkody, P. 2006, ApJ, 642, 1018

Godon, P., Sion, E. M., Cheng, F. H., Szkody, P., Long, K. S., & Froning, C. S. 2004, ApJ, 612, 429

Hamilton, R., & Sion, E. M. 2006, PASP, 116, 926

Hamilton, R. T., & Sion, E. M. 2008, PASP, 120, 165

Hamilton, R. T., Harrison, T. E., Tappert, C., & Howell, S. B. 2011, ApJ, 728, 16

Harrison, T. E. 2016, ApJ, 832, 14

Hartley, L. E., Long, K. S., Froning, C. S., & Drew, J. E. 2005, ApJ, 623, 425

Heinonen, R. A., Saumon, D., Daligault, J., Starrett, C. E., Baalrud, S. D., & Fontaine, G. 2020, ApJ, 896, 2

Heise, J., & Verbunt, F. 1988, A&A, 189, 112

Hertfelder, M., Kley, W., Suleimanov, V., & Werner, K. 2013, A&A, 560A, 56

Hertfelder, M., & Kley, W.2017, A&A, 605, 24

Hillman, Y., Shara, M. M., Prialnik, D., & Kovetz, A. 2020, NatAs, 4, 886

Hillman, Y., Prialnik, D., Kovetz, A., & Shara, M. M. 2015, MNRAS, 446, 192

Hillman, Y., Prialnik, D., Kovetz, A., & Shara, M. M. 2016, ApJ, 819, 168

Hillman, Y., et al. 2020, NatureAst, 4, 886

Hillman, Y. 2021, MNRAS, 505, 3260

Hillman, Y., & Kashi, A. 2021, MNRAS, 501, 201

Hillebrandt, W., & Niemeyer, J. C. 2000, ARA&A, 38, 191

Howell, S. B., Gänsicke, B. T., Szkody, P., & Sion, E. M. 2002, ApJ, 575, 419

Howell, S. B., Nelson, L. A., & Rappaport, S. 2001, ApJ, 550, 897

Hoard, D. W., Linnell, A. P., Szkody, P., et al. 2004, ApJ, 604, 346

Hubeny, I. 1988, CoPhC, 52, 103

Hubeny, I., & Lanz, T. 1995, ApJ, 439, 875

Iben, I. Jr. 1982, ApJ, 259, 244

José, J. 2017, Stellar Explosions (Series in Astronomy and Astrophysics, CRC Press)

Kley, W. 1989, A&A, 222, 141

Kley, W. 1991, A&A, 247, 95

Knigge, C., Baraffe, I., & Patterson, J. 2011, ApJS, 194, 28

Knigge, C., Long, K. S., Hoard, D. W., Szkody, P., & Dhillon, V. S. 2000, ApJ, 539, L49

Ko, T., et al. 1996, ApJ, 457, 363

Koester, D. 2013, Planets, Stars and Stellar Systems, ed. T. D. Oswalt, & M. A. Barstow (Dordrecht: Springer)

Kraft, R. P., & Luyten, W. J. 1965, ApJ, 142, 1041

Kromer, M., Nagel, T., & Werner, K. 2007, A&A, 475, 301

La Dous, C. 1991, A&A, 252, 100

Leibundgut, B. 2000, A&A Rev, 10, 179

Leibundgut, B. 2001, ARA&A, 39, 67

Liu, D., Wang, B., Ge, H., Chen, X., & Han, Z. 2019, A&A, 622, A35

Livio, M., & Pringle, J. 1998, ApJ, 503, 339

Long, K. S., & Gilliland, R. L. 1999, ApJ, 511, 916

Long, K. S., Blair, W. P., Bowers, C. W., et al. 1993, ApJ, 405, 327

Long, K. S., Blair, W. P., Hubeny, I., & Raymond, J. 1996, ApJ, 466, 964

Long, K. S., Brammer, G., & Froning, C. S. 2006, ApJ, 648, 541

Long, K. S., Gnsicke, B., Knigge, C., Froning, C. S., & Monard, B. 2009, ApJ, 697, 1528

Long, K. S., Sion, E. M., Huang, M., & Szkody, P. 1994, ApJL, 424, L49

Long, K. S., et al. 1994, ApJ, 424, L49

Long, K. S., et al. 2005, ApJ, 630, 511

Luyten, W. J., & Hughes, H. S. 1965, Pub. Univ. Minnesota, No. 36

Marks, P. B., Sarna, M. J., & Prialnik, D. 1999, MNRAS, 290, 283

Marks, P. B., & Sarna, M. J. 1999, MNRAS, 301, 691

Mateo, M., & Szkody, P. 1984, AJ, 89, 863

McAllister, M., Littlefair, S. P., Parsons, S. G., et al. 2019, MNRAS, 486, 5535

Medvedev, D., & Menou, K. 2002, ApJ, 565, L39

Medvedev, M. V., & Narayan, R. 2001, ApJ, 554, 1255

Menou, K., Perna, R., & Raymond, J. C. 2001, ApJ, 549, 509

Mestel, L. 1952, MNRAS, 112, 398

Mestel, L. 1952, MNRAS, 112, 583

Mikolajewska, J, & Shara, M 2018, ApJ, 877, 99

Muchmore, D. 1984, ApJ, 278, 769

Mukai, K. 2017, PASP, 129, 062001

Mukai, K., & Charles, P. A. 1987, MNRAS, 226, 209

Mukadam, A. S., Townsley, D. M., Gänsicke, B. T., et al. 2010, ApJ, 714, 1702

Nilsson, R., Uthas, H., Ytre-Eide, M., Solheim, J.-E., & Warner, B. 2006, MNRAS, 370, 56

Nomoto, K, Saio, H, Kato, M, & Hachisu, I 2007, ApJ, 663, 1269

Pala, A. F., Gänsicke, B. T., Townsley, D., et al. 2017, MNRAS, 466, 2855

Pala, A. F., Gänsicke, B. T., Belloni, D., et al. 2022, MNRAS, 510, 6110

Pala, A, Gaensicke, B, Marsh, T, et al. 2022, MNRAS, 600, 510

Panek, R., & Holm, A. V. 1984, ApJ, 277, 700

Patat, F, Chandra, P, Chevalier, R, et al. 2007, Science, 317, 924

Paquette, C., Pelletier, C., Fontaine, G., & Michaud, G. 1986, ApJS, 61, 177

Patterson, J. R., Thorstensen, J., & Kemp, 2005, PASP, 117, 427

Patterson, J., Thorstensen, J. R., Armstrong, E., Henden, A. A., & Hynes, R. I. 2005, PASP, 117, 922

Pavlenko, E. 2009, JPhCS, 172, 012071

Pavlenko, E., Malanushenko, V., Tovmassian, G., et al. 2012, MmSAI, 83, 520

Piro, A., Arras, P., & Bildsten, L. 2005, ApJ, 628, 401

Piro, A., & Bildsten, L. 2004a, ApJ, 610, 977

Piro, A., & Bildsten, L. 2004b, ApJ, 616, 155

Prialnik, D 1986, ApJ, 310, 222

Prialnik, D, & Kovetz, A 1995, ApJ, 445, 789

Pringle, J. E. 1988, MNRAS, 230, 587

Piersanti, L., Gagliardi, S., Iben, I. Jr., & Tornamb, A. 2003, ApJ, 583, 885

Popham, R., & Narayan, R. 1993, Nature, 362, 820

Podsiadlowski, Ph., Han, Z., & Rappaport, S. 2001, arXiv:astro-ph/0109171

Podsiadlowski, Ph., Han, Z., & Rappaport, S. 2003, MNRAS, 340, 1214

Popham, R., & Narayan, R. 1995, ApJ, 442, 337

Ritter, H, Politano, M, Livio, M, & Webbink, R F 1991, ApJ, 376, 177

Rose, W K 1968, ApJ, 152, 245

Rosen, S. R., Rainger, J. F., Burleigh, M. R., et al. 2001, MNRAS, 322, 631

Rothschild, R. E., Gruber, D. E., Knight, F. K., et al. 1981, ApJ, 250, 723

Saslaw, W 1968, MNRAS, 138, 337

Savoury, C. D. J., Littlefair, S. P., Dhillon, V. S., et al. 2011, MNRAS, 415, 2025

Scaringi, S, Groot, P J, Knigge, C, et al. 2022a, Nature, 604, 447

Scaringi, S, Groot, P J, Knigge, C, et al. 2022b, MNRAS, 514, L11

Schaefer, B 2010, ApJS, 187, 275

Schatzman, E 1950, Ann. Astrophys, 12, 81

Schwarzschild, M, & Harm, R 1965, ApJ, 142, 855

Schenker, K., King, A. R., Kolb, U., Wynn, G. A., & Zhang, Z. 2002, MNRAS, 337, 1105

Shafter, A. W. 1983, PhD thesis, UCLA

Shafter, A., Szkody, P., Liebert, J., et al. 1985, ApJ, 290, 707

Shara, M M 1982, ApJ, 261, 649

Shara, M., Drissen, L., Martin, T., Alarie, A., & Stephenson, F. R. 2017, MNRAS, 465, 739

Shara, M, Prialnik, D, Hillman, Y, & Kovetz, A 2018, ApJ, 860, 110

Shen, K, & Quataert, E 2022, ApJ, 938, 31

Shaw, A. W., et al. 2020, MNRAS, 498, 3457

Sion, E. 1991, AJ, 102, 295

Sion, E. 1999, PASP, 111, 532

Sion, E. M., & Sparks, W. M. 2014, ApJ, 796, 10

Sion, E. M., Long, K. S., Szkody, P., & Huang, M. 1994, ApJ, 430, L53

Sion, E. M. 1985, ApJ, 297, 538

Sion, E. M. 1995, ApJ, 438, 876

Sion, E. M. 2002, AIP Conference Proceedings, 637, 21

Sion, E. M. 2003, in White Dwarfs, NATO Sci. Ser. Vol. 105, ed. D. deMartino, R. Silvotti, J. E.
    Solheim, & R. Kalytis (Berlin: Springer), 303

Sion, E. M., Cheng, F. H., Long, K. S., et al. 1995, ApJ, 439, 957

Sion, E. M., Cheng, F. H., Szkody, P., et al. 1998, ApJ, 496, 449

Sion, E. M., Szkody, P., Gaensicke, B., et al. 2001, ApJ, 555, 834

Sion, E. M., Cheng, F., Huang, M., Hubeny, I., & Szkody, P. 1996, ApJ, 471, L41,

Sion, E. M., Cheng, F., Godon, P., Urban, J. A., & Szkody, P. 2004, AJ, 128, 1834

Sion, E. M., Cheng, F. H., Sparks, W. M., Szkody, P., Huang, M., & Hubeny, I. 1997, ApJ, 480, L17

Sion, E. M., Gänsicke, B. T., Long, K. S., Szkody, P., Knigge, C., Hubeny, I., deMartino, D., &
    Godon, P. 2008, ApJ, 681, 543

Sion, E. M., Godon, P., Cheng, F., & Szkody, P. 2007, AJ, 134, 886

Sion, E., Godon, P., Myzcka, J., & Blair, W. P. 2010, ApJ, 716, L157

Sion, E. M., Long, K. S., Szkody, P., & Huang, M. 1994, ApJ, 430, L53

Sion, E. M., Szkody, P., Cheng, F., Gänsicke, B. T., & Howell, S. B. 2003, ApJ, 583, 907

Sion P, E. M., et al. 1995a, ApJ, 439, 957

Sion, E. M., et al. 1995b, ApJ, 444, 97

Sion, E. M., et al. 1997, ApJ, 403, 987

Slevinsky, R. J., Stys, D., West, S., Sion, E. M., & Cheng, F. H. 1999, PASP, 111, 1292

Sparks, W. M., & Sion, E. M. 2022, ApJ, 914, 5

Starrfield, S G 1971, MNRAS, 152, 307

Starrfield, S., Bose, M., Iliadis, C., et al. 2020a, ApJ, 895, 70

Starrfield, S., Bose, M., Iliadis, C., Hix, R., Woodward, C. E., & Wagner, R. M. 2020b, ApJ, 895, 70

Starrfield, S G, Truran, J W, & Sparks, W M 1972, ApJ, 176, 169

Starrfield, S, et al. 2012, BaltA, 21, 76

Starrfield, S 2014, AIP Adv., 4, 041007

Starrfield, S, Iliadis, C, & Hix, W R 2016, PASP, 128, 1001

Starrfield, S, et al. 2020, ApJ, 895, 70

Steeghs, D., Howell, S. B., Knigge, C., et al. 2007, ApJ, 667, 442

Strope, R J, Schaefer, B E, & Henden, A A 2010, AJ, 140, 34

Suleimanov, V. F., Doroshenko, V., & Werner, K. 2019, MNRAS, 482, 3622

Suleimanov, V., Revnivtsev, M., & Ritter, H. 2005, A&A, 435, 191

Szkody, P., Gänsicke, B. T., Sion, E. M., & Howell, S. B. 2002a, ApJ, 574, 950

Szkody, P., Gänsicke, B. T., Howell, S. B., & Sion, E. 2002b, ApJ, 575, L79

Szkody, P., Fraser, O., Silvestri, N., et al. 2003, AJ, 126, 1499

Szkody, P., Mukadam, A., Gänsicke, B. T., et al. 2007, ApJ, 658, 1188

Szkody, P., Mukadam, A., Gänsicke, B. T., et al. 2010, ApJ, 710, 64

Szkody, P., Mukadam, A. S., Gänsicke, B. T., et al. 2012, ApJ, 753, 158

Szkody, P., Mukadam, A. S., Gänsicke, B. T., et al. 2016, ApJ, 152, 48

Szkody, P., Raymond, J. C., & Capps, R. W. 1982, ApJ, 257, 686

Szkody, P., Godon, P., Gänsicke, B. T., et al. 2021, ApJ, 914, 40

Szkody, P. 2022a, in Frontiers (Stuttgart: Holtzbrinck Publishing Group)

Szkody, P. 2022b, Invited Mini-Review of Pulsating Accreting White Dwarfs, in Frontiers (Stuttgart: Holtzbrinck Publishing Group)

Thorstensen, J. R., Charles, P. A., Margon, B., & Bowyer, S. 1978, ApJ, 223, 260

Townsley, D., & Bildsten, L. 2003, ApJ, 596, L227

Townsley, D., & Bildsten, L. 2004, ApJ, 600, 390

Townsley, D, & Bildsten, L 2004, ApJ, 600, 510

Townsley, D., & Gänsicke, B. 2009, ApJ, 693, 100

Urban, J., & Sion, E. M. 2006, ApJ, 642, 1029

Uthas, H., Knigge, C., Long, K. S., Patterson, J., & Thorstensen, J. 2011, MNRAS, 414, 85

Uthas, H., Patterson, J., Kemp, J., et al. 2012, MNRAS, 420, 329

van Zyl, L., Warner, B., O'Donoghue, D., et al. 2004, MNRAS, 350, 307

Vennes, S., et al. 1991, ApJ, 372, L37

Verbunt, F. 1987, A&A, 71, 339

Warner, B. 1995, Cataclysmic Variables (Cambridge: Cambridge Univ. Press)

Warner, B., & van Zyl, L. 1998, New Eyes to See inside the Sun and Stars, ed. F. L. Deubner, J. Christensen-Dalsgaard, & D. Kurtz (Dordrecht: Springer)

Whelan, J, & Iben, I 1973, ApJ, 186, 1007

Wood, M. A. 1995, LNP, 443, 41

Woudt, P. A., & Warner, B. 2004, MNRAS, 348, 599

Woudt, P., & Warner, B. 2011, ApSS, 333, 119

Xu, X.-J., Yu, Z.-l., & Li, X.-d. 2019, ApJ, 878, 53
Yaron, O., Prialnik, D., Shara, M. M., & Kovetz, A. 2005, ApJ, 623, 398
Yoon, S.-C., & Langer, N. 2004, A&A, 419, 645
Yuasa, T., Nakazawa, K., Makishima, K., et al. 2010, A&A, 520, A25
Zorotovic, M., Schreiber, M., & Gänsicke, B. T. 2011, A&A, 536, 42

Accreting White Dwarfs
From exoplanetary probes to classical novae and Type 1a supernovae
Edward M Sion

Appendix A

Masses and Effective Temperatures of CV White Dwarfs

doi:10.1088/2514-3433/ac930cch8   A-1

System	P_{orb}	M_{wd}	T_{eff}	ref(M_{wd}/T_{eff})
V485 Cen	59.03		15200 ± 655	Pala et al. (2017)
		0.696 + 0.117/-0.108	16188 ± 313	Godon & Sion (2023a)
SDSSJ150772.30 + 5	66.61	0.90 + 0.10/-0.14	14207 + 356/- 403	Pala et al. (2022)
	66.611	0.892 ± 0.008	11300 ± 1000	Savoury et al. (2011)
SDSSJ074531.91+ 453829.5	76.0	0.75 + 0.18/-0.20	15447 + 556/-654	Pala et al. (2022)
GW Lib	76.78	0.83 + 0.08/-0.12	16166 + 253/-350	Pala et al. (2022)
SDSSJ143544.02 + 233638.7	78.00	0.84 + 0.07/-0.06	11997 + 99/-160	Pala et al. (2022)
OT J213806.6 + 261957	78.10	0.57 + 0.06/-0.07	15317 + 216/-228	Pala et al. (2022)
SDSS1433 + 1011	78.1	0.865 +/-0.005		Zorotovic et al. (2011)
	78.106	1.098 ± 0.024	15500 ± 2400	Savoury et al. (2011)
BW Scl	78.23	1.007 + 0.010/-0.012	15145 + 51/57	Pala et al. (2022)
		1.047 ± 0.051	15830 ± 261	Godon & Sion (2023b)
V844 Her	78.69		14850 ± 622	Pala et al. (2017)
LL And	79.28	0.7 + 0.4/-0.3	14353 + 1210/-743	Pala et al. (2022)
SDSS J013701.06-091234.8	79.71		14547 ± 594	Pala et al. (2017)
		0.615 + 0.088/-0.083	14430 ± 276	Godon & Sion (2023a)
SDSS J1610-0102	80.40		14500 ± 1500	Townsley & Gänsicke (2009)
SDSS J123813.73-033932.9	80.52		18358 ± 922	Pala et al. (2017)
EF Eri	81.00		9500 ± 500	Townsley & Gänsicke (2009)
HS2331 + 3905	81.06		11500 ± 750	Townsley & Gänsicke (2009)
WZ Sge	81.63	0.8 ± 0.02	13190 + 115/-105	Pala et al. (2022)
		0.85 ± 0.04		Steeghs et al. (2007)
AL Com	81.66	0.9 + 0.5/-0.3	15840 +1453/1243	Pala et al. (2022)
SW UMa	81.81	0.61 + 0.061/0.04	13854 +189/-131	Pala et al. (2022)
		0.510 + 0.118/-0.108	13875 ± 480	Godon & Sion (2023b)
SDSS 1501	81.851	0.767 ± 0.027	10800 ± 1500	Savoury et al. (2011)
V1108 Her	81.87	0.88 + 0.09/0.11	13943 + 185/-226	Pala et al. (2022)
ASAS J002511 + 1217.2	82.00	1.02 ± 0.04	13208 + 154/-112	Pala et al. (2022)
SDSS J103533.02 + 055158.4	82.08	0.94 ± 0.01	10500 ± 1000	Littlefair et al. (2006)
	82.089	0.835 ± 0.009	10000 ± 1100	Savoury et al. (2011)
	82.1	0.830 + 0.102/-0.118	11475 ± 188	Godon & Sion (2022)

HV Vir	82.18	0.87 + 0.13/0.17	12958 + 257/-291	Pala et al. (2022)
SDSS J103533.02 + 055158.4	82.22	1.00 + 0.08/-0.10	11876 + 108/-115	Pala et al. (2022)
WX Cet	83.94	1.10 + 0.17/-0.21	15186 + 1460/1468	Pala et al. (2022)
CC Sci	84.10		16855 ± 801	Pala et al. (2017)
SSS100615	84.53	0.88 ± 0.03		McAllister et al. (2019)
EG Cnc	84.60		12300 ± 700	Townsley & Gänsicke (2009)
SDSS J075507.70 + 143547.6	84.76	0.91 + 0.09/0.12	16193 + 280/-357	Pala et al. (2022)
SDSS1502+ 3334	84.8	0.723 + 0.017/-0.013		McAllister et al. (2019)
	84.829	0.709 ± 0.004	11800 ± 1200	Savoury et al. (2011)
SDSS J0804434.20 + 510349.2	84.97	1.01 + 0.03/-0.04	13715 + 55/-79	Pala et al. (2022)
SDSS0903 + 3300	85.065	0.872 ± 0.011	13300 ± 1700	Savoury et al. (2011)
	85.1	0.872 + /-0.011		Zorotovic et al. (2011)
CSS080623	85.79	0.710 ± 0.019	15500 ± 1700	McAllister et al. (2019)
EK TrA	86.36	0.87 + 0.09/-0.13	17608 + 269/-481	Pala et al. (2022)
EG Cnc	86.36	1.03 + 0.04/-0.05	12295 + 56/-57	Pala et al. (2022)
XZ Eri	88.069	0.769 ± 0.017	15300 ± 1900	Savoury et al. (2011)
1RXS J105010.8-140431	88.56	0.77 + 0.02/-0.03	11523 + 29/-47	Pala et al. (2022)
DP Leo	89.82		13500	Townsley & Gänsicke (2009)
V347 Pav	90.06		11800 ± 600	Townsley & Gänsicke (2009)
BC UMa	90.16	0.48 + 0.08/-0.09	14378 + 272/327	Pala et al. (2022)
		0.449 + 0.086/-0.076	14538 ± 351	Godon & Sion (2023b)
SDSS1227+ 5139	90.661	0.796 ± 0.018	15900 ± 1400	Savoury et al. (2011)
VY Aqr	90.85	1.06 ± 0.06	14453 + 316/-366	Pala et al. (2022)
OY Car	90.89	0.882 + 0.011/-0.015	18600 + 2800/-0.1600	McAllister et al. (2019)
		0.84 + /- 0.04		Zorotovic et al. (2011)
MR UMa	91.17		15182↓ ± 654	Pala et al. (2017)
QZ Lib	92.36	0.82 + 0.14/-0.19	11 419 + 175/-229	Pala et al. (2022)
SDSS J153817.35+ 512338.0	93.11	0.97 + 0.09/-0.11	35284 +600/-688	Pala et al. (2022)
		1.073 + 0.074/-0.066	35743 ± 576	Godon & Sion (2022)
UV Per	93.44	0.7 ± 0.2	14040 +539/-645	Pala et al. (2022)
CTCV J2354-4700	94.392	0.935 ± 0.031	14800 ± 700	Savoury et al. (2011)
SSS130413	94.71	0.84 ± 0.03	24,000 +/- 3000	McAllister et al. (2019)

(Continued)

(*Continued*)

System	P_{orb}	M_{wd}	T_{eff}	ref(M_{wd}/T_{eff})
IRXS J023238.8-371812	95.04	1.15 ± 0.04	14457 +118/-135	Pala et al. (2022)
CSS110113	95.11	1.00 + 0.04/-0.01	14500 ± 2200	McAllister et al. (2019)
SDSS J1152+4049	97.518	0.560 ± 0.028	12400 ± 1400	Savoury et al. (2011)
	97.56	0.62 ± 0.04	15900 +/-2000	McAllister et al. (2019)
RZ Sge	98.32	0.78 + 0.13/-0.18	15197 + 419/-507	Pala et al. (2022)
CY UMa	100.18	0.57 + 0.08/-0.06	14692 + 471/-394	Pala et al. (2022)
VV Pup Polar	100.44		11900 ± 600	Townsley & Gänsicke (2009)
BB Ari	101.20		14948↓ ± 632	Pala et al. (2017)
V834 Cen	101.52		14300 ± 900	Townsley & Gänsicke (2009)
GD 552	102.73	0.78 + 0.03/-0.04	10761 +37/-43	Pala et al. (2022)
OU Vir	104.696	0.703 ± 0.012	22300 ± 2100	Savoury et al. (2011)
HT Cas	106.08	0.61 + 0.04/-0.04 Hor91	14000 ± 1000	Townsley & Gänsicke (2009)
IY UMa*	106.43	0.99 + 0.04/-0.03	17057 +179/-79	Pala et al. (2022)
		0.903 + 0.088/-0.090	17130 ± 249	Godon & Sion (2022)
		0.79 +/- 0.04		Zorotovic et al. (2011)
VW Hyi	106.98	0.71 + 0.18/-0.26	20000 ± 1000	Townsley & Gänsicke (2009)
		0.869 + 0.065/-0.064	21538 ± 235	Godon & Sion (2023b)
Z Cha	107.28	0.803 ± 0.014	14900 +/- 2000	McAllister et al. (2019)
		0.84 +/-0.09		Zorotovic et al. (2011)
SDSS J100515.38+191107.9	107.60	0.44 + 0.15/-0.09	14483 + 520/-430	Pala et al. (2022)
RZ Leo	110.17	0.97 + 0.13/-0.16	15573 + 437/-424	Pala et al. (2022)
SDSS 0901	112.15	0.752 + 0.024/-0.018	14900 +/-2000	McAllister et al. (2019)
CU Vel	113.04	0.47 + 0.04/-0.05	14174 +117/-169	Pala et al. (2022)
AX For	113.04	0.76 ± 0.07	16571↓ ± 777	Pala et al. (2017)
		1.09 + 0.08/-0.09		Pala et al. (2022)
MR Ser	113.46		14200 ± 900	Townsley & Gänsicke (2009)
SDSS J164248.52+134751.4	113.60		17710: ± 871	Townsley & Gänsicke (2009)
BL Hyi	113.64		13300 ± 900	Townsley & Gänsicke (2009)
ST LMi	113.88		10800 ± 500	Townsley & Gänsicke (2009)
QZ Ser	119.75		14481 ± 587	Pala et al. (2017)

		0.766 + 0.090/-0.079	15893 ± 225	Godon & Sion (2023a)
EF Peg	120.53	0.8 + 0.16/-0.17	16644 + 448/-570	Pala et al. (2022)
V713 Cep	123.00	0.703 + 0.012/-0.015	17000 + 6000/-2000	McAllister et al. (2019)
DV UMa	123.62	0.96 + 0.07/-01.	19410 + 244/-400	Pala et al. (2022)
		0.968 + 0.127/-0.116	19325 ± 481	Godon & Sion (2022)
		1.098 +/- 0.024		Zorotovic et al. (2011)
HU Aqr	125.04		14000	Townsley & Gänsicke (2009)
IR Com	125.34	1.03 + 0.07/-0.09	17531 + 236/-271	Pala et al. (2022)
		1.001 + 0.066/-0.055	17860 ± 180	Godon & Sion (2022)
CTCV J1300-3052	128.07	0.717 ± 0.019	11000 ± 1000	McAllister et al. (2019)
	128.074	0.736 ± 0.014	11100 ± 800	Savoury et al. (2011)
QS Tel	139.92		17500 ± 1500	Townsley & Gänsicke (2009)
SDSS1702+322	144.1	0.91 ± 0.03		Zorotovic et al. (2011)
SDSS 1702	144.118	0.91 ± 0.03	15200 ± 1200	Savoury et al. (2011)
SDSS J1702+3229	144.12	0.94 ± 0.01	17000 ± 500	Townsley & Gänsicke (2009)
SDSS J001153.08-064739.2	144.40		13854 ± 525	Pala et al. (2017)
TU Men	168.77	0.77 + 0.16/-0.13	27750 ± 1000	Godon & Sion (2021)
AM Her Polar	185.65	0.63 + 0.1/-0.08	19248 + 486/-450	Pala et al. (2022)
		0.78 +/- 0.15		Zorotovic et al. (2011)
MV Lyr	190.56		47000	Townsley & Gänsicke (2009)
DW UMa	196.71	0.82 + 0.03/-0.04	56760 + 146/-259	Pala et al. (2022)
		0.7-0.84	45000 - 60000	Godon & Sion (2023b)
		0.87 +/- 0.19		Zorotovic et al. (2011)
TT Ari	198.06		39000	Townsley & Gänsicke (2009)
IP Peg	227.8	1.16 +/-0.02		Zorotovic et al. (2011)
SDSS J040714.78-064425.1	245.04		20885 ± 1104	Pala et al. (2017)
V1043 Cen	251.40		15000	Townsley & Gänsicke (2009)
GY Cnc	252.64	0.881 ± 0.016	25900 ± 2300	McAllister et al. (2019)
		0.99 +/-0.12		Zorotovic et al. (2011)
WW Cet	253.20		26000 ± 1000	Townsley & Gänsicke (2009)

(Continued)

(Continued)

System	P_{orb}	M_{wd}	T_{eff}	ref(M_{wd}/T_{eff})
U Gem	254.76	1.160 ± 0.013	33070 + 648/-616	Pala et al. (2022)
		1.20 +/- 0.09		Zorotovic et al. (2011)
HS 2214+2845	258.02		26443↓ ± 1436	Pala et al. (2017)
		0.798 + 0.086/-0.108	27643 ± 268	Godon & Sion (2022)
BD Pav	258.19		17775 ± 876	Pala et al. (2017)
		0.683 + 0.083/-0.087	19330 ± 312	Godon & Sion (2022)
SS Aur	263.23	0.98 + 0.09/-0.11	28627 + 190/-269	Pala et al. (2022)
		1.045 + 0.21/0.28	33375 ± 1875	Godon & Sion (2021)
SDSS J100658.41+233724.4	267.71	0.78 ± 0.12	16000 ± 1000	Zorotovic et al. (2011)
	267.71	0.82 ± 0.11		McAllister et al. (2019)
DQ-Her	278.8	0.60 + /-0.07		Zorotovic et al. (2011)
V895 Cen	285.90		14000 ± 900	Townsley & Gänsicke (2009)
RX And	302.25	0.81 + 0.06/-0.08	33900 + 634/-995	Pala et al. (2022)
		0.83 + 0.10/-0.08	33813 +/- 869	Godon & Sion (2023b)
V347 Pup	334.0	0.63 + /-0.04		Zorotovic et al. (2011)
TT Crt	386.423	0.663 0.049	26990 ± 557	Godon & Sion (2022)
EM Cyg	418.9	1.00 +/-0.06		Zorotovic et al. (2011)
HS 0218+3229	428.02		17990 ± 893	Pala et al. (2017)
		0.502 + 0.078/-0.072	17980 ± 177	Godon & Sion (2023a)
AC Cnc	432.7	0.76 +/-0.03		Zorotovic et al. (2011)
V363 Aur	462.6	0.90 +/- 0.06		Zorotovic et al. (2011)
AE Aqr	592.8	0.63 +/-0.05		Zorotovic et al. (2011)
EY Cyg	661.4	1.026 + 0.174/-0.090	28040 ± 1173	Godon & Sion (2023a)
V442 Cen	662.4	0.64 + 0.06/-0.05	29802 + 211/-247	Pala et al. (2022)
		0.671 ± 0.057	31032 ± 308	Godon & Sion (2022)

Notes from reference Pa22 are given here.

* = A precise mass measurement is available.

↓ = Represent upper limits for the temperature of the systems.

: = Represent unreliable effective temperature.

References

Godon, P., & Sion, E. M. 2021, ApJ, 908, 173

Godon, P., & Sion, E. M. 2022, ApJ, 928, 26

Godon, P., & Sion, E. M. 2023a, ApJ, 950, 139

Godon, P., & Sion, E. M. 2023b, ApJ, accepted

Littlefair, S. P., Dhillon, V. S., Marsh, T. R., Gänsicke, B. T., Southworth, J., & Watson, C. A. 2006, Science, 314, 1578

McAllister, M, Littlefair, S. P., Parsons, S. G., et al. 2019, MNRAS, 486, 5535

Pala, A. F., et al. 2017, MNRAS, 466, 2855

Pala, A. F., et al. 2022, MNRAS, 510, 6110

Savoury, C. D. J., Littlefair, S., Dhillon, V. S., et al. 2011, MNRAS, 415, 2025

Steeghs, D., Howell, S. B., Knigge, C., Gänsicke, B. T., Sion, E. M., & Welsh, W. F. 2007, ApJ, 667, 442

Townsley, D. M., & Gänsicke, B. T. 2009, ApJ, 693, 100

Zorotovic, M., Schreiber, M., & Gaensicke, B. T. 2011, A&A, 536, 42

Accreting White Dwarfs
From exoplanetary probes to classical novae and Type 1a supernovae
Edward M Sion

Appendix B

Masses and X-Ray Temperatures, T_max, of Magnetic White Dwarfs in Intermediate Polars

doi:10.1088/2514-3433/ac930cch9   B-1

Object Name	P_{orb} Minutes	WD Mass de Martino et al. (2020)	WD Mass Xu et al. (2019) C	WD Mass Xu et al. (2019). D	Method	WD Mass Shaw et al. (2020)	WD Mass Suleimanov et al. (2019)	Temperature keV	Temperature Xu et al. (2019). A keV	Temperature Xu et al. (2019) B keV
IGRJ19552+0044/ SWIFTJ1955.2+0077	83.6	0.77 + 0.02/−0.03								
V1025 Cen	84.6	0.6 + 0.06/−0.03				0.61 ± 0.14				
IGR G1817-2508	91.872					0.54 ± 0.08				
IGRJ18173-2509/ SWIFTJ1817.4-2510	91.9	0.96 ± 0.05								
EX Hya	98.3	0.78 ± 0.03	0.77 ± 0.10	0.76 ± 0.07		0.70 ± 0.04		14.5 ± 0.3	16.7 ± 2.3	
Paloma	156									
V515 And	163.9	0.79 ± 0.07			0.73 ± 0.06	0.67 ± 0.07	18.2 + 2.3/−2.0			
BG CMi	194.1	0.67 ± 0.19	0.75 ± 0.12	0.75 ± 0.07	0.78 + 0.08/−0.07	0.63 ± 0.07	20.7 + 3.1/−2.6	32.5 ± 6.0	24.1 ± 0.12	
BY Cam	201.244 32				0.76 ± 0.06		21.0 + 2.4/−2.0			
V1223 Sgr	201.9	0.75 ± 0.02	0.81 ± 0.06	0.75 ± 0.03		0.72 ± 0.02		32.4 ± 2.5	35.4 + 1.7/−1.5	
V1432 Aql	201.939 84				0.76 + 0.09/−0.08		20.8 + 3.9/−3.1			
V2400 Oph	205.8	0.81 ± 0.10	0.75 ± 0.04	0.74 ± 0.04	0.67 + 0.06/−0.05	0.72 ± 0.05	16.6 + 2.0/−1.7	30.9 ± 2.0	30.2 ± 3.9	
IGRJ19267+1325	206.9									
V647 Aur	207.9	0.74 ± 0.06				0.85 ± 0.16				
AO Psc	215.5	0.55 ± 0.06	0.58 ± 0.04	0.53 ± 0.03	0.55 ± 0.05	0.53 ± 0.05	12.4 + 1.7/−1.4	17.9 ± 1.2	17.5 ± 3.4	
IGRJ16500-3307/ SWIFTJ1649.9-3307	217	0.92 ± 0.06								
IGR J1649-3307	217.008					0.82 ± 0.09				
IGR J16547-1916/ SWIFTJ1654.7-1917	222.9	0.85 ± 0.15								
IGR J16547-1916	222.912				0.74 + 0.09/−0.08		18.9 + 3.4/−2.7			
IGR J1654-1916	222.912					0.83 ± 0.10				
YY Dra	238.1	0.75 ± 0.02	0.80 ± 0.14	0.82 ± 0.04				37.8 ± 4.0	33.1 ± 7.6	
DO Dra	238.1388					0.76 ± 0.08				
IGRJ1719-4100/ SWIFTJ1719.6-4102	240.3	0.86 ± 0.06								
IGR J17195-4100	240.336				0.84 + 0.08/−0.07		22.9 + 3.6/−2.9			
J1719-4100	240.336		0.85 + 0.29/−0.15	0.82 ± 0.04				37.8 ± 4.8	30.2 ± 5.3	
IGR J1719-4100	240.336					0.72 ± 0.06				

SWIFTJ2113.5+5422	241.2	0.81 + 0.16/−0.10							
V1033 Cas	242	0.91 + 0.14/−0.16				1.02 ± 0.15			
IGRJ142576117/ 4PBCJ1425.1-6118	243	0.6 ± 0.2							
V405 Aur	249.6	0.89 ± 0.13			0.75 + 0.07/−0.06	0.73 ± 0.03	19.1 + 2.6/−2.2		
V2306 Cyg	261	0.77 ± 0.16				0.7 ± 0.10			
V418 Gem	262.8	0.5 ± 0.2				0.74 ± 0.23			
SWIFTJ0927.7-6945	281	0.58 ± 0.1							
MU Cam	283.1	0.74 ± 0.13	1.17 + 0.08/−0.33	0.86 ± 0.10		0.67 ± 0.08		41.9 ± 11.6	26.7 ± 5.8
FO Aqr	290.9	0.61 ± 0.05	0.61 ± 0.11	0.61 ± 0.03	0.61 + 0.06/−0.05	0.57 ± 0.03	14.8 + 1.8/−1.7	22.3 ± 1.5	20.8 ± 1.4
PQ Gem	311.6	0.65 ± 0.09	0.78 + 0.28/−0.23	0.84 ± 0.04	0.71 + 0.06/−0.05	0.77 ± 0.07	18.0 + 2.3/−1.9	40.5 ± 3.8	34.6 ± 5.7
V709 Cas	320	0.88 + 0.05/−0.04	0.94 + 0.16/−0.28	0.88 ± 0.05		0.83 ± 0.04		43.5 ± 6.8	50.0 + 4.6/−3.9
IGRJ18308-1232/ SWIFTJ1830.8-1253	322.4	0.85 ± 0.06							
TV Col	329.2	0.78 ± 0.06	0.72 ± 0.06	0.73 ± 0.02		0.79 ± 0.03		30.1 ± 2.1	36.0 ± 3.5
V667 Pup	336.2	0.79 ± 0.11			0.83 + 0.11/−0.08	0.86 ± 0.08	22.8 + 4.3/−3.4		
TX Col	343.2	0.67 ± 0.10	0.67 ± 0.11	0.60 ± 0.05		0.52 ± 0.07		22.4 ± 3.5	17.7 ± 3.4
AXJ1740.2-2903	343.3								
IGR J15094-6649	352.8				0.73 ± 0.06		18.5 + 2.4/−2.0		
IGR J1509-6649	353.376					0.85 ± 0.09			
IGRJ1509-6649/ SWIFTJ1509.4-6649	353.4	0.89 ± 0.08							
XY Ari	363.9	0.96 ± 0.12	0.91 ± 0.27	0.99 ± 0.06		1.06 ± 0.10		57.5 ± 7.9	67.0 + 19.5/−14.8
SWIFTJ0457.1+4528	371.3	1.12 ± 0.06							
EI UMa	386.1	9.1 + 0.017e/−0.07							
RXJ2133.7+5107	428.300 64				0.96 + 0.08/−0.07	0.95 ± 0.04	28.2 + 4.6/−3.7		
J2133.7+5107	428.302 08		0.87 ± 0.10	0.91 ± 0.07				47.3 ± 7.5	52.0 ± 5.5
IGRJ21335+5105/ SWIFTJ2133.6+5105	431.6	0.93 ± 0.04							
IGR J0457+4527	432					0.87 ± 0.14			
V2069 Cyg	448.8	0.82 ± 0.08			0.73 + 0.09/−0.08	0.83 ± 0.10	18.5 + 2.8/−3.6		
IGR J0838-4831	475.2					1.27 ± 0.15			

(*Continued*)

(*Continued*)

Object Name	P_{orb} Minutes	WD Mass de Martino et al. (2020)	WD Mass Xu et al. (2019) C	WD Mass Xu et al. (2019). D	Method	WD Mass Shaw et al. (2020)	WD Mass Suleimanov et al. (2019)	Temperature keV	Temperature Xu et al. (2019). A keV	Temperature Xu et al. (2019) B keV
IGRJ08390-4833/ SWIFTJ0838.8-4832	480	0.95 ± 0.08								
IGR J08390-4833	480				0.81 + 0.13/−0.11		12.1 + 5.8/−4.2			
SWIFTJ0746.3-1608	562.8	0.78 ± 0.13								
NY Lup	591.8	1.16 + 0.04/−0.02	1.17 + 0.07/−0.30	0.94 ± 0.04		1.05 ± 0.04		53.2 ± 5.1	55.5 + 6.9/−3.4	
El Uma	591.84				0.91 + 0.15/−0.13	0.91 ± 0.08	26.0 + 8.1/−5.6			
V1062 Tau	598.9	0.72 ± 0.17			0.72 + 0.07/−0.06	0.81 ± 0.10	18.7 + 2.4/−2.0			
Nova Sco AD 1437	769	1.16 ± 0.12								
V2731 Oph	925.3	1.16 ± 0.05				1.06 ± 0.03				
GK Per	2875.4	0.87 ± 0.08								
SWIFTJ0525.6+2416		1.01 ± 0.06								
2PBCJ0801.2-4625		1.18 ± 0.10								
SWIFTJ0958.0-4208		0.74 ± 0.11								
IGRJ14091-6108/Swift J1408.26113		1.2–1.3								
SWIFTJ1832.5-0863/ AXJ1832.3-0840										
IRXS J052523.2+241331					0.81 + 0.13/−0.10		19.6 + 5.2/−0.07			

References

de Martino, D, Bernardini, F, Mukai, K, Falanga, M, & Masetti, N 2020, AdSpR, 66, 1209
Shaw, A W, et al. 2020, MNRAS, 498, 3457
Suleimanov, V F, Doroshenko, V, & Werner, K 2019, MNRAS, 482, 3622
Xu, X-J, Yu, Z-l, & Li, X-d 2019, ApJ, 878, 53